W0245718

Lecture Notes in Physics

Editorial Board

R. Beig, Wien, Austria
J. Ehlers, Potsdam, Germany
U. Frisch, Nice, France
K. Hepp, Zürich, Switzerland
W. Hillebrandt, Garching, Germany
D. Imboden, Zürich, Switzerland
R. L. Jaffe, Cambridge, MA, USA
R. Kippenhahn, Göttingen, Germany
R. Lipowsky, Golm, Germany
H. v. Löhneysen, Karlsruhe, Germany
I. Ojima, Kyoto, Japan
H. A. Weidenmüller, Heidelberg, Germany
J. Wess, München, Germany
J. Zittartz, Köln, Germany

Springer
Berlin
Heidelberg
New York
Barcelona
Hong Kong
London
Milan
Paris
Singapore
Tokyo

Physics and Astronomy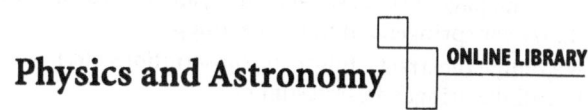

http://www.springer.de/phys/

Editorial Policy

The series *Lecture Notes in Physics* (LNP), founded in 1969, reports new developments in physics research and teaching – quickly, informally but with a high quality. Manuscripts to be considered for publication are topical volumes consisting of a limited number of contributions, carefully edited and closely related to each other. Each contribution should contain at least partly original and previously unpublished material, be written in a clear, pedagogical style and aimed at a broader readership, especially graduate students and nonspecialist researchers wishing to familiarize themselves with the topic concerned. For this reason, traditional proceedings cannot be considered for this series though volumes to appear in this series are often based on material presented at conferences, workshops and schools (in exceptional cases the original papers and/or those not included in the printed book may be added on an accompanying CD ROM, together with the abstracts of posters and other material suitable for publication, e.g. large tables, colour pictures, program codes, etc.).

Acceptance

A project can only be accepted tentatively for publication, by both the editorial board and the publisher, following thorough examination of the material submitted. The book proposal sent to the publisher should consist at least of a preliminary table of contents outlining the structure of the book together with abstracts of all contributions to be included. Final acceptance is issued by the series editor in charge, in consultation with the publisher, only after receiving the complete manuscript. Final acceptance, possibly requiring minor corrections, usually follows the tentative acceptance unless the final manuscript differs significantly from expectations (project outline). In particular, the series editors are entitled to reject individual contributions if they do not meet the high quality standards of this series. The final manuscript must be camera-ready, and should include both an informative introduction and a sufficiently detailed subject index.

Contractual Aspects

Publication in LNP is free of charge. There is no formal contract, no royalties are paid, and no bulk orders are required, although special discounts are offered in this case. The volume editors receive jointly 30 free copies for their personal use and are entitled, as are the contributing authors, to purchase Springer books at a reduced rate. The publisher secures the copyright for each volume. As a rule, no reprints of individual contributions can be supplied.

Manuscript Submission

The manuscript in its final and approved version must be submitted in camera-ready form. The corresponding electronic source files are also required for the production process, in particular the online version. Technical assistance in compiling the final manuscript can be provided by the publisher's production editor(s), especially with regard to the publisher's own Latex macro package which has been specially designed for this series.

Online Version/ LNP Homepage

LNP homepage (list of available titles, aims and scope, editorial contacts etc.):
http://www.springer.de/phys/books/lnpp/
LNP online (abstracts, full-texts, subscriptions etc.):
http://link.springer.de/series/lnpp/

P. A. Vermeer S. Diebels W. Ehlers
H. J. Herrmann S. Luding E. Ramm (Eds.)

Continuous and Discontinuous Modelling of Cohesive-Frictional Materials

Springer

Editors

Pieter A. Vermeer
Institute of Geotechnical Engineering
University of Stuttgart
Pfaffenwaldring 35
70569 Stuttgart, Germany

Wolfgang Ehlers
Stefan Diebels
Institute of Applied Mechanics
University of Stuttgart
Pfaffenwaldring 7
70569 Stuttgart, Germany

Hans J. Herrmann
Stefan Luding
Institute of Computer Applications 1
University of Stuttgart
Pfaffenwaldring 27
70569 Stuttgart, Germany

Ekkehard Ramm
Institute of Structural Mechanics
University of Stuttgart
Pfaffenwaldring 27
70569 Stuttgart, Germany

Cover picture: see D'Addetta et al. in this volume

Library of Congress Cataloging-in-Publication Data applied for.
Die Deutsche Bibliothek - CIP-Einheitsaufnahme

Continuous and discontinuous modelling of cohesive frictional
materials / P. A. Vermeer ... (ed.). - Berlin ; Heidelberg ; New York
; Barcelona ; Hong Kong ; London ; Milan ; Paris ; Singapore ; Tokyo :
Springer, 2001
  (Lecture notes in physics ; Vol. 568)
  (Physics and astronomy online library)
  ISBN 3-540-41525-4
ISSN 0075-8450
ISBN 3-540-41525-4 Springer-Verlag Berlin Heidelberg New York

This work is subject to copyright. All rights are reserved, whether the whole or part of the material is concerned, specifically the rights of translation, reprinting, reuse of illustrations, recitation, broadcasting, reproduction on microfilm or in any other way, and storage in data banks. Duplication of this publication or parts thereof is permitted only under the provisions of the German Copyright Law of September 9, 1965, in its current version, and permission for use must always be obtained from Springer-Verlag. Violations are liable for prosecution under the German Copyright Law.

Springer-Verlag Berlin Heidelberg New York
a member of BertelsmannSpringer Science+Business Media GmbH

© Springer-Verlag Berlin Heidelberg 2001
Printed in Germany

The use of general descriptive names, registered names, trademarks, etc. in this publication does not imply, even in the absence of a specific statement, that such names are exempt from the relevant protective laws and regulations and therefore free for general use.

Typesetting: Camera-ready by the authors/editors
Cover design: *design & production*, Heidelberg

Printed on acid-free paper
SPIN: 10792073   55/3141/du - 5 4 3 2 1 0

Preface

A knowledge of the mechanical behaviour of both naturally occurring materials such as soils and rocks, and artificial materials such as concrete and industrial granular matter is of fundamental importance to their proper use in engineering and scientific applications. The research activities in this broad area of applied mechanics have attracted scientists and engineers with a variety of backgrounds ranging from physics to civil engineering. For simulating the mechanical behaviour of the cohesive granular materials considered, two different frameworks of modelling and analysis have emerged. On the one hand, continuum-based models and, on the other hand, discrete particle methods, or in other words "Continuous and Discontinuous Modelling" as referred to in the title of this volume, are successfully applied to cohesive-frictional materials. In addition, the *micro-to-macro* or homogenization approaches, respectively, are used to relate microscopic discontinuum models to macroscopic continuum models.

This volume contains contributions to the International Symposium on "Continuous and Discontinuous Modelling of Cohesive-Frictional Materials", as organized at the University of Stuttgart by the research group "Modelling of Cohesive-Frictional Materials". This research group was established in May 1998 with the full support of the German Science Foundation (DFG). Four different institutes of the University of Stuttgart participate in the research group, namely

- Institute of Applied Mechanics (Prof. W. Ehlers, Dr. S. Diebels),
- Institute of Geotechnical Engineering (Prof. P.A. Vermeer),
- Institute of Structural Mechanics (Prof. E. Ramm),
- Institute of Computer Applications 1 (Prof. H.J. Herrmann,
  Dr. S. Luding).

The research group focuses on the development of a multilevel approach for the modelling of cohesive-frictional materials. Within this framework, the main research areas can be found in the enhancement of the discontinuous (particle level) and continuum based modelling with an emphasis on the transition between these two approaches.

After two years of research, it was considered appropriate to organize this International Symposium. For doing so, we obtained support from the German Science Foundation (DFG). We are grateful for the sponsoring provided by this organization. The success of this symposium largely rested on the efforts of a small Organizing Committee within our research group, namely:

- Dipl.-Ing. G.A. D'Addetta, Institute of Structural Mechanics,
- Dipl. Phys. M. Lätzel, Institute of Computer Applications 1,
- Dipl.-Ing. T. Marcher, Institute of Geotechnical Engineering,
- Dr. T. Michelitsch, Institute of Applied Mechanics.

We would like to thank this Organizing Committee for the work in preparing and coordinating this meeting of researchers. They invested a lot of time and energy to guarantee a successful meeting for about 110 participants. Most of them came from Europe, but some had to travel much further as they came from America, Australia or South Africa.

The highlight of the symposium was a series of lectures of outstanding speakers. International experts in targeted research areas lectured on current developments and problems in the numerical modelling of cohesive-frictional materials and provided a deeper understanding of the microscopic and macroscopic description of geomaterials. We are grateful for their willingness to prepare and present their lectures. Their contributions are published in this proceedings volume. This book will prove not only helpful for specialist researchers in the fields of physics and engineering but also for students who want to gain experience in the fascinating field of cohesive-frictional materials.

In conclusion, we are convinced that this International Symposium on "Continuous and Discontinuous Modelling of Cohesive-Frictional Materials" has fulfilled its objective as a vehicle for the cross-fertilization of ideas between engineers and scientists engaged in research on continuous and discontinuous modelling of cohesive-frictional materials.

Stuttgart, November 2000

P.A. Vermeer
S. Diebels
W. Ehlers
H.J. Herrmann
S. Luding
E. Ramm

Contents

Computational models for failure in cohesive-frictional materials with stochastically distributed imperfections

M. A. Gutiérrez, R. de Borst

Koiter Institute Delft/Faculty of Aerospace Engineering, Delft University of Technology, P.O. Box 5058, 2600 GB Delft, The Netherlands

Abstract. The finite element reliability method is applied to the analysis of strain localisation phenomena. Properties like the Young's modulus, the softening modulus and the yield stress are considered to be random fields. The most probable configurations of imperfections leading to failure are sought by means of an optimisation algorithm. This allows for evaluation of the significance of different modes and of the global probability of failure. The symmetry/asymmetry of shear band patterning in a biaxial test is studied for a viscoplastic material.

1 Introduction

Failure in most engineering materials is preceded by the emergence of narrow zones of intense straining. During this phase of so-called strain localisation, the deformation pattern in a body rather suddenly evolves from relatively smooth into one in which thin zones of highly strained material dominate. In fact, these so-called zones of strain localisation act as a precursor to ultimate fracture and failure. Thus, in order to accurately and properly describe the failure behaviour of materials it is of pivotal importance that the strain localisation phase is modelled in a physically and mathematically correct manner, and that proper numerical tools are utilised to actually solve strain localisation phenomena in boundary value problems.

In recent years the study of strain localisation in solids has received an increasing amount of attention, even though typical localisation phenomena like Lüders bands and rock faults have been known and studied for many decades. Until the mid-1980s analyses of localisation phenomena in materials were commonly carried out for standard, rate-independent continuum models. This is sufficient when the principal aim is to determine the behaviour in the pre-localisation regime and some properties at incipient localisation, such as the direction of shear bands in tension tests, and in biaxial and triaxial devices. However, there is a major difficulty in the post-localisation regime, since localisation in standard, rate-independent solids is intimately related to a possible change of the character of the governing set of partial differential equations. In the static case the elliptic character of the set of partial differential equations can be lost, while, on the other hand, in the dynamic case we typically observe a change of a hyperbolic set into an elliptic set. In both cases the rate boundary value problem becomes ill-posed and numerical solutions suffer from spurious mesh sensitivity.

The inadequacy of the standard, rate-independent continuum to model zones of localised straining correctly can be viewed as a consequence of the fact that force-displacement relations measured in testing devices are simply mapped onto stress-strain curves by dividing the force and the elongation by the original load-carrying area and the original length of the specimen, respectively, without taking into account the changes in the micro-structure. Therefore, the mathematical description ceases to be a meaningful representation of the physical reality.

To remedy this problem one must either introduce additional terms in the continuum description which reflect the changes in the micro-structure, or one must take into account the inherent viscosity of most engineering materials [4]. The effect is that the governing equations do not change type during the loading process and that physically meaningful solutions are obtained for the entire loading range. A more mathematical way to look at the introduction of additional terms in the continuum description is that the Dirac distributions for the strain at failure are replaced by continuous strain distributions, which lend themselves for description by standard numerical schemes. Although the strain gradients are now finite, they are very steep and the concentration of strain in a small area can still be referred to as strain localisation or localisation of deformation.

Until recently, analyses on strain localisation were carried out either for perfect specimens where localisation was triggered by injecting a bifurcation mode that results from an eigenvalue analysis of the tangent stiffness matrix [18], or by arbitrarily imposed imperfections [2]. The latter approach is numerically more convenient, but suffers from a certain degree of inobjectivity: the position and size of the imperfections may strongly influence the failure mode and the collapse load. To enable a more objective assessment of localisation and failure in cohesive-frictional materials, a procedure has been developed where, starting from a given imperfection pattern, the most critical pattern is found in the sense that the probability that this particular pattern leads to failure is a local maximum [8][9][10][11]. Thus, by starting from different imperfection patterns a fairly accurate assessment can be made on the probability that different imperfection patterns lead to failure. This procedure will be elaborated for an elasto-viscoplastic material and will be applied to the classical example of a biaxial test.

2 The finite element reliability method

2.1 Introduction to the reliability method

The reliability method is essentially a computational tool for the approximation of the probability of failure. This probability is defined as

$$
\begin{aligned}
\mathcal{P}_f &= \Pr(Z < 0) \\
&= \int_{z<0} p_Z(\theta)\, d\theta,
\end{aligned}
\tag{1}
$$

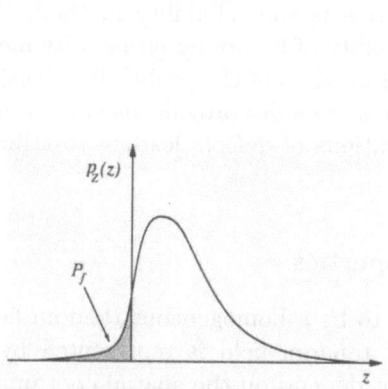

Fig. 1. Probability content of the region $Z < 0$

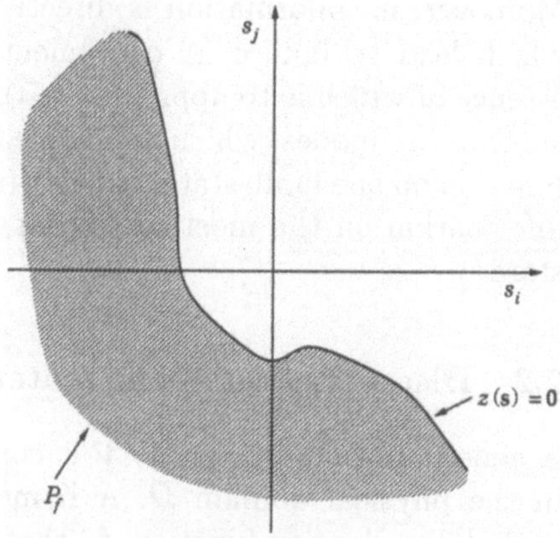

Fig. 2. Probability content of the region $z(\mathbf{s}) < 0$ in the standard normal space

where Z represents a function of the response of the structure such that positive realisations of Z denote serviceability and negative realisations denote failure. This function is referred to as the limit-state function. The function p_Z represents the probability density function of Z and is represented in Figure 1. However, this function is not known. Distribution information is available on the material properties, represented by a random field $\mathcal{V}$. Since the transformation from this field to Z is of a functional nature, an analytical expression of $\mathcal{P}_f$ in terms of distributions of $\mathcal{W}$ is seldom available and a numerical approximation must be used. For this purpose the original field of properties must be discretised into a set of n random variables $\mathbf{S}$ which, through suitable transformations, can be considered to be uncorrelated and standard normally distributed, that is,

$$
\begin{aligned}
\mathbf{ES} &= \mathbf{0}; \\
\mathrm{Cov}(\mathrm{S_i}, \mathrm{S_j}) &= \delta_{ij}.
\end{aligned}
\tag{2}
$$

The probability of failure (1) can thus be recast as

$$
\mathcal{P}_f = \int_{z(\mathbf{s})<0} \phi_n(\varsigma)\, d\varsigma,
\tag{3}
$$

which is represented in Figure 2. The function ϕ_n represents the n-dimensional uncorrelated standard normal function. The shape of the domain $z(\mathbf{s}) < 0$ is often complicated, so that the integral (3) cannot be computed exactly. The most accurate method to carry out this computation is the well-known Monte Carlo method, in which different realisations of the random variables are generated according to their probability distribution. The relation between the number of realisations leading to failure and the total number of realisations considered provides an estimation of the probability of failure. When this probability is small, the number of simulations required to obtain accurate estimates increases enormously, which makes the Monte Carlo method unaffordable in practice.

Moreover, no information is directly provided on the most critical realisations which lead to failure. A convenient alternative is the reliability method, the essence of which is to approximate the probability of failure by probability measures of the modes, which are defined as local maxima of the probability density function on the limit-state surface $z(\mathbf{s}) = 0$. These modes provide also individual information on the most significant configurations of defects leading to a limit state.

2.2   Discretisation of the material properties

A generic material property $\mathcal{V}$ is considered to be a homogeneous random field in the physical domain Ω. A homogeneous random field is represented by a probability density function $f_{\mathcal{V}}$ that does not depend on the spatial coordinate and a correlation function ρ that gives the correlation coefficient between two different points $\mathbf{x}_1$ and $\mathbf{x}_2$ in Ω and is exclusively dependent on the distance $|\mathbf{x}_1 - \mathbf{x}_2|$. This function is usually scaled by a length parameter l_c that is referred to as the correlation length. In order to perform numerical simulations, this random field must be discretised into a set of random variables characterised by marginal probability density functions and a correlation matrix [14]. Without loss of generality, the midpoint method has been adopted in this study, in which the domain is discretised in a finite number of subdomains Ω_i. The random field within Ω_i is then represented by a variable V_i such that

$$V_i = \mathcal{V}(\mathbf{x}_i^c), \tag{4}$$

where $\mathbf{x}_i^c$ is the central point of Ω_i (Figure 3). The marginaldistribution of V_i equals that of $\mathcal{V}$ and the correlation coefficient between V_i and V_j is given by $\rho(\mathbf{x}_i, \mathbf{x}_j)$. To a certain extent, this discretisation technique resembles the usual practice in computational localisation analysis of imposing an imperfection by reducing the strength within a patch of finite elements.

The vector $\mathbf{V}$ of random variables can be converted into a vector of uncorrelated, standard normally distributed variables $\mathbf{S}$ through the Nataf distribution model [6].

2.3   Response as a function of the imperfections

The relation between the (discretised) random field of material properties and the response in quasi-static conditions, neglecting body forces, is governed by the boundary value problem

$$\begin{aligned}
\nabla \cdot \boldsymbol{\sigma} &= 0 & &\text{in } \Omega; \\
\mathbf{u} &= \bar{\mathbf{u}}(t) & &\text{on } \partial\Omega_1; \\
\boldsymbol{\sigma} \cdot \mathbf{n} &= \bar{\boldsymbol{\sigma}}(t) & &\text{on } \partial\Omega_2,
\end{aligned} \tag{5}$$

at each instant t, where $\boldsymbol{\sigma}$ is the stress tensor, $\mathbf{u}$ is the displacement field, $\partial\Omega_1 \cup \partial\Omega_2 = \partial\Omega$, $\mathbf{n}$ is the outward normal vector to $\partial\Omega$ and $\bar{\mathbf{u}}(t)$ and $\bar{\boldsymbol{\sigma}}(t)$ are the

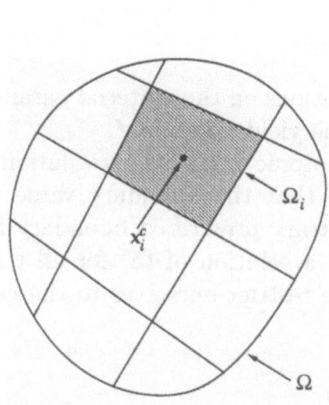

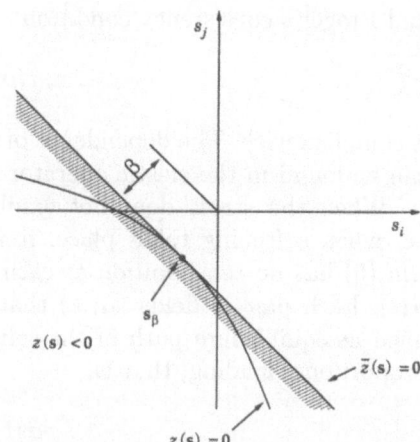

Fig. 3. Schematic representation of the midpoint discretisation technique for random fields

Fig. 4. Schematic representation of the single hyperplane approximation procedure

prescribed boundary displacements and loading. The relation between the stress tensor, the displacements field and the material properties is formalised by means of a Duvaut-Lions elasto-viscoplastic material model. For such material, the constitutive equations in rate form are written as

$$\dot{\sigma} = \mathbf{D}_e \dot{\varepsilon} - \frac{1}{\eta}(\sigma - \sigma_p);$$
$$\dot{\kappa} = -\frac{1}{\eta}(\kappa - \kappa_p), \tag{6}$$

where ε is the strain tensor, κ is the hardening (softening) parameter, $\mathbf{D}_e$ is the linear elastic tangent operator, η is a fluidity-like parameter and the subindex p signifies that these quantities belong to a rate-independent elasto-plastic problem. Equations (6) are valid when the yield function is non-negative

$$f(\sigma, \kappa) \geq 0 \tag{7}$$

otherwise the material is in the elastic regime,

$$\dot{\sigma} = \mathbf{D}_e \dot{\varepsilon}. \tag{8}$$

The underlying rate-independent elasto-plastic problem (denoted by p above) follows the usual relation

$$\dot{\sigma}_p = \mathbf{D}_e \left(\dot{\varepsilon} - \dot{\kappa}_p \left. \frac{\partial f}{\partial \sigma_p} \right|_{\kappa_p} \right) \tag{9}$$

when the yield function vanishes

$$f(\sigma_p, \kappa_p) = 0 \tag{10}$$

and Prager's consistency condition

$$\dot{f}(\sigma_p, \kappa_p) = 0 \tag{11}$$

is complied with. The dependence of these equations on the material parameters can be found in the elastic operator $\mathbf{D}_e$ and the yield function f.

When the stress does not exhibit a monotonic increasing evolution law, i.e. when softening takes place, it is possible that the boundary value problem (5) has no real solution at each t for arbitrary prescribed boundary forces $\bar{\sigma}(t)$. Each pair of fields $(\mathbf{u}, \bar{\sigma})$ that furnishes a solution of (5) for all t is defined as equilibrium path of the solid Ω. If we restrict ourselves to the case of proportional loading, that is

$$\bar{\sigma}(t) = \lambda(t)\sigma_d \tag{12}$$

where λ is a scalar valued multiplier and σ_d is a fixed, design load, we can represent the equilibrium path corresponding to σ_d as $(\mathbf{u}, \lambda)$. Functionals can be defined of the equilibrium path to represent characteristics of practical interest. For instance,

$$\lambda_c = \max_t \lambda \tag{13}$$

represents the maximal load multiplier, which is typically attained when $\dot{\lambda} = 0$. The critical load is then defined as

$$\sigma_c = \lambda_c \sigma_d. \tag{14}$$

Since λ_c is coupled to the random material properties, it can be considered as a random variable, so that λ_c is a realisation of Λ_c. A random measure of the capacity of the structure to resist the design load σ_d is then given by

$$Z(\Lambda_c) = \Lambda_c - 1. \tag{15}$$

According to this equation, positive realisations of Z represent a safe structure whereas negative realizations reflect failure. In a similar fashion, energy functionals can be defined to study the degree of brittleness.

2.4   Approximation of the probability of failure

As commented upon in Section 2.1, the shape of the failure domain $z(\mathbf{s}) < 0$ is usually complicated and the integral (1) cannot be evaluated exactly. To remedy this, the surface $z(\mathbf{s}) = 0$ is approximated by low-order surfaces like hyperplanes (first-order approximation) or quadratic surfaces (second-order approximation). In this study only hyperplane approximations are considered. This particular case is referred to as the *first-order reliability method*. The relative positions of these hyperplanes provide a valuable insight into the mechanical behaviour of the solid. When second-order surfaces are used, the approximation technique is referred to as the *second-order reliability method*.

If the failure domain is convex, the surface $z(\mathbf{s}) = 0$ is approximated by the hyperplane

$$
\begin{aligned}
\bar{z}(\mathbf{s}) &= \boldsymbol{\alpha}^T \mathbf{s} + \beta \\
&= 0
\end{aligned}
\tag{16}
$$

where $\|\boldsymbol{\alpha}\| = 1$ and β is the distance from the hyperplane to the origin. This is represented in Figure 4. The approximation point is chosen as the closest point of $z(\mathbf{s}) = 0$ to the origin. This point, with coordinates $-\beta\boldsymbol{\alpha}$, is referred to as design point or β-point and represents a maximum of the probability density function of $\mathbf{S}$ on $z(\mathbf{s}) = 0$. The β index is referred to as reliability index and the probability content of the approximated failure domain is given by

$$
\mathcal{P}_f = \varPhi(-\beta),
\tag{17}
$$

where $\varPhi$ is the one-dimensional standard normal cumulative distribution function.

When the surface $z(\mathbf{s}) = 0$ exhibits several β-points, as often happens in localisation analysis, multiple approximation hyperplanes $\bar{z}_i(\mathbf{s}) = 0$, $i = 1, \ldots n$, must be considered. The probability of failure is then expressed as [13]

$$
\begin{aligned}
\mathcal{P}_f &\approx \Pr(\bigcup_{i=1}^{n} \{\bar{z}_i(\mathbf{s}) < 0\}) \\
&= 1 - \varPhi_n(\boldsymbol{\beta}, \boldsymbol{\rho}^{\beta}),
\end{aligned}
\tag{18}
$$

where $\boldsymbol{\beta}$ represents a vector containing the β-indices of each linearisation point and $\boldsymbol{\rho}^{\beta}$ represents the correlation matrix of the approximating domains that is obtained from

$$
\rho_{ij}^{\beta} = \boldsymbol{\alpha}_i^T \boldsymbol{\alpha}_j.
\tag{19}
$$

The computation of the multinormal cumulative distribution function $\varPhi_n$, i.e., the integration of (18) within an n-cell, is numerically prohibitive as n increases. As an alternative, narrow bounds of the probability (18) have been derived [5].

The reliability method is not only useful to evaluate the probability of failure, but also to identify the most critical configurations of imperfections leading to failure. These imperfections, characterised by the β-points, provide a measure of the significance of the corresponding localisation patterns. For symmetric solids subject to symmetric loading, the limit-state surface is also symmetric. However, a symmetric β-point can be much more separated from the origin than other, non-symmetric points, and can even furnish a local minimum of the probability density on the limit-state function. It is also observed that each asymmetric β-point has a homologous point with respect to the symmetry hyperplane (Figure 5). The relative spatial position of both asymmetric points and the symmetric point provides also information on the spread of the main contributions to the probability of failure. If these points are relatively close to each other, which situation is quantified by the correlation coefficient (19) of the corresponding β-indices, the regions that contribute most to the probability of failure overlap each other (Figure 6).

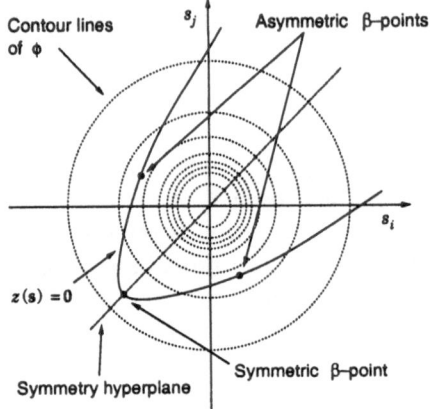
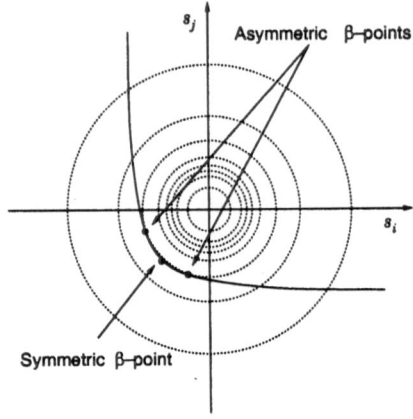

Fig. 5. Symmetric limit-state surface with symmetric and asymmetric β-points

Fig. 6. Symmetric limit-state surface with highly correlated symmetric and asymmetric β-points

2.5 Computation of the β-points

The computation of the β-points, which furnish the local modes on the limit-state surface, is of capital importance for an accurate approximation of the marginal and global probabilities of failure and constitutes the main task in the reliability method. Since the uncorrelated standard normal probability density function decays monotonically as the distance to the origin increases, it follows immediately that β-points are locally the closest points to the origin. Therefore, the computation of the β-points can be stated as a constrained optimisation problem

$$\text{minimise} \quad \|\mathbf{s}\|; \tag{20}$$
$$\text{subject to} \quad z(\mathbf{s}) = 0.$$

Several algorithms exist to solve this problem, and have been reviewed by [15]. These algorithms are gradient-based and usually consist of a recursive sequence of points in the form

$$\mathbf{s}^{(k+1)} = \mathcal{F}\left(\mathbf{s}^{(k)}, z^{(k)}, \frac{\partial z^{(k)}}{\partial \mathbf{s}}\right), \tag{21}$$

where k accounts for the iteration number.

In this study a simple iterative algorithm has been adopted that was introduced by [12] and later improved by [17]. This algorithm is usually referred to as the HL-RF algorithm and attains the form

$$\mathbf{s}^{(k+1)} = \frac{1}{\left\|\dfrac{\partial z^{(k)}}{\partial \mathbf{s}}\right\|^2} \left[\frac{\partial z^{(k)}}{\partial \mathbf{s}}\mathbf{s}^{(k)} - z^{(k)}\right] \left(\frac{\partial z^{(k)}}{\partial \mathbf{s}}\right)^T \tag{22}$$

This algorithm requires the evaluation of the gradient of the sensitivity $\frac{\partial z}{\partial s}$ of the limit-state function with respect to the discretised imperfection field s. This is elaborated in the next section. When the algorithm (22) has converged to a local β-point there is no way to force convergence to another point. The traditional methodology ([16]) to cope with the multiplicity problem is to select different initial approximations for algorithm (22). Engineering judgement and experience often provide an adequate estimation of the starting points. For the particular case of localisation problems, the usual practice of placing an initial imperfection within a patch of finite elements so as to trigger a symmetric or a non-symmetric localisation pattern has proved to be a meaningful choice for the starting point of (22).

3 Computation of the mechanical transformation

The mechanical transformation defined in Section 2.3 must be computed by discretising u in finite elements,

$$u = \sum_i a_i N_i, \tag{23}$$

where a is defined as the nodal displacement vector and N represents the shape functions. The discretised version of (5) with the load parameter λ as defined in (12) is then integrated in time with an Euler-forward scheme according to the expressions

$$\begin{aligned} {}^{t+\Delta t}a &= {}^t a + \Delta a; \\ {}^{t+\Delta t}\lambda &= {}^t\lambda + \Delta\lambda, \end{aligned} \tag{24}$$

where the notation ${}^t\gamma = \gamma(t)$ has been used for the dependence of any object on time. Instead, the material equations (6)–(11) are solved at each integration point with an Euler-backward scheme. The gradient of the mechanical transformation is obtained from a careful differentiation of the global as well as the local algorithms. A study on the existence, behaviour and computation of these derivatives in softening solids can be found in [7].

3.1 Computation of the equilibrium path

Global equilibrium. Elaborating a weak form of equation (5) for the discretisation (4) of the material properties and the discretization (23) of the displacements field, the non-linear system of algebraic equations

$$^t f_{int}(a, v) = {}^t\lambda f_d \tag{25}$$

is obtained, where the nodal internal force vector is defined as

$$^t f_{int} = \int_\Omega B^T {}^t\sigma \, d\Omega, \tag{26}$$

B is the strain-nodal displacement matrix, $\mathbf{f}_d$ accounts for the nodal design forces and $\boldsymbol{\sigma}$ represents the stress tensor, which is related to the nodal displacement vector **a** and the material properties **v** through the constitutive equations. Using equations (23) the evolution of the equilibrium path can be written incrementally as

$$\int_\Omega \mathbf{B}^T \Delta\boldsymbol{\sigma}\, d\Omega = \Delta\lambda \mathbf{f}_d, \tag{27}$$

assuming that equilibrium is complied with at time t. A load control procedure, in which $\Delta\lambda$ is supplied by the analyst, cannot be used because this increment can be incompatible with the equilibrium path (see Section 2.3). Instead, a selective arc-length control procedure is used [1]. In the selective arc-length method, both the nodal incremental displacements and external forces are parametrized by the length of a projection of the nodal incremental displacement vector. This is formally done by introducing a constraint equation

$$\Delta\mathbf{a}^T \boldsymbol{\Psi} \Delta\mathbf{a} = \Delta l^2, \tag{28}$$

with $\boldsymbol{\Psi}$ a projection matrix to select the degrees-of-freedom which contribute to the increment of arc-length Δl, which, in a context of quasi-static loading, can be considered congruent with the time increment Δt. However, such a restriction is not necessary and in general it can be considered that Δl is any function of t. Combining (27) and (28) we obtain

$$\int_\Omega \mathbf{B}^T \Delta\boldsymbol{\sigma}\, d\Omega = \Delta\lambda \mathbf{f}_d;$$
$$\Delta\mathbf{a}^T \boldsymbol{\Psi} \Delta\mathbf{a} = \Delta l^2, \tag{29}$$

which system is solved with the Newton-Raphson method

$$\begin{pmatrix} \Delta\mathbf{a} \\ \Delta\lambda \end{pmatrix}^{(k+1)} = \begin{pmatrix} \Delta\mathbf{a} \\ \Delta\lambda \end{pmatrix}^{(k)} - \begin{bmatrix} \mathbf{K} & -\mathbf{f}_d \\ 2\Delta\mathbf{a}^{T(k)}\boldsymbol{\Psi} & 0 \end{bmatrix} \mathbf{r}(\Delta\mathbf{a}, \Delta\lambda)^{(k)} \tag{30}$$

where **K** is the consistent tangent stiffness matrix, which will be discussed in the next section, and the residual **r** is defined as

$$\mathbf{r}(\Delta\mathbf{a}, \Delta\lambda) = \begin{pmatrix} \int_\Omega \mathbf{B}^T \Delta\boldsymbol{\sigma}\, d\Omega - \Delta\lambda \mathbf{f}_d \\ \Delta\mathbf{a}^T \boldsymbol{\Psi} \Delta\mathbf{a} - \Delta l^2 \end{pmatrix} \tag{31}$$

From this procedure and equations (24) a time-discretised equilibrium path is obtained. However, a point such that $\dot\lambda = 0$, which is associated with λ_c, does not generally exist in the context of spatially discretised systems with softening behaviour. In this study it has been considered that the peak load multiplier λ_c can be interpolated between $^{t-\Delta t}\lambda$, $^t\lambda$ and $^{t+\Delta t}\lambda$ when $^t\lambda$ is larger than both $^{t-\Delta t}\lambda$ and $^{t+\Delta t}\lambda$.

Integration of the constitutive relation. When the yield criterion (7) is satisfied at a generic integration point of the finite element discretization, equations (9)–(11) for the underlying rate-independent elasto-plastic problem must be integrated in time to obtain a time discretised increment of stress $\Delta\sigma$. This is done with an Euler-backward return-mapping algorithm [3]. Without loss of generality, a Von Mises material in plane-stress conditions with a linear hardening (softening) rule is considered,

$$(\mathbf{I} - \mathbf{A})^t\sigma_p + \mathbf{D}_e\Delta\varepsilon = \mathbf{A}\Delta\sigma_p;$$
$$f(^t\sigma_p + \Delta\sigma_p, {}^t\kappa_p + \Delta\kappa_p) = 0,$$

(32)

where

$$\mathbf{A} = \mathbf{I} + \frac{3\Delta\kappa_p}{\bar{\sigma}(^t\kappa_p + \Delta\kappa_p)}\mathbf{D}_e\mathbf{P};$$

$$f(\sigma, \kappa) = \sqrt{3J_2} - \bar{\sigma}(\kappa);$$

(33)

$$\bar{\sigma}(\kappa) = \bar{\sigma}_0 + h\kappa;$$

$$J_2 = \tfrac{1}{2}\sigma^T\mathbf{P}\sigma.$$

In these equations, h is the hardening (softening) modulus, $\bar{\sigma}$ is the yield stress

and

$$\mathbf{P} = \begin{bmatrix} 2/3 & -1/3 & 0 \\ -1/3 & 2/3 & 0 \\ 0 & 0 & 2 \end{bmatrix}.$$

(34)

Equations (32) are solved with any suitable iterative procedure for a given increment of strain $\Delta\varepsilon$. The consistent tangential relation between $\Delta\sigma_p$ and $\Delta\varepsilon$ is given by the matrix [3]

$$\mathbf{D}_{cp} = \mathbf{H} - \frac{\mathbf{H}\dfrac{\partial f}{\partial\sigma}\dfrac{\partial f}{\partial\sigma}^T\mathbf{H}}{h + \dfrac{\partial f}{\partial\sigma}^T\mathbf{H}\dfrac{\partial f}{\partial\sigma}}$$

(35)

where

$$\mathbf{H} = \left(\mathbf{I} + \Delta\kappa_p\mathbf{D}_e\frac{\partial^2 f}{\partial\sigma^2}\right)^{-1}\mathbf{D}_e.$$

(36)

Equations (6) are integrated with an Euler-backward algorithm as well, resulting in the explicit formulæ

$$\Delta\sigma = \tfrac{1}{1+\Delta t/\eta}\left(\mathbf{D}_e\Delta\varepsilon + \Delta t/\eta\,\Delta\sigma_p - \Delta t/\eta(^t\sigma - {}^t\sigma_p)\right);$$

$$\Delta\kappa = \tfrac{1}{1+\Delta t/\eta}\left(\Delta t/\eta\,\Delta\kappa_p - \Delta t/\eta(^t\kappa - {}^t\kappa_p)\right).$$

(37)

The consistent tangential relation between $\Delta\sigma$ and $\Delta\varepsilon$ has been derived by [19],

$$\mathbf{D}_c = \frac{\mathbf{D}_e + \Delta t/\eta\,\mathbf{D}_{cp}}{1 + \Delta t/\eta}$$

(38)

This consistent tangent operator is assembled in the stiffness matrix in (30) in order to ensure a quadratic rate of convergence of the Newton-Raphson procedure. Moreover, the accurate evaluation of the response gradient requires the use of the consistent tangent operator (see Section 3.2).

3.2   Computation of the gradient of the equilibrium path

A general methodology has been presented in [7] to evaluate the gradient of the equilibrium path in softening solids. The particular case for a Duvaut-Lions elasto-viscoplastic solid will be given here. Adopting the notation

$$\partial_i = \frac{\partial}{\partial v_i} \tag{39}$$

for the partial derivative with respect to a single coefficient of the discretisation of the material property, differentiation of equation (29) with respect to v_i leads to

$$\int_\Omega \mathbf{B}^T \partial_i \Delta\boldsymbol{\sigma} \, d\Omega - \partial_i \Delta\lambda \hat{\mathbf{f}} = 0;$$

$$2\Delta\mathbf{a}^T \boldsymbol{\Psi} \partial_i \Delta\mathbf{a} = 0. \tag{40}$$

The derivative of the stress increment $\Delta\boldsymbol{\sigma}$ with respect to v_i can be decomposed into two contributions

$$\partial_i \Delta\boldsymbol{\sigma} = \partial_i \Delta\boldsymbol{\sigma}|_{\Delta\boldsymbol{\varepsilon}} + \frac{\partial\Delta\boldsymbol{\sigma}}{\partial\Delta\boldsymbol{\varepsilon}} \partial_i \Delta\boldsymbol{\varepsilon}, \tag{41}$$

where the term $\frac{\partial\Delta\boldsymbol{\sigma}}{\partial\Delta\boldsymbol{\varepsilon}}$ is the consistent tangent operator (38). Substituting (41) into (40), the system

$$\int_\Omega \mathbf{B}^T \partial_i \Delta\boldsymbol{\sigma}|_{\Delta\boldsymbol{\varepsilon}} \, d\Omega + \int_\Omega \mathbf{B}^T \frac{\partial\Delta\boldsymbol{\sigma}}{\partial\Delta\boldsymbol{\varepsilon}} \partial_i \Delta\boldsymbol{\varepsilon} \, d\Omega - \partial_i \Delta\lambda \hat{\mathbf{f}} = 0;$$

$$2\Delta\mathbf{a}^T \boldsymbol{\Psi} \partial_i \Delta\mathbf{a} = 0 \tag{42}$$

results, which, using $\partial_i \Delta\boldsymbol{\varepsilon} = \mathbf{B}^T \partial_i \Delta\mathbf{a}$, can be recast as

$$\begin{bmatrix} \mathbf{K} & -\hat{\mathbf{f}} \\ 2\Delta\mathbf{a}^{T(k)}\boldsymbol{\Psi} & 0 \end{bmatrix} \begin{pmatrix} \partial_i \Delta\mathbf{a} \\ \partial_i \Delta\lambda \end{pmatrix} = \begin{pmatrix} -\int_\Omega \mathbf{B}^T \partial_i \Delta\boldsymbol{\sigma}|_{\Delta\boldsymbol{\varepsilon}} \, d\Omega \\ 0 \end{pmatrix}. \tag{43}$$

This is a system of linear simultaneous equations with a matrix identical to that in equation (30) at a converged state.

The computation of the right-hand side of (43) is explained next. Differentiation of equations (32) with respect to v_i yields

$$\partial_i \mathbf{A}^{t+\Delta t}\boldsymbol{\sigma}_p + \mathbf{A}\partial_i \Delta\boldsymbol{\sigma}_p = (\mathbf{I} - \mathbf{A})\partial_i {}^t\boldsymbol{\sigma}_p + \partial_i \mathbf{D}_e \Delta\boldsymbol{\varepsilon} + \mathbf{D}_e \partial_i \Delta\boldsymbol{\varepsilon};$$

$$\frac{3 {}^{t+\Delta t}\boldsymbol{\sigma}_p^T \mathbf{P}(\partial_i {}^t\boldsymbol{\sigma}_p + \partial_i \Delta\boldsymbol{\sigma}_p)}{2\,\bar{\sigma}({}^{t+\Delta t}\kappa_p)} = \partial_i \bar{\sigma}({}^t\kappa_p + \Delta\kappa_p), \tag{44}$$

where the terms $\partial_i {}^t\sigma_p$ and $\partial_i {}^t\kappa_p$ are known from the previous time increment. Developing terms in (44) for constant $\Delta\varepsilon$, a linear system of four simultaneous equations is obtained which is solved for $\partial_i \Delta\sigma_p|_{\Delta\varepsilon}$ and $\partial_i \Delta\kappa_p|_{\Delta\varepsilon}$. Differentiation of (37) then gives

$$\partial_i \Delta\sigma = \tfrac{1}{1+\Delta t/\eta}\Big(\partial_i \mathbf{D}_e \Delta\varepsilon + \mathbf{D}_e \partial_i \Delta\varepsilon + \Delta t/\eta \partial_i \Delta\sigma_p - \Delta t/\eta(\partial_i {}^t\sigma - \partial_i {}^t\sigma_p)\Big);$$

$$\partial_i \Delta\kappa = \tfrac{1}{1+\Delta t/\eta}\Big(\Delta t/\eta \partial_i \Delta\kappa_p - \Delta t/\eta(\partial_i {}^t\kappa - \partial_i {}^t\kappa_p)\Big).$$

$$(45)$$

For constant $\Delta\varepsilon$, equations (45) then yield $\partial_i \Delta\sigma|_{\Delta\varepsilon}$ and $\partial_i \Delta\kappa|_{\Delta\varepsilon}$ which furnish the right-hand side of (41). Solving (41) $\partial_i \Delta\mathbf{a}$ and $\partial_i \Delta\lambda$ are obtained. The derivatives of the strain increment are obtained from $\partial_i \Delta\varepsilon = \mathbf{B}\partial_i \Delta\mathbf{a}$ and those of $\Delta\sigma_p$, $\Delta\kappa_p$, $\Delta\sigma$ and $\Delta\kappa$ are solved from (44) or (45). Finally, the derivatives at $t + \Delta t$ are updated as in (24).

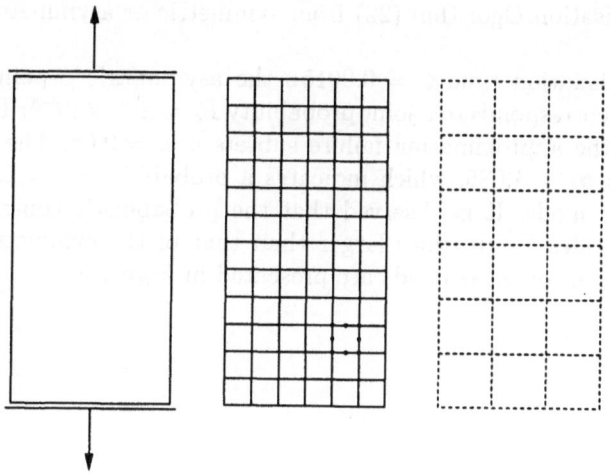

Fig. 7. Biaxial tension test: Static scheme, finite element mesh and random field mesh

4   Numerical simulation

The localisation behaviour has been analysed for the biaxial tension test of Figure 7. The specimen has a size of $120 \times 60\,\mathrm{mm}^2$ and a thickness of $5\,\mathrm{mm}$. The domain is discretised into a structured mesh of 28×14 eight-noded plane-stress finite elements with a four-point Gauss-Legendre integration quadrature. Linear constraint equations are used at the top and the bottom to ensure that both boundaries remain parallel during the loading process, which is controlled by a selective arc-length procedure such that the distance between both ends increases with a velocity of $1\,\mathrm{mm/s}$.

The Young's modulus and the initial yield stress are logarithmic-normally distributed according to the medians

$$Y_m = 20\,000\,\mathrm{N/mm^2};$$
$$\bar{\Sigma}_{0m} = 100\,\mathrm{N/mm^2}. \tag{46}$$

The hardening (softening) modulus obeys a normal distribution with

$$\mu_H = -500\,\mathrm{N/mm^2}. \tag{47}$$

The coefficient of variation is $C = 0.1$ for all properties. An exponential correlation function is considered with the correlation length $l_c = 20\,\mathrm{mm}$. Each random field is discretised into 98 subdomains corresponding with patches of 2×2 finite elements. The Poisson's ratio has been taken $\nu = 0.2$.

The energy dissipation in the time interval $[0\,\mathrm{s}, 0.75\,\mathrm{s}]$ is studied. A generic energy dissipation threshold $\psi_0 = 12\,000\,\mathrm{N\,mm}$ is considered and the adopted time step is $\Delta t = 0.0125\,\mathrm{s}$. Symmetric or asymmetric β-points are found by starting the optimisation algorithm (22) from symmetric or asymmetric realisations respectively.

For the relaxation time $\zeta = 0.0015\,\mathrm{s}$ the asymmetric β-points have $\beta_a = 3.6693$, which corresponds to a joint probability $P_a = 2.29 \times 10^{-4}$. The correlation coefficient of the approximating failure subsets is $\rho = 0.68$. The symmetric β-point exhibits $\beta_s = 4.085$, which indicates a probability $P_s = 2.20 \times 10^{-5}$ for the symmetric mode. It is observed that the probabilistic contribution of the asymmetric modes is ten times larger than that of the symmetric mode. The respective profiles of shear bands are presented in Figure 8.

 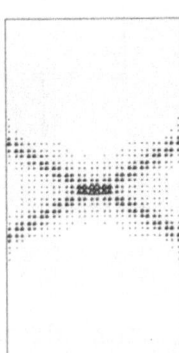

Fig. 8. Analysis of energy dissipation: Profiles of equivalent plastic strain at the asymmetric (left) and the symmetric (right) β-points with $\zeta = 0.0015\,\mathrm{s}$

If a larger relaxation time ($\zeta = 0.0045\,\mathrm{s}$) is considered, the computed β-points are close to each other. The reliability indices now read $\beta_a = 4.4598$ and

Fig. 9. Analysis of energy dissipation: Profiles of equivalent plastic strain at the asymmetric (left) and the symmetric (right) β-points with $\zeta = 0.0045\,\mathrm{s}$

$\beta_s = 4.4578$ and the correlation coefficient of the homologous asymmetric points is $\rho = 0.93$. The probability contribution of the asymmetric modes is reduced to $P_a = 6.56 \times 10^{-6}$ and that of the symmetric mode is $P_s = 4.14 \times 10^{-6}$. The resulting localisation patterns are shown in Figure 9.

It is observed that larger relaxation times lead to lower probabilistic contributions to the energy dissipation. This result could be expected, since a larger relaxation time usually corresponds to a larger energy dissipation during failure.

5   Conclusions

The influence of stochastically distributed imperfections in the failure behaviour of cohesive-frictional materials can be evaluated with the finite element reliability method. In order to carry out reliability computations the derivatives of the equilibrium path with respect to the discretised material properties and the boundary constraints must be computed. This is successfully achieved by direct differentiation of the discretised governing equations. The results obtained by applying the developed procedure to the analysis of a biaxial test are in agreement with the experimental observations and engineering practice.

References

1. R. de Borst: Comp. Struct. **25**, 211 (1987)
2. R. de Borst: Ingenieur-Archiv **59**, 160 (1989)
3. R. de Borst, P. H. Feenstra: Int. J. Numer. Methods Eng. **29**, 315 (1990)
4. R. de Borst, L. J. Sluys, H. -B. Mühlhaus, J. Pamin: Eng. Comp. **10**, 99 (1993)
5. O. Ditlevsen, H. O. Madsen: *Structural reliability methods* (Wiley, Chichester 1996)
6. A. Der Kiureghian, P. L. Liu: J. Eng. Mech. ASCE **112**, 85 (1986)

16   M. A. Gutiérrez, R. de Borst

7. M. A. Gutiérrez, R. de Borst: Comp. Meth. Appl. Mech. Eng. **162**, 337 (1998)
8. M. A. Gutiérrez, R. de Borst: Int. J. Numer. Methods Eng. **44**, 1823 (1999)
9. M. A. Gutiérrez, R. de Borst: Arch. Appl. Mech. **69**, 655 (1999)
10. M. A. Gutiérrez: Objective simulation of failure in heterogeneous softening solids. PhD Thesis, Delft University of Technology (1999)
11. M. A. Gutiérrez, R. de Borst: Int. J. Solids Struct. in press
12. A. M. Hasofer, N. C. Lind: J. Eng. Mech. ASCE **100**, 111 (1974)
13. M. Hohenbichler, R. Rackwitz: Structural Safety **1**, 177 (1983)
14. C. C. Li, A. Der Kiureghian: J. Eng. Mech. ASCE **119**, 161 (1993)
15. P. L. Liu, A. Der Kiureghian: Structural Safety **9**, 161 (1991)
16. H. O. Madsen, S. Krenk, N. C. Lind: *Methods of structural safety* (Prentice-Hall, Englewood Cliffs, 1986)
17. R. Rackwitz, B. Fiessler: Comp. Struct. **9**, 489 (1978)
18. H. I. van der Veen: The significance and use of eigenvalues and eigenvectors in the numerical analysis of elasto-plastic soils. PhD Thesis, Delft University of Technology (1998)
19. W. M. Wang, L. J. Sluys, R. de Borst: Int. J. Numer. Methods Eng. **40**, 3839 (1997)

Modeling of localized damage and fracture in quasibrittle materials

M. Jirásek

Laboratory of Structural and Continuum Mechanics (LSC), Swiss Federal Institute of Technology at Lausanne (EPFL), CH-1015 Lausanne, Switzerland

Abstract. This paper focuses on strain and damage localization due to propagation and coalescence of microcracks in quasibrittle materials such as concrete or rock. The mathematical description and numerical simulation of such phenomena can be based on an indirect representation of the cracking-induced deformation by inelastic strain, or on a direct representation by discontinuities in the displacement field. The first part of the paper gives a general overview and classification of modeling approaches and numerical techniques related to both types of representations and discusses their possible combination. The second part proposes a new technique for computational resolution of localization zones in regularized softening continua, based on special enrichments of the standard finite element approximations.

1 Representation of localized deformation

The behavior of quasibrittle materials (such as concrete, rock, tough ceramics, or ice) subjected to a high level of mechanical solicitations is characterized by the localization of strain and damage in relatively narrow zones and by a gradual development of macroscopic stress-free cracks. Despite a considerable progress in the past two decades, theoretical modeling and computational resolution of the localization process up to structural failure still remains a challenging issue of contemporary solid mechanics.

Narrow zones of highly concentrated evolving microdefects can be represented in many different ways. To establish a systematic classification, we will look at the (i) kinematic description, (ii) material models, and (iii) numerical approximation techniques. From the discussion it will transpire that all these aspects are closely related and a successful modeling approach must combine them in an appropriate manner.

1.1 Kinematic description

Depending on the regularity of the displacement field, $u(x)$, we can distinguish three types of kinematic descriptions. The first one incorporates *strong discontinuities*, i.e., jumps in displacements across a discontinuity curve (in two dimensions) or discontinuity surface (in three dimensions). The strain field, $\varepsilon(x)$, then consists of a regular part, obtained by standard differentiation of the displacement field, and a singular part, having the character of a multiple of the Dirac

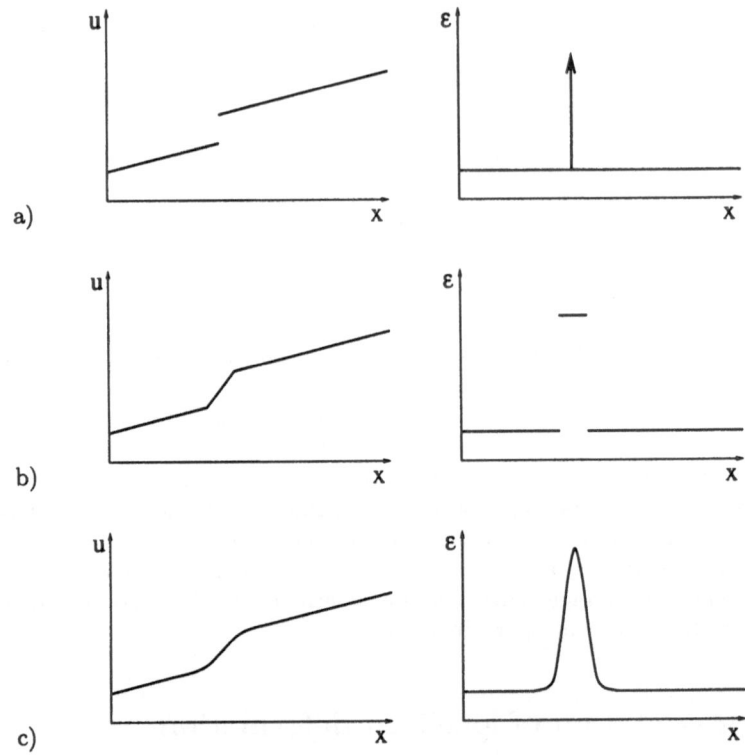

Fig. 1. Kinematic description with a) one strong discontinuity, b) two weak discontinuities, c) no discontinuities

delta distribution. This is schematically shown for the one-dimensional case in
Fig. 1a. In physical terms, the strong discontinuity corresponds to a sharp crack.

Another possible kinematic description represents the region of localized deformation by a band of a small but finite thickness, separated from the remaining
part of the body by two *weak discontinuities*, i.e., curves or surfaces across which
certain strain components have a jump but the displacement field remains continuous. This is illustrated in Fig. 1b. Since the displacement is continuous, the
strain components in the plane tangential to the discontinuity surface must remain continuous as well, and only the out-of-plane components can have a jump.
In physical terms, the band between the weak discontinuities corresponds to a
damage process zone with an almost constant density of microdefects.

Finally, the most regular description uses a continuously differentiable displacement field, and the strain field remains *continuous*. Strain localization is
manifested by high strains in a narrow band, with a continuous transition to
much lower strains in the surrounding parts of the body. A typical strain profile
of this type is shown in Fig. 1c. In physical terms, this corresponds to a damage
process zone with a higher concentration of defects around its center.

1.2   Constitutive models

Each of the kinematic descriptions discussed in the preceding section requires a different type of constitutive law for the cracking material. Of course, one could also consider the material law as the primary component of the model, and say that the regularity of the displacement field obtained as the solution of the corresponding boundary value problem depends on the choice of that law.

In the simplest case, the strong discontinuity can be considered as stress-free, which necessarily leads to a stress singularity at the discontinuity front (crack tip). The criterion for crack propagation is then formulated in terms of the stress intensity factors characterizing the singular part of the stress field. This is the linear elastic fracture mechanics approach, applicable at very large scales, when the process zone is negligible with respect to the characteristic dimensions of the body (such as the crack length and ligament size).

With the exception of huge quasibrittle structures such as concrete dams, it is usually necessary to take into account the finite size of the process zone at least in the direction tangential to the discontinuity surface. In the direction normal to the discontinuity, all inelastic effects are lumped and replaced by the equivalent displacement jump (crack opening), $[[u]]$. Before the growing microdefects coalesce and form a true physical crack, the process zone still carries some stresses, which are represented in the model as cohesive tractions transmitted by the discontinuity. The usual assumption of *cohesive crack models* or *cohesive zone models* is that the cohesive traction depends only on the displacement jump (separation vector), even though the evolution of the process zone is in general affected also by strains (or stresses) in the tangential plane. This means that the strong discontinuity is governed by its own constitutive law, formulated as a traction-separation law, which complements the stress-strain law valid for the continuous part of the body; see Fig. 2a. The area under the traction-separation curve is the energy needed to create a stress-free crack of a unit area, and it is usually called the Mode-I fracture energy, G_F. Even though it is questionable whether this is a true material property, it is certainly a fundamental parameter of every cohesive crack model.

Models with localization bands bounded by weak discontinuities can be considered as simple regularizations of models with strong discontinuities. Instead of lumping all the inelastic effects into a surface, it is possible to distribute them uniformly across the width of a band of a finite thickness h. This naturally leads to the *smeared crack models*, which transform the traction-separation law into a law that links the stress transmitted by the localization band to the average inelastic strain in that band; see Fig. 2b. If $[[u]]$ is the normal component of the displacement jump, the corresponding inelastic normal strain is obviously $\varepsilon_i = [[u]]/h$, and the area under the softening curve, G_F/h, has now the meaning of energy spent per unit volume of the localization band. Instead of splitting the constitutive law into the elastic and inelastic parts (more precisely, into the basic part and the part activated after localization), one could use a law that directly links the stress to the total strain, as is the case, e.g., in continuum dam-

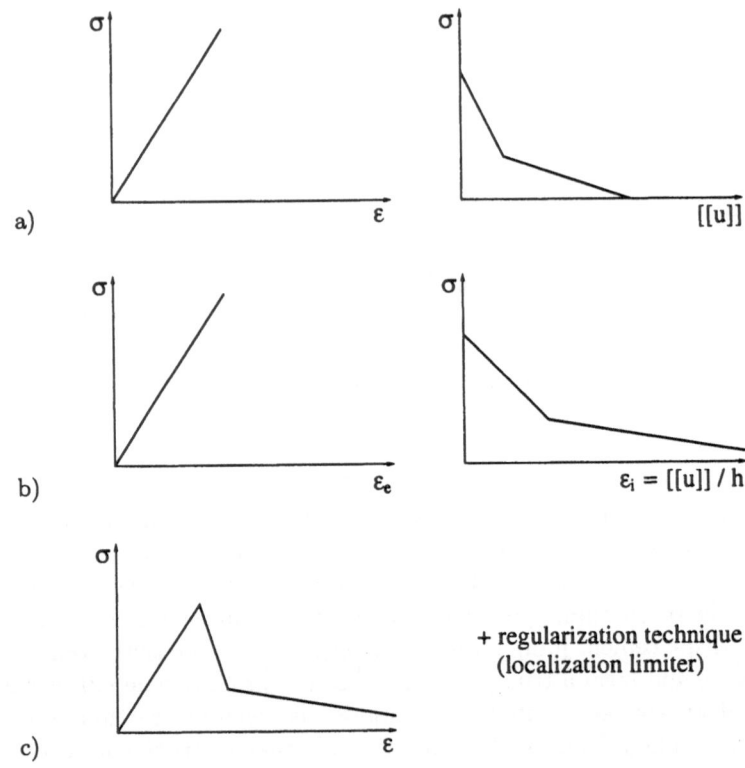

Fig. 2. Schematic representation of the constitutive description: a) cohesive crack model, b) smeared crack model, c) regularized continuum model with softening

age mechanics. The adjustment of the fracture energy is then more complicated but still possible.

Strain fields that remain continuous even after the onset of localization can be obtained with more sophisticated regularization techniques that serve as localization limiters. They are usually based on various forms of *enriched continuum theories*, e.g., on nonlocal or higher-order gradient continua. Such enrichments typically introduce a parameter defining a characteristic length of the material and related to the size and spacing of heterogeneities that control the width of the localized zone. The standard "local" form of the constitutive model corresponds to a deformation process during which the strain field remains uniform. However, the actual solution is nonuniform (either from the very beginning, or at least after bifurcation from a uniform solution), and the relation between the stress and local strain does not follow the same curve at every point. The energy needed to create a completely stress-free process zone depends on the area under the local stress-strain curve and on the characteristic length introduced by the localization limiter. One could define an equivalent dissipation length, l_d, equal to the thickness of a localization band that would give the same energy dissipation when the corresponding local model is used. The dissipation length is roughly proportional to the characteristic length but is also affected by the ductility of the constitutive law and by specific details of the nonlocal formulation; see [1].

1.3   Numerical approximations

As is clear from the foregoing discussion, each class of constitutive models gives rise to solutions that are characterized by a certain degree of regularity of the kinematic fields (displacements and strains). This must be taken into account by the numerical technique used for an approximate solution of the boundary value problem. We will focus here on the finite element method, with standard as well as enriched finite elements. The idea common to all types of models is that the details of the kinematic representation of the localized process zone can be resolved either by standard finite elements, which often requires fine meshes or frequent remeshing, or by special enrichments used in conjunction with a relatively coarse mesh.

For example, a strong discontinuity can be captured by *interface elements* inserted between two- or three-dimensional elements into which the body is divided; see Fig. 3a left. This is relatively straightforward if the discontinuity curve or surface is known in advance and is taken into account by the mesh generation procedure. In a general case, however, the discontinuity propagates across the body along an unknown path, and a good approximation of that path by interfaces between elements in a fixed mesh is not always possible. One solution is frequent remeshing, at least in the vicinity of the propagating discontinuity tip. As an elegant alternative, the recently emerged *elements with embedded strong discontinuities* [2–4] allow capturing the displacement jump across a segment that has an arbitrary position and orientation with respect to the basic finite element; see Fig. 3a right. Such formulations are based on enrichments of the standard shape functions by special discontinuous functions.

A similar discussion applies to models with localization bands of a finite thickness (Fig. 3b). They can be approximated by *standard finite elements* (this time only by the basic mesh, without the need for introducing interface elements), but if the band is not aligned with a layer of elements, it must be approximated by a zig-zag band. The thickness of the band, h, is dictated by the size of the finite elements, and the softening part of the stress-strain law must be adjusted accordingly, as explained in Section 1.2. The solution is usually biased by the mesh orientation and, for many constitutive models, the misalignment between the band direction and the mesh lines generates spurious stresses (this pathological behavior is sometimes called the stress locking). Again, special *elements with embedded localization bands* [5, 6] permit the modeling of bands of an arbitrary direction with respect to the basic mesh, and also of an arbitrary thickness. So, an additional advantage of this approach is that the thickness of the computationally resolved localization band does not depend on the element size any more, and it can be considered as a part of the constitutive model.

Finally, regularized models lead to localized strain profiles that remain continuous. To capture steep strain gradients with standard approximation tech-

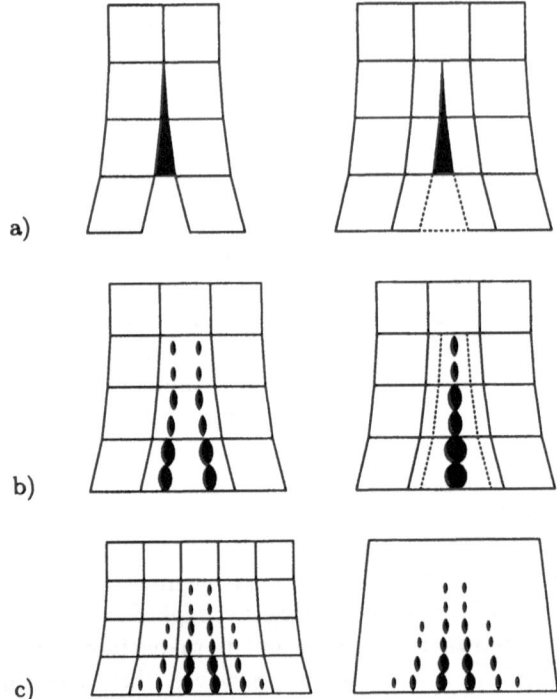

Fig. 3. Finite element models of the a) strong discontinuity, b) localization band, c) localization zone in a regularized softening continuum

niques, it is necessary to use a sufficiently fine discretization (Fig. 3c left). Alternatively, one can develop special enrichments such as the spectral overlays proposed in [7, 8] for shear bands modeled by softening viscoplasticity. One of such techniques, specifically designed for the resolution of evolving process zones in quasibrittle materials modeled by nonlocal damage theories, will be examined in Section 2. As shown in the right part of Fig. 3c, the finite elements can then be larger than the thickness of the process zone, which makes the computation much more efficient.

1.4   Combined continuous-discontinuous description

Enhanced formulations dealing with displacement or strain discontinuities embedded in finite elements are potentially very powerful but, in order to avoid pathological locking effects, the kinematic description of such elements and the internal equilibrium conditions must be constructed carefully, and appropriate criteria for the positioning of newly initiated discontinuities must be used. Recently it has been shown [9] that the mesh-induced directional bias can be dramatically reduced by models combining the discontinuous description with a continuous one. The early stage of cracking (at each material point) is represented by a stress-strain law with softening, either in the local form (with adjustment

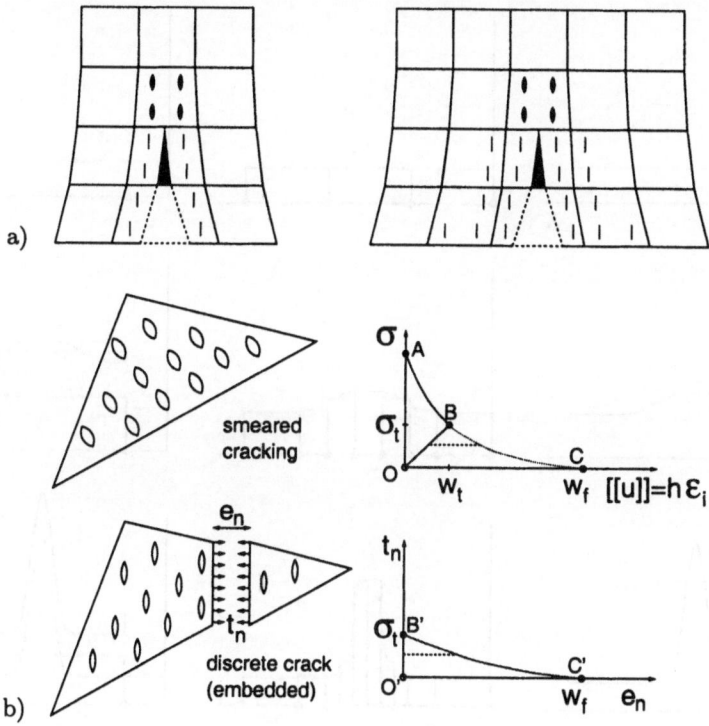

Fig. 4. Combined description: a) kinematic representation, b) constitutive model

of the softening curve according to the element size), or regularized by a localization limiter. When the damage at some point reaches a certain critical level, a discrete crack is introduced and modeled as a strong discontinuity embedded in the corresponding finite element; see Fig. 4a. The material laws describing the continuous and discrete parts of the model must be matched such that the desired global behavior is obtained; see Fig. 4b.

A related idea of a transition from an embedded localization band to an embedded strong discontinuity was explored in [10].

2   Elements with embedded localization zones

2.1   Motivation

The overview of numerical approximation techniques in Section 1.3 has shown that, in analogy to finite elements with embedded discontinuity lines or bands, it may be worthwhile to develop elements with special enrichments that are tailored for regularized continuum models. This idea is illustrated in Fig. 5, which shows from top to bottom three stages of a typical evolution of strain in the direction perpendicular to a propagating process zone. The left column corresponds to the development of the "exact" strain field, from a uniform profile before the onset

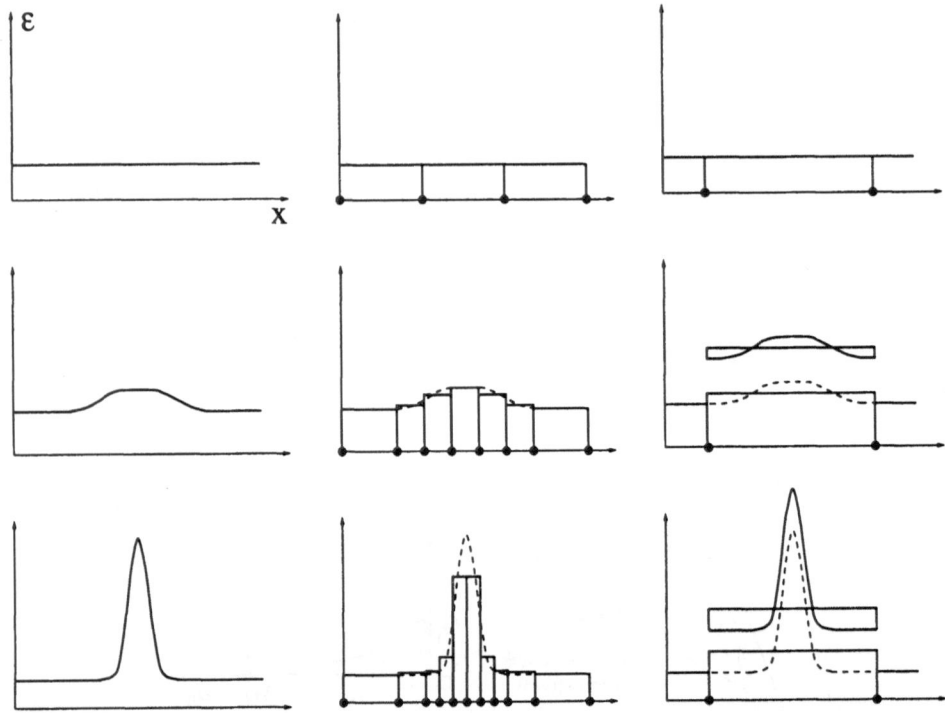

Fig. 5. Evolving strain profile and its approximation using standard low-order finite elements or embedded localization zones

of localization to a highly localized one at later stages of the loading process. The middle column in Fig. 5 illustrates the resolution of the evolving strain profile with low-order (constant-strain) finite elements. If the initial mesh is relatively coarse with respect to the characteristic material length, an adaptive refinement of the mesh is needed as the localization profile evolves. Even then, the approximation is not smooth and can substantially deviate from the actual profile, unless an extremely fine mesh is used. The right column shows that a good resolution can be hoped for if the standard finite element approximation on a coarse mesh is combined with special wave-like enrichments that are adaptively adjusted to the characteristic "wave length" of the evolving localization profile.

An additional argument motivating the development of special elements for regularized continuum models is that, after the onset of localization, standard low-order elements are often incapable of reproducing the state of constant stress. Of course, when linear elements with one integration point are used in a one-dimensional localization problem, the stress profile always remains perfectly uniform. However, this is rather an exception. In multiple dimensions, this would be the case only for underintegrated bilinear elements of a rectangular shape, perfectly aligned with the localization zone. As soon as several integration points per element are used, or the elements have an arbitrary shape and orientation, stress oscillations arise. For nonlocal integral models, the problem is caused by the imbalanced quality of approximations of the local and nonlocal strains. The local strain approximations are discontinuous across element interfaces, while the nonlocal strain approximations are constructed using a discretized integral operator and thus are by definition continuous. A similar problem arises for gradient-type models discretized by mixed finite elements [11], if the approximations of the displacement field and of the additional Lagrange-multiplier field are of the same order.

2.2 Low-order elements

To illustrate the stress oscillations, Fig. 6 presents the strain and stress profiles obtained with a one-dimensional version of the nonlocal damage model [12] using linear finite elements with two integration points per element. As already explained, one-dimensional elements with one integration point would give perfectly uniform stress profiles, but in multiple dimensions this is no longer the case, even with constant-strain elements using one-point integration. One-dimensional elements with two-point integration indicate what would happen for bilinear elements with the usual 2×2 integration, so the problem documented here is not artificially caused by an unusual integration scheme. To simplify the plotting procedure, the profiles were constructed by connecting the values at integration points by straight segments, which gives the false impression of continuity. In reality, the strain approximation is piecewise constant, and the stress approximation varies continuously inside each element, with jumps at nodes that connect the elements. The figure presents the results for 4 different meshes and two different states—one just before the onset of localization and the other after softening to about 80% of the peak load. The spatial coordinate x on the horizontal axis is normalized by the characteristic length of the nonlocal continuum,

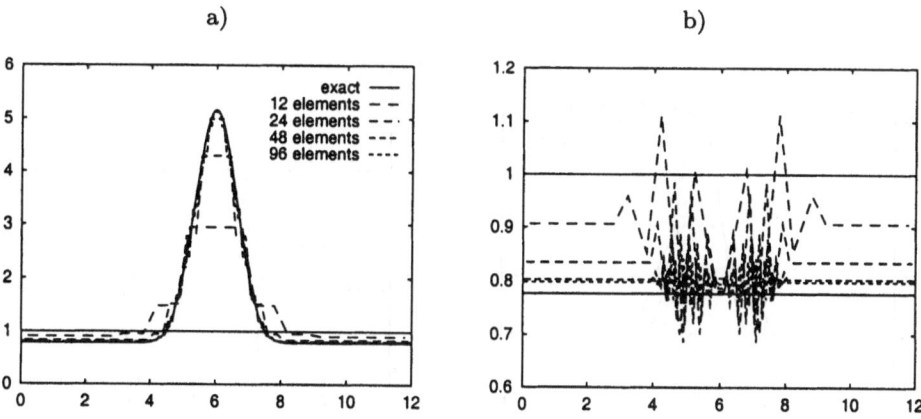

Fig. 6. Localization in a one-dimensional nonlocal continuum approximated by low-order finite elements: a) strain profiles, b) stress profiles

and the stresses and strains on the vertical axes are normalized by their values at the peak of the load-displacement diagram. The profiles before localization are uniform and they are reproduced exactly by each of the meshes. For the localized state we can see that the strain profiles converge to the exact solution as the mesh is refined. The stress profiles also converge, but they exhibit wild oscillations, even on a fine mesh with more than 30 elements across the thickness of the localized zone (the material in the left and right thirds of the bar is unloading elastically). The weak form of the equilibrium conditions implies that the average stress in each element must be the same, but even this average stress level is still quite far from the exact solution. Note that on the coarser meshes the stresses at some integration points are above the peak stress.

2.3 Higher-order elements

At first one may think that the simplest remedy is to increase the polynomial order of the finite element approximation. Higher-order polynomials should better correspond to the smooth character of the exact solution and reduce the imbalance between the quality of the local and nonlocal strain approximations. This is only partially true. If the mesh is sufficiently fine, higher-order elements give indeed better results than linear ones (of course, a fair comparison must be done for the same number of nodes). However, if the basic mesh is relatively coarse compared to the characteristic material length, it is not sufficient to increase the order of polynomial approximation, keeping the element size constant. Fig. 7 shows that the stress oscillations remain large even for a quartic displacement interpolation. Only three finite elements across the entire bar length were used, with a large number of integration points (16 in the middle element and 6 in each of the other elements). The results are somewhat disappointing, because the size and position of the middle element matches very well with the size and

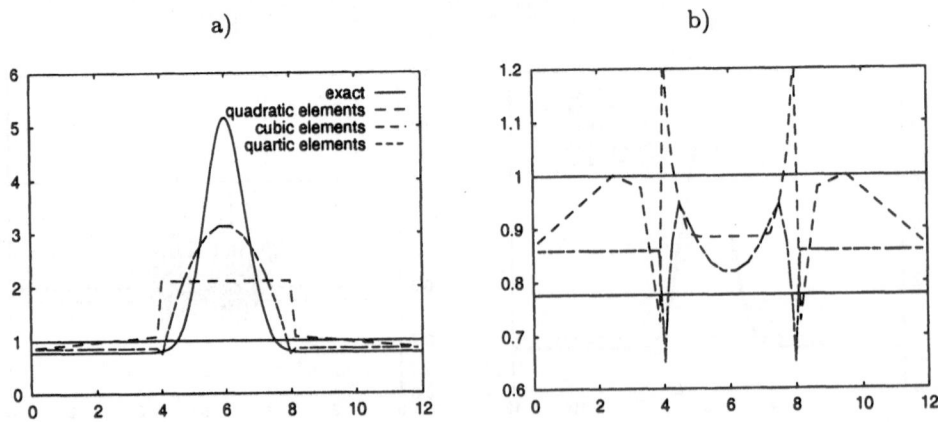

Fig. 7. Localization in a one-dimensional nonlocal continuum approximated by higher-order finite elements: a) strain profiles, b) stress profiles

position of the actual localization zone, so the element has in principle a good chance to closely approximate the exact solution. Note that, due to symmetry of the present problem, there is no difference between the quartic and the cubic interpolation.

2.4   Enriched elements

Finally, Fig. 8 shows the results obtained with special enrichment functions incorporated into the finite element approximation. Again, only three elements across the entire bar length were used, with the same integration scheme as before. The enrichment (applied only in the middle element) was based on truncated harmonic functions

$$\hat{G}_k(\xi) = \chi(\xi; \xi_L, \xi_R) \, \sin \frac{k\pi(\xi - \xi_L)}{\xi_R - \xi_L}, \qquad k = 1, 2, \ldots M \tag{1}$$

where ξ is the natural element coordinate varying between -1 and 1, ξ_L and ξ_R are parameters defining the position of the localization zone with respect to the element, and $\chi(\xi; \xi_L, \xi_R)$ is the characteristic function of the interval $[\xi_L, \xi_R]$, equal to 1 for ξ inside that interval and 0 for all other ξ. Functions $\hat{G}_k(\xi)$ are harmonic "pulses" added to the strain interpolation using the enhanced assumed strain technique [13]. To satisfy the compatibility condition, they are first transformed into functions

$$G_k(\xi) = \hat{G}_k(\xi) - \frac{1}{2} \int_{-1}^{1} \hat{G}_k(\xi) \, d\xi = \tag{2}$$

$$= \chi(\xi; \xi_L, \xi_R) \, \sin \frac{k\pi(\xi - \xi_L)}{\xi_R - \xi_L} - \frac{1 - (-1)^k}{2k\pi}(\xi_R - \xi_L), \qquad k = 1, 2, \ldots M$$

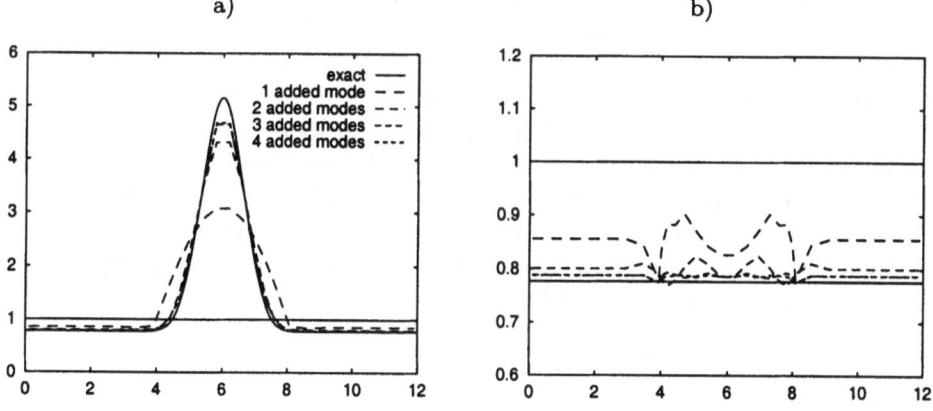

Fig. 8. Localization in a one-dimensional nonlocal continuum approximated by enriched finite elements: a) strain profiles, b) stress profiles

that have zero mean on the interval $[-1, 1]$.

The results in Fig. 8 indicate that already one added mode ($M = 1$) leads to a strain approximation comparable to cubic or quartic elements and, at the same time, the stress oscillations are substantially reduced. With three added modes ($M = 3$), the strain profile is approximated very closely and the stress oscillations become almost negligible. In fact, the agreement with the exact strain profile is better than it appears in the figure, where the values at integration points are connected by straight segments. The actual approximation has a sinusoidal shape and its value at the center of the localization zone is above the chord shown in Fig. 8a, much closer to the peak of the exact profile. Also, note that each added strain mode corresponds to one additional degree of freedom. So the complete model with three added strain modes in the middle element has as few as 5 degrees of freedom, while the finest mesh consisting of standard linear elements needed 95 degrees of freedom (or at least 33 degrees of freedom, if only the localization zone is refined).

3   Concluding remarks

This paper has briefly sketched the idea of finite elements with embedded localization zones and illustrated it with a simple one-dimensional example. Of course, much additional work is needed to make the proposed concept useful in practical applications.

The most difficult step is no doubt the extension to multiple dimensions. In one dimension, the weak compatibility condition (vanishing integral of the enhanced strain mode over the element) is equivalent to the strong compatibility condition (continuity of displacements). In multiple dimensions, this is no longer the case, and the exact way in which weak compatibility is enforced has an important influence on the kinematic properties of the enriched model. For instance, the elements with embedded discontinuities that are derived in a variationally consistent manner from the stress continuity condition are inherently incapable of reproducing a clean separation with no stress transfer across a widely open crack; see [9] for a discussion.

Other issues to be addressed in the future include the optimal choice of enrichment functions and efficient integration schemes. Finally, it is desirable to develop the technique into a truly adaptive approach, in which the parameters of the enhanced modes (e.g., ξ_L and ξ_R in the one-dimensional example) are continuously adapted to the evolving shape of the localization profile. These topics are the subject of an ongoing research.

Acknowledgment

Financial support of the Swiss Commission for Technology and Innovation under grant CTI 4424.1 is gratefully acknowledged.

References

1. M. Jirásek: Comparison of Nonlocal Models for Damage and Fracture. LSC Report 98/02, Swiss Federal Institute of Technology, Lausanne (1998)
2. E. N. Dvorkin, A. M. Cuitiño and G. Gioia: Comp. Meth. Appl. Mech. Eng. **90**, 829 (1990)
3. M. Klisinski, K. Runesson and S. Sture: J. Eng. Mech. **117**, 575 (1991)
4. J. C. Simo and J. Oliver: 'A new approach to the analysis and simulation of strain softening in solids'.
   In: *Fracture and Damage in Quasibrittle Structures*, ed. by Z. P. Bažant et al. (E.&F. N. Spon, London, 1994) pp. 25-39
5. T. Belytschko, J. Fish and B. E. Engelmann: Comp. Meth. Appl. Mech. Eng. **70**, 59 (1988)
6. L. J. Sluys: 'Discontinuous modeling of shear banding'. In: *Computational Plasticity: Fundamentals and Applications*, ed. by D. R. J. Owen, E. Oñate and E. Hinton (Int. Center for Numerical Methods in Engineering, Barcelona, 1997) pp. 735-744
7. J. Fish and T. Belytschko: Comp. Meth. Appl. Mech. Eng. **78**, 181 (1990)
8. T. Belytschko, J. Fish and A. Bayliss: Comp. Meth. Appl. Mech. Eng. **81**, 71 (1990)
9. M. Jirásek: 'Embedded crack models for concrete fracture'. In: *Computational Modelling of Concrete Structures*, ed. R. de Borst et al. (Balkema, Rotterdam, 1998) pp. 291-300.
10. J. Oliver, M. Cervera and O. Manzoli: 'On the Use of J-2 Plasticity Models for the Simulation of 2D Strong Discontinuities in Solids'. In: *Computational Plasticity: Fundamentals and Applications*, ed. by D. R. J. Owen, E. Oñate and E. Hinton (Int. Center for Numerical Methods in Engineering, Barcelona, 1997) pp. 38-55
11. J. Pamin: Gradient-Dependent Plasticity in Numerical Simulation of Localization Phenomena. PhD Thesis, Delft University of Technology, Delft, The Netherlands (1994)
12. G. Pijaudier-Cabot and Z. P. Bažant: J. Eng. Mech. **113**, 1512 (1987)
13. J. C. Simo and M. S. Rifai: Int. J. Num. Meth. Eng. **29**, 1595 (1990)

Microplane modelling and particle modelling of cohesive-frictional materials

E. Kuhl, G. A. D'Addetta, M. Leukart, E. Ramm

Institute of Structural Mechanics, University of Stuttgart
Pfaffenwaldring 7, 70569 Stuttgart, Germany

Abstract. This paper aims at comparing the microplane model as a particular representative of the class of continuous material models with a discrete particle model which is of discontinuous nature. Thereby, the constitutive equations of both approaches will be based on VOIGT's hypothesis defining the strain state on the individual microplanes as well as the relative particle displacements. Through an appropriate constitutive assumption, the microplane stresses and the contact forces can be determined. In both cases, the equivalence of microscopic and macroscopic virtual work yields the overall stress strain relation. An elastic and an elasto-plastic material characterization of the microplane model and the particle model are derived and similarities of both approaches are illustrated.

1 Motivation

The constitutive description of the mechanical behavior of granular systems is of great interest to the fields of geotechnics and various other related applications. By taking into account the discrete nature of the microstructure of a granular assembly, numerous different discrete models have been developed, compare for example CUNDALL & STRACK [10] or BATHURST & ROTHENBURG [1]. Most of them are based on a finite number of discrete, semi-rigid spherical or polygon-shaped particles interacting by means of contact forces. The specification of an appropriate contact law is probably the most significant part of the discrete model. For example, the contact behavior can be formulated in a linear or non-linear elastic fashion according to the classical HERTZ model [13] or include frictional effects along the line of COULOMB's friction law. In order to compare the results of the discrete element simulation with macroscopic measurements, different averaging techniques can be applied to derive homogenized quantities characterizing the overall behavior of the assembly. Thus, the definition of the macroscopic stress tensor has been studied intensively in the beginning of the 80s and can now be considered as well-established, see ROTHENBURG & SELVADURAI [22] or CHRISTOFFERSEN, MEHRABADI & NEMAT-NASSER [9]. During the last decade, various studies have been dedicated to the derivation of explicit expressions for the overall strain tensor, compare KRUYT & ROTHENBURG [15], and, accordingly, for the macroscopic constitutive moduli WALTON [24], CAMBOU, DUBUJET, EMERIAULT & SIDOROFF [5], EMERIAULT & CAMBOU [12], CHANG [8] and LIAO, CHANG, YOUNG & CHANG [19]. While these homogenization strategies are based on the assumption of an elastic material

behavior, dislocation and plastic flow which have been found experimentally by
DRESCHER & DE JOSSELIN DE JONG [11] are incorporated in more advanced
studies, see CHANG [7] for example. Even damage laws can be applied to characterize the degradation of the contact stiffness, see SUIKER, ASKES & SLUYS [23].
While discrete models take into account the individual behavior of each single
particle, continuum-based approaches can only describe the material behavior
in an average sense. Although, in most cases, the choice of the specific constitutive formulation is motivated by microstructural considerations, the material
response is characterized exclusively in terms of stresses or strains and a set
of internal variables, which represent microstructural effects in a phenomenological fashion. The microplane plasticity model is a classical representative of
this class of continuum–based constitutive models. It is based on the early ideas
of MOHR [21], who suggested to characterize the response of a material point
by describing its behavior in various representative directions in space. Similar
to the particle models, the choice of the constitutive assumption relating the
corresponding stress and strain vector of each direction can be considered as
the most important feature of the model. The overall response of the material
point is obtained by integrating the resulting stress vectors over the entire solid
angle. The first application of the microplane model to rock–like geomaterials
was presented at the end of the 70s by ZIENKIEWICZ & PANDE [25]. A decade
later, a microplane model for cementitious materials was proposed by BAŽANT
& GAMBAROVA [2] and BAŽANT & PRAT [4], which serves as the basis for most
existing microplane formulations, compare CAROL, BAŽANT & PRAT [6], KUHL
& RAMM [16] or KUHL, D' ADDETTA, HERRMANN & RAMM [17].
Although derived from two completely different fields, both models show significant similarities from a theoretical point of view. This contribution aims at
highlighting the equivalences of the two different material formulations. Therefore, chapter 2 summarizes the basic ideas of the microplane model followed by
a basic introduction to discrete particle modelling in chapter 3. Particular interest is dedicated to the fact, that the initially continuous microplane model
has to be *discretized* for computational reasons while the initially discrete particle model must be *'continuumized'* in order to relate its material parameters to
macroscopic quantities. Finally, in chapter 4, the two different approaches are
compared and advantages and disadvantages of both formulations are discussed.
The following derivations are based on the assumption of small displacements
and small strains, restricting the models to linear kinematics.

2   Continuum-based microplane models

The microplane model belongs to the class of continuous material models. These
models are based on the assumption that the displacement field is continuous
throughout the whole domain of consideration which is associated with one integration point in a finite element simulation. With a given displacement field
u, the corresponding strain field ϵ can be determined as the symmetric part of

the displacement gradient.

$$\epsilon = \nabla^{sym} u \tag{1}$$

Although alternative homogenization strategies could be thought of, we will restrict ourselves to a kinematic constraint, assuming that the strains are distributed equally in space. Consequently, the strain vector t_ϵ associated with the direction characterized through the normal n is given in the following form.

$$t_\epsilon(n) = \epsilon \cdot n = \epsilon_N\, n + \epsilon_T \tag{2}$$

This strain vector can be additively decomposed into a normal and a tangential contribution ϵ_N and ϵ_T as illustrated in figure 1, left, whereby the normal strain component and the tangential strain vector are given as follows.

$$
\begin{aligned}
\epsilon_N(n) &= \quad n \cdot \epsilon \cdot n = N : \epsilon \\
\epsilon_T(n) &= \epsilon \cdot n - \epsilon_N n = T : \epsilon
\end{aligned}
\tag{3}
$$

For sake of transparency, we have introduced the second and third order projection tensors N and T as functions of the characteristic direction n and the symmetric fourth order unit tensor $\mathcal{I}$.

$$
\begin{aligned}
N(n) &= n \otimes n \\
T(n) &= n \cdot \mathcal{I} - n \otimes n \otimes n
\end{aligned}
\tag{4}
$$

By making use of the following analytical integration formulae of the zeroth, second and fourth order fabric tensor as given by LUBARDA & KRAJCINOVIC [20] or KANATANI [14],

$$
\begin{aligned}
\tfrac{3}{4\pi} \int_\Omega & & d\Omega &= \quad 3 \\
\tfrac{3}{4\pi} \int_\Omega & n \otimes n & d\Omega &= \quad 1 \\
\tfrac{3}{4\pi} \int_\Omega & n \otimes n \otimes n \otimes n \; d\Omega &= \tfrac{3}{5}\mathcal{I}^{vol} + \tfrac{2}{5}\mathcal{I}
\end{aligned}
\tag{5}
$$

the fourth order products of the projection tensors N and T can by integrated analytically over the solid angle Ω.

$$
\begin{aligned}
\tfrac{3}{4\pi} \int_\Omega N \otimes N \; d\Omega &= \quad \tfrac{3}{5}\mathcal{I}^{vol} + \tfrac{2}{5}\mathcal{I} \\
\tfrac{3}{4\pi} \int_\Omega T^T \cdot T \; d\Omega &= -\tfrac{3}{5}\mathcal{I}^{vol} + \tfrac{3}{5}\mathcal{I}
\end{aligned}
\tag{6}
$$

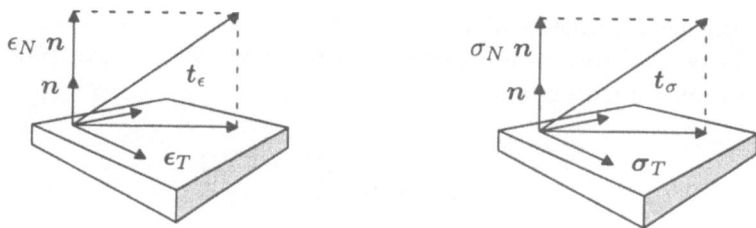

Fig. 1. Normal and tangential strains and stresses on microplane

Herein, $\mathbf{1}$ denotes the second order unit tensor. Its dyadic product defines the volumetric fourth order unit tensor $\mathcal{I}^{vol} = 1/3\ \mathbf{1} \otimes \mathbf{1}$. Note, that by the transpose of the third order tensor $\mathbf{T}$ denoted by $\mathbf{T}^T$, we will understand the expression $\mathbf{T}^T = \mathcal{I} \cdot \mathbf{n} - \mathbf{n} \otimes \mathbf{n} \otimes \mathbf{n}$.

2.1   Microplane elasticity

First, we will assume a linear elastic material behavior. Consequently, the microplane stresses σ_N and σ_T can be expressed exclusively in terms of the normal strain ϵ_N, the tangential strain vector ϵ_T and the normal and tangential elasticity modulus of the corresponding plane $\mathcal{E}_N$ and $\mathcal{E}_T$.

$$\begin{aligned} \sigma_N\,(\mathbf{n}) &= \mathcal{E}_N\ \epsilon_N \\ \boldsymbol{\sigma}_T\,(\mathbf{n}) &= \mathcal{E}_T\ \boldsymbol{\epsilon}_T \end{aligned} \tag{7}$$

The resulting traction vector $\mathbf{t}_\sigma$ of the plane can thus be expressed in terms of its normal and tangential components as indicated in figure 1, right.

$$\mathbf{t}_\sigma(\mathbf{n}) = \sigma_N \mathbf{n} + \boldsymbol{\sigma}_T \tag{8}$$

By making use of the equivalence of the macroscopic and the microplane–based virtual work $\delta W^{mac} = \delta W^{mic}$ with

$$\begin{aligned} \delta W^{mac} &= \boldsymbol{\sigma} : \delta\boldsymbol{\epsilon} \\ \delta W^{mic} &= \frac{3}{4\pi} \int_\Omega \mathbf{t}_\sigma \cdot \delta\mathbf{t}_\epsilon\ d\Omega \end{aligned} \tag{9}$$

we can determine the overall stress tensor. Together with equation (2) and the kinematic constraint condition of equation (3), such that $\delta\mathbf{t}_\epsilon = \delta\boldsymbol{\epsilon} \cdot \mathbf{n}$, equation (9) yields the definition of the macroscopic stress tensor as a function of the stress vector and the plane's normal.

$$\boldsymbol{\sigma} = \frac{3}{4\pi} \int_\Omega [\mathbf{t}_\sigma \otimes \mathbf{n}]^{sym}\ d\Omega \tag{10}$$

Note, that $(\bullet)^{sym} = \left[(\bullet) + (\bullet)^T\right]/2$ extracts only the symmetric part of the corresponding quantity. By making use of the definition of the stress vector (8) and the algebraic equivalence of $[\boldsymbol{\sigma}_T \otimes \mathbf{n}]^{sym} = \mathbf{T}^T \cdot \boldsymbol{\sigma}_T$, we obtain the definition of the macroscopic stress tensor in terms of the microscopic stress components and the two projection tensors.

$$\boldsymbol{\sigma} = \frac{3}{4\pi} \int_\Omega \mathbf{N}\sigma_N + \mathbf{T}^T \cdot \boldsymbol{\sigma}_T\ d\Omega \tag{11}$$

The overall macroscopic constitutive relation can be derived by inserting the definition of the microplane stresses (7) into (11) yielding the definition of the *continuous* fourth order tensor of constitutive moduli relating the macroscopic stresses and strains as $\sigma = \mathcal{E} : \epsilon$.

$$\mathcal{E} = \frac{3}{4\pi} \int_\Omega \mathcal{E}_N \boldsymbol{N} \otimes \boldsymbol{N} + \mathcal{E}_T \boldsymbol{T}^T \cdot \boldsymbol{T} \, d\Omega \tag{12}$$

If we assume an isotropic material behavior, the elastic moduli of the microplane are independent of the orientation and can thus be written in front of the integral. By applying the integration formulae (6) to evaluate the integrals of the projection tensors, we can compare the result of the analytical integration with the generalized form of HOOKE's law,

$$\mathcal{E} = \left[\frac{3}{5}\mathcal{E}_N - \frac{3}{5}\mathcal{E}_T\right] \boldsymbol{\mathcal{I}}^{vol} + \left[\frac{2}{5}\mathcal{E}_N + \frac{3}{5}\mathcal{E}_T\right] \boldsymbol{\mathcal{I}}$$

$$\mathcal{E} = \frac{3\,E\,\nu}{[1+\nu][1-2\nu]} \boldsymbol{\mathcal{I}}^{vol} + \frac{E}{1+\nu} \quad \boldsymbol{\mathcal{I}} \tag{13}$$

yielding the following relation for YOUNG's modulus and POISSON's ratio in terms of the elastic microplane moduli $\mathcal{E}_N$ and $\mathcal{E}_T$.

$$E = \frac{\mathcal{E}_N\,[2\mathcal{E}_N + 3\mathcal{E}_T]}{4\mathcal{E}_N + \mathcal{E}_T} \qquad \text{and} \qquad \nu = \frac{\mathcal{E}_N - \mathcal{E}_T}{4\mathcal{E}_N + \mathcal{E}_T} \tag{14}$$

Note, that this particular microplane model is restricted to values of POISSON's ratio lying in the range of $-1 \leq \nu \leq 1/4$. In order to cover the whole range of POISSON's ratio, an additional split of the normal microplane component into normal volumetric and normal deviatoric contributions has been proposed by BAŽANT & GAMBAROVA [2]. The values of the elastic microplane moduli for varying POISSON's ratios are depicted in figure 5, whereby the vertical axis is scaled by YOUNG's modulus $E^* = E$. For this linear elastic isotropic material model, the integration over the solid angle of equation (12) can be carried out analytically. For an inelastic material behavior, however, this analytical integration becomes nearly impossible. Consequently, the integral expression is commonly evaluated numerically by replacing the integral by a discrete sum evaluated at a certain number of integration points, $c = 1, .., n_{mp}$, and weighted by the weighting coefficients w^c.

$$\int_\Omega F(n)d\Omega \approx \sum_{c=1}^{n_{mp}} F(n^c)w^c \tag{15}$$

Thus, the definition of the macroscopic stress tensor of equation (10) can be approximated by the following sum,

$$\sigma \approx \sum_{c=1}^{n_{mp}} [t_\sigma^c \otimes n^c]^{sym} \, w^c \tag{16}$$

whereas the *discrete* fourth order constitutive tensor defined in equation (12) can be approximated as follows.

$$\mathcal{E} \approx \sum_{c=1}^{n_{mp}} \left[\mathcal{E}_N^c \, \boldsymbol{N}^c \otimes \boldsymbol{N}^c + \mathcal{E}_T^c \, \boldsymbol{T}^{cT} \cdot \boldsymbol{T}^c \right] w^c \tag{17}$$

Obviously, the number of integration points determines the order of accuracy of the approximation. By comparing some of the various different integration formulae for the integration over a solid angle, BAŽANT & OH [3] have found, that the integration with 42 integration points as depicted in figure 2 yields a sufficiently accurate approximation at an acceptable level of effort.

2.2   Microplane elasto-plasticity

The microplane–based plasticity model is based on the additive decomposition of the microplane strains into elastic and plastic parts.

$$\begin{aligned} \epsilon_N(\boldsymbol{n}) &= \epsilon_N^{el} + \epsilon_N^{pl} \\ \epsilon_T(\boldsymbol{n}) &= \epsilon_T^{el} + \epsilon_T^{pl} \end{aligned} \tag{18}$$

The microplane stresses are assumed to depend only on the elastic parts of the strain components.

$$\begin{aligned} \sigma_N(\boldsymbol{n}) &= \mathcal{E}_N \, \epsilon_N^{el} \\ \boldsymbol{\sigma}_T(\boldsymbol{n}) &= \mathcal{E}_T \, \boldsymbol{\epsilon}_T^{el} \end{aligned} \tag{19}$$

Moreover, a yield function Φ of DRUCKER–PRAGER type is introduced. It can be expressed as the difference of an equivalent stress $\|\boldsymbol{\sigma}_T\| - \tan\varphi\sigma_N$ and the yield stress Y.

$$\Phi(\boldsymbol{n}) = \|\boldsymbol{\sigma}_T\| - \tan\varphi\,\sigma_N - Y \leq 0 \tag{20}$$

Herein, φ denotes the angle of internal friction whereas the normals to the yield function will be denoted by ν_N and $\boldsymbol{\nu}_T$.

$$\begin{aligned} \nu_N(\boldsymbol{n}) &:= \partial\Phi \,/\, \partial\sigma_N \\ \boldsymbol{\nu}_T(\boldsymbol{n}) &:= \partial\Phi \,/\, \partial\boldsymbol{\sigma}_T \end{aligned} \tag{21}$$

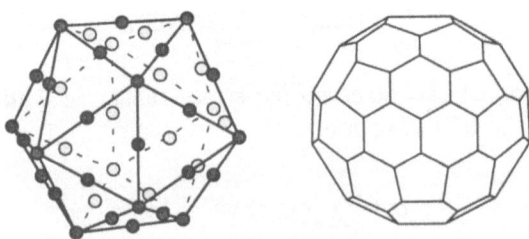

Fig. 2. Discrete model – polyhedron with 42 microplanes

The evolution of the plastic strains is determined by the plastic multiplier γ and the normal and tangential flow directions μ_N and μ_T which can be understood as normals to the plastic potential Φ^*.

$$
\begin{aligned}
\dot{\epsilon}_N^{pl}(\boldsymbol{n}) &= \dot{\gamma}\,\mu_N \qquad \text{with} \qquad \mu_N(\boldsymbol{n}) := \partial\Phi^* \,/\, \partial\sigma_N \\
\dot{\epsilon}_T^{pl}(\boldsymbol{n}) &= \dot{\gamma}\,\mu_T \qquad \text{with} \qquad \mu_T(\boldsymbol{n}) := \partial\Phi^* \,/\, \partial\boldsymbol{\sigma}_T
\end{aligned}
\tag{22}
$$

The loading–unloading process is governed by the KUHN–TUCKER conditions and the consistency condition.

$$
\Phi \leq 0 \qquad \dot{\gamma} \geq 0 \qquad \Phi\dot{\gamma} = 0 \qquad \dot{\Phi}\dot{\gamma} = 0 \tag{23}
$$

The evolution of the plastic multiplier can be directly determined from the evaluation of the consistency condition (23.4).

$$
\dot{\gamma}(\boldsymbol{n}) = \frac{1}{h}\left[\nu_N \mathcal{E}_N \boldsymbol{N} + \boldsymbol{\nu}_T \cdot \mathcal{E}_T \boldsymbol{T}\right] : \dot{\boldsymbol{\epsilon}} \quad \text{with} \quad h(\boldsymbol{n}) := \nu_N \mathcal{E}_N \mu_N + \boldsymbol{\nu}_T \cdot \mathcal{E}_T \boldsymbol{\mu}_T \tag{24}
$$

According to equation (10) the macroscopic stress tensor is given as the symmetric part of the dyadic product of the traction vector with the corresponding normal,

$$
\boldsymbol{\sigma} = \frac{3}{4\pi}\int_\Omega [\boldsymbol{t}_\sigma \otimes \boldsymbol{n}]^{sym}\,d\Omega \tag{25}
$$

whereby the components of the traction vector $\boldsymbol{t}_\sigma = \sigma_N \boldsymbol{n} + \boldsymbol{\sigma}_T$ are defined in equations (19). Moreover, the continuous elasto–plastic tangent operator $\mathcal{E}_{tan}^{ep}$ relating the macroscopic stress and strain rates according to $\dot{\boldsymbol{\sigma}} = \mathcal{E}_{tan}^{ep} : \dot{\boldsymbol{\epsilon}}$ can be expressed as follows.

$$
\mathcal{E}_{tan}^{ep} = \mathcal{E}^{el} - \frac{3}{4\pi}\int_\Omega \frac{1}{h}\left[\boldsymbol{N}\mathcal{E}_N \nu_N + \boldsymbol{T}^T \cdot \mathcal{E}_T \boldsymbol{\nu}_T\right] \otimes \left[\mu_N \mathcal{E}_N \boldsymbol{N} + \boldsymbol{\mu}_T \cdot \mathcal{E}_T \boldsymbol{T}\right] d\Omega \tag{26}
$$

2.3   Example

The features of the elasto–plastic microplane model are demonstrated by means of a simulation of the two classical model problems of uniaxial compression and simple shear. The loading is applied in the horizontal direction in both cases. A plane strain situation is assumed. In accordance with most existing particle models from the literature, the yield stress has been set to $Y = 0$ N/mm^2, the friction angle has been chosen to $\varphi = 30^0$ and a non-associated flow rule with $\mu_N = 0$ and $\boldsymbol{\mu}_T = \boldsymbol{\nu}_T$ has been assumed. Figure 3 depicts the distribution of the plastic multiplier γ characterizing the amount of plastic sliding associated with the normal of the corresponding plane. For each problem, three different ratios of the normal and the tangential elastic contact stiffness have been studied, manifesting themselves in three different values of the macroscopic POISSON's ratio with YOUNG's modulus constant at $E = 30000$ N/mm^2.

As expected, for the case of uniaxial compression depicted in figure 3, top, maximum sliding takes place normal to the loading direction. Obviously, the

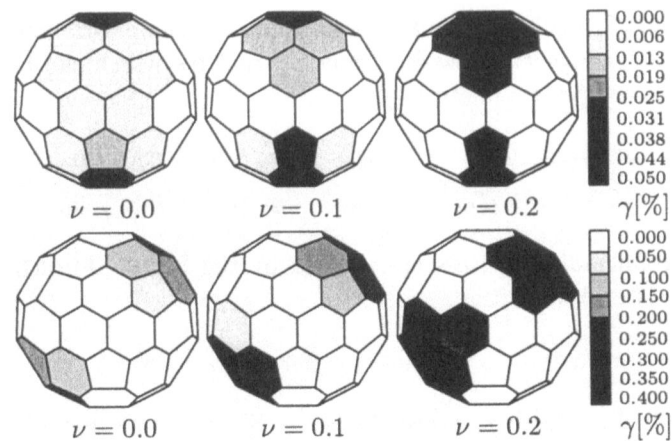

Fig. 3. Uniaxial compression & simple shear – distribution of plastic multiplier

amount of sliding increases with increasing POISSON ratios. The response under simple shear is characterized by the REYNOLD's effect introduced through the lateral confinement due to the boundary conditions. Consequently, plastic sliding tends to concentrate under an angle of 45^0 towards the loading axis, figure 3, bottom. Again, the amount of sliding increases with increasing POISSON ratios. This example provides inside into the anisotropic nature of plastic slip in granular media.

3   Discrete particle models

According to VOIGT's hypothesis, we will assume that the strain ϵ is distributed uniformly in the granular assembly. Consequently, every particle within the packing displaces in accordance with the uniform strain as the mean displacement field. Thus, any vector l connecting two arbitrary points of the assembly is strained by the amount Δl given as the scalar product of the normalized displacement ϵ, which will be interpreted as the strain tensor in the following, and the vector l itself.

$$\Delta l = \epsilon \cdot l \tag{27}$$

In particular, this relation holds for the relative displacement Δl^c of the contact vector l^c which connects the centers of mass of the two corresponding particles in contact. By expressing the contact vector in terms of its length $||l^c|| = \sqrt{l^c \cdot l^c}$ and the unit normal to the contact plane $n^c = l^c/||l^c||$, we can rewrite equation (27) in the following form.

$$\Delta l^c(n^c) = ||l^c||\; \epsilon \cdot n^c = \Delta l_N^c n + \Delta l_T^c \tag{28}$$

Figure 4, left, indicates that the contact displacement Δl^c can be additively decomposed into the displacements normal and tangential to the contact plane, denoted by Δl_N^c and Δl_T^c, respectively. Both components can be directly related to the overall strain tensor through the projection tensors N and T defined in equation (4).

$$\begin{aligned} \Delta l_N^c(n^c) &= \qquad ||l^c|| \, n^c \cdot \epsilon \cdot n^c = ||l^c|| \, N^c : \epsilon \\ \Delta l_T^c(n^c) &= ||l^c|| \, \epsilon \cdot n^c - \Delta l_N^c n^c = ||l^c|| \, T^c : \epsilon \end{aligned} \qquad (29)$$

3.1   Elastic particles

The normal and the tangential contact forces f_N^c and f_T^c are related to the normal and the tangential contact displacements through the constitutive assumption at the contact. For sake of simplicity, we will assume a simplified version of the elastic contact model along the lines of HERTZ [13] which, in the linear case, takes the following form.

$$\begin{aligned} f_N^c(n^c) &= k_N \, \Delta l_N^c \\ f_T^c(n^c) &= k_T \, \Delta l_T^c \end{aligned} \qquad (30)$$

Herein, k_N and k_T denote the normal and the tangential contact stiffness, respectively. Moreover, we have neglected the rotations of the particles, which is a reasonable assumption for dense packings, compare BATHURST & ROTHENBURG [1]. As indicated in figure 4, right, the normal and the tangential contact forces represent the components of the contact force vector f^c.

$$f^c(n^c) = f_N^c n^c + f_T^c \qquad (31)$$

In analogy to the microplane model, the contact forces can be related to the macroscopic stress tensor through the principle of virtual work as demonstrated by CHRISTOFFERSEN, MEHRABADI & NEMAT – NASSER [9] and more recently by CHANG [8]. Again, the equivalence of the overall macroscopic virtual work and the virtual work of the granular assembly $\delta W^{mac} = \delta W^{mic}$ with

$$\begin{aligned} \delta W^{mac} &= \qquad \sigma : \delta\epsilon \\ \delta W^{mic} &= \frac{1}{V} \sum_{c \in V} f^c \cdot \delta\Delta l^c \end{aligned} \qquad (32)$$

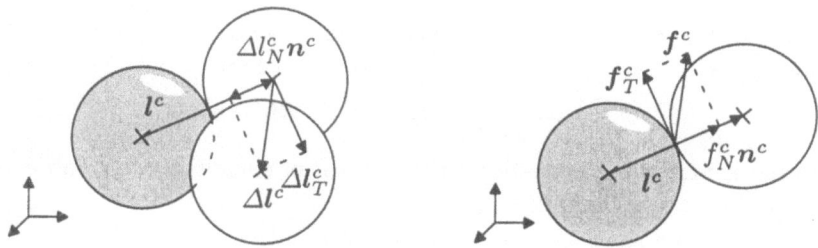

Fig. 4. Normal and tangential components of contact vector and contact force

serves as starting point for the derivation of the macroscopic stress tensor. Equation (32) together with the kinematic constraint condition $\delta \Delta l^c = \delta \epsilon \cdot l^c$ as postulated in (27) yield the following expression for the *discrete* macroscopic stress tensor.

$$\sigma = \frac{1}{V} \sum_{c \in V} [f^c \otimes l^c]^{sym} \tag{33}$$

Note, that this derivation based on the equivalence of virtual work differs from the derivation presented by LÄTZEL, LUDING & HERRMANN [18] which, in general, leads to a non-symmetric stress tensor. By combining the above definition with the definition of the contact force vector (31), and making use of the identities $l^c = \|l^c\|\, n^c$ and $[f_T^c \otimes n^c]^{sym} = T^{cT} \cdot f_T^c$ we can rewrite the definition of the stress tensor in the following form.

$$\sigma = \frac{1}{V} \sum_{c \in V} \|l^c\| \left[N^c f_N^c + T^{cT} \cdot f_T^c \right] \tag{34}$$

The combination of the particle contact law (30) and the kinematic constraint (29) with equation (34) yields the following relation between the overall stress and strain tensor $\sigma = \mathcal{E} : \epsilon$ in terms of the *discrete* fourth order constitutive tensor $\mathcal{E}$ of the granular material.

$$\mathcal{E} = \frac{1}{V} \sum_{c \in V} \|l^c\|^2 \left[k_N N^c \otimes N^c + k_T T^{cT} \cdot T^c \right] \tag{35}$$

For certain specific classes of particle models, an analytical solution can be derived. We will restrict ourselves to granular materials with an isotropic packing structure, for which the contact points are distributed with uniform probability over all possible directions in space. Moreover, the assembly is assumed to be composed of equally sized particles, for which $l^c = 2r\, n^c$ with $r = \text{const.}$ Consequently, the above definitions for the *discrete* stress tensor and the *discrete* constitutive moduli can be transformed into an integral form, if a suitably large representative volume V with a large number of particles is considered. According to LIAO, CHANG, YOUNG & CHANG [19], the summation of an arbitrary function F over all contacts $c = 1, .., N$ can be expressed through its integral over the solid angle Ω weighted by the number of contacts N divided by 4π.

$$\sum_{c=1}^{N} F(n^c) = \frac{N}{4\pi} \int_{\Omega} F(n) d\Omega \tag{36}$$

Consequently, the *continuous* form of the stress definition of equation (33) can be expressed as

$$\sigma = \frac{Nr}{2V\pi} \int_{\Omega} [f \otimes n]^{sym} d\Omega \tag{37}$$

whereas the *continuous* counterpart of the tensor of constitutive moduli defined in equation (35) takes the following form.

$$\mathcal{E} = \frac{Nr^2}{V\pi} \int_{\Omega} k_N N \otimes N + k_T T^T \cdot T \, d\Omega \tag{38}$$

If we postulate a linear elastic contact law which is identical for each contact, the contact stiffnesses k_N and k_T are independent of the direction n and can therefore be written in front of the integral. It remains to apply the integration formulae of the fourth order products of the projection tensors of equation (6). By comparing the result of the integration with the generalized form of HOOKE's law for a linear elastic material,

$$\mathcal{E} = \frac{4Nr^2}{5V}[k_N - k_T]\,\boldsymbol{\mathcal{I}}^{vol} + \frac{4Nr^2}{15V}[2k_N + 3k_T]\,\boldsymbol{\mathcal{I}}$$

$$\mathcal{E} = \frac{3E\nu}{[1+\nu][1-2\nu]}\,\boldsymbol{\mathcal{I}}^{vol} + \frac{E}{1+\nu}\,\boldsymbol{\mathcal{I}}$$

(39)

we can easily express YOUNG's modulus and POISSON's ratio as functions of the contact stiffnesses k_N and k_T.

$$E = \frac{4Nr^2}{3V}\frac{k_N\,[2k_N + 3k_T]}{4k_N + k_T} \qquad \text{and} \qquad \nu = \frac{k_N - k_T}{4k_N + k_T} \qquad (40)$$

Similar to the microplane model, this particle model is restricted to a material with POISSON's ratio of $-1 \leq \nu \leq 1/4$, as can be concluded from equation (40). A larger value for POISSON's ratio can only be obtained, if the tangential stiffness becomes negative, which is a non-reasonable assumption from a physical point of view. Figure 5 illustrates this drawback of the model. It shows the normal and tangential contact stiffnesses for different POISSON ratios. Their values are scaled by the modified YOUNG's modulus $E^* = [3V/4Nr^2]\,E$.

3.2   Elasto-plastic particles

The existence of irreversible contact displacements was verified experimentally by DRESCHER & DE JOSSELIN DE JONG [11]. Frictional contact laws for granular assemblies are based on the idea that the normal and tangential components of the contact displacement can be decomposed into reversible elastic parts denoted

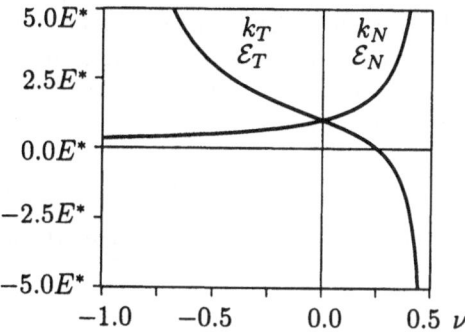

Fig. 5. Contact stiffnesses as a function of POISSON's ratio

by Δl_N^{el} and Δl_T^{el} and irreversible plastic contributions Δl_N^{pl} and Δl_T^{pl}. For sake of simplicity the superscript c has been omitted in the sequel.

$$\Delta l_N(n) = \Delta l_N^{el} + \Delta l_N^{pl}$$
$$\Delta l_T(n) = \Delta l_T^{el} + \Delta l_T^{pl} \tag{41}$$

The normal and tangential components of the contact force can be expressed exclusively in terms of the elastic contact displacements.

$$f_N(n) = k_N \Delta l_N^{el} = k_N \Delta l_N - f_N^{pl} \qquad f_N^{pl} := k_N \Delta l_N^{pl}$$
$$f_T(n) = k_T \Delta l_T^{el} = k_T \Delta l_T - f_T^{pl} \qquad f_T^{pl} := k_T \Delta l_T^{pl} \tag{42}$$

The frictional sliding of the particles is characterized through COULOMB's friction law, motivating the introduction of a yield function Φ of the following form.

$$\Phi(n) = \| f_T \| - \tan\varphi\, f_N - Y \le 0 \tag{43}$$

Note, that a coupling of the normal and the tangential components of the contact force is introduced through the friction angle φ while Y accounts for the influence of cohesion. The normals to the yield function can thus be expressed as follows.

$$\nu_N(n) := \partial\Phi \,/\, \partial f_N$$
$$\nu_T(n) := \partial\Phi \,/\, \partial f_T \tag{44}$$

The evolution of the plastic contact displacements can be expressed in terms of the plastic multiplier γ and the directions of plastic displacement μ_N and μ_T as normals to the plastic potential Φ^*.

$$\Delta l_N^{pl}(n) = \dot\gamma\, \mu_N \qquad \text{with} \qquad \mu_N(n) := \partial\Phi^*/\partial f_N$$
$$\Delta l_T^{pl}(n) = \dot\gamma\, \mu_T \qquad \text{with} \qquad \mu_T(n) := \partial\Phi^*/\partial f_T \tag{45}$$

Only in case of an associated friction model, these directions correspond to the normals to the yield function such that $\mu_N = \nu_N$ and $\mu_T = \nu_T$. In general, however, the directions of irreversible displacement can be chosen independently yielding a non-associated friction law. In most existing formulations in the literature, the following choice is applied, $\mu_N = 0$ and $\mu_T = \nu_T$. The sliding process is further characterized through KUHN-TUCKER conditions and the consistency condition.

$$\Phi \le 0 \qquad \dot\gamma \ge 0 \qquad \Phi\dot\gamma = 0 \qquad \dot\Phi\dot\gamma = 0 \tag{46}$$

The evaluation of the consistency condition (46.4), yields the evolution equation for the plastic multiplier γ.

$$\dot\gamma(n) = \frac{\|l\|}{h}[\nu_N k_N N + \nu_T \cdot k_T T] : \dot\varepsilon \quad \text{with} \quad h(n) := \nu_N k_N \mu_N + \nu_T \cdot k_T \mu_T \tag{47}$$

Consequently, for the case of frictional sliding with the flow directions $\mu_N = 0$ and $\mu_T = \nu_T$ the rate of the normal and tangential plastic contact forces of equation (42) take the following form.

$$
\begin{aligned}
\dot{f}_N^{pl}(n) &= 0 \\
\dot{f}_T^{pl}(n) &= \left[\| \dot{f}_T \| - \tan\varphi \dot{f}_N\right] \mu_T
\end{aligned}
\tag{48}
$$

Although derived in a different way, these evolution equations correspond to the ones given in the literature, compare for example CHANG [7]. Finally, we can specify the macroscopic stress tensor, which can again be derived by applying the principle of virtual work. It can be expressed in accordance with equation (33),

$$
\sigma = \frac{1}{V} \sum_{c \in V} \|l^c\| \left[f^c \otimes n^c \right]^{sym}
\tag{49}
$$

whereby $f^c = f_N^c n + f_T^c$ denotes the contact vector whose components are specified in equation (42). Moreover, we obtain the definition of the overall tangent operator $\mathcal{E}_{tan}^{ep}$ relating the macroscopic stress rates to the macroscopic strain rates as $\dot{\sigma} = \mathcal{E}_{tan}^{ep} : \dot{\varepsilon}$.

$$
\mathcal{E}_{tan}^{ep} = \mathcal{E}^{el} - \frac{1}{V} \sum_{c \in V} \frac{\|l^c\|^2}{h} \left[N^c k_N \nu_N^c + T^{cT} \cdot k_T \nu_T^c \right] \otimes \left[\mu_N^c k_N N^c + \mu_T^c \cdot k_T T^c \right]
\tag{50}
$$

4   Comparison

Although derived from two completely different fields, the constitutive equations of the discrete particle model and the continuum–based microplane model show various similarities. In both cases, the kinematic relation between the microscopic and the macroscopic quantities was assumed in accordance with VOIGT's hypothesis. While the particle model is formulated in terms of relative displacements and contact forces, the microplane model is based on strain and stress vectors. Consequently, the material parameters of the particle model can be interpreted as normal and tangential contact stiffnesses, whereas the related microplane parameters can be understood as normal and tangential elastic moduli. In both cases, the overall response is derived through the principle of virtual work. A comparison of the most important equations of both models is summarized in table 1.

Even for frictional materials, both models show a similar behavior. The particle model is usually associated with COULOMB's friction law, whereas the yield function of the microplane model can be introduced in a more general DRUCKER–PRAGER based fashion. If both material models are written in a similar notation, the existing particle models for frictional sliding can be interpreted as a special case of a non–associated plasticity formulation. Based on the introduction of

Table 1. Comparison of linear elastic models

microplane model	particle model
$t_\epsilon = \epsilon_N n + \epsilon_T$	$\Delta l^c = \Delta l_N^c n + \Delta l_T^c$
$\epsilon_N = N : \epsilon$	$\Delta l_N^c = \|l^c\| N : \epsilon$
$\epsilon_T = T : \epsilon$	$\Delta l_T^c = \|l^c\| T : \epsilon$
$t_\sigma = \sigma_N n + \sigma_T$	$f^c = f_N^c n + f_T^c$
$\sigma_N = \mathcal{E}_N \epsilon_N$	$f_N^c = k_N \Delta l_N^c$
$\sigma_T = \mathcal{E}_T \epsilon_T$	$f_T^c = k_T \Delta l_T^c$
$\sigma = \frac{3}{4\pi} \int [t_\sigma \otimes n]^{sym} \, d\Omega$	$\sigma = \frac{Nr}{2V\pi} \int [f^c \otimes n]^{sym} \, d\Omega$
$\mathcal{E}_N = 2\mu + 3\lambda$	$k_N = \frac{3V}{4Nr^2} [2\mu + 3\lambda]$
$\mathcal{E}_T = 2\mu - 2\lambda$	$k_T = \frac{3V}{4Nr^2} [2\mu - 2\lambda]$
$\mathcal{E} = \frac{3}{4\pi} \int [\mathcal{E}_N N \otimes N + \mathcal{E}_T T^T \cdot T] \, d\Omega$	$\mathcal{E} = \frac{Nr^2}{V\pi} \int [k_N N \otimes N + k_T T^T \cdot T] \, d\Omega$

plastic multipliers, a homogenized tangent operator for the particle model can be derived in the same fashion as for the microplane plasticity model. The remarkable similarity of both formulations is documented in table 2.

Finally, it should be mentioned, that although there are numerous similarities between both formulations, each of them is extremely valuable for its own fields of application. While microplane–based continuum models are usually applied to simulate the behavior of larger structures, particle models are believed to provide further inside into complex microstructural phenomena. The comparison presented in this paper was restricted to granular assemblies of equally sized, spherical

Table 2. Comparison of elasto-plastic models

microplane model	particle model
$t_\epsilon = \epsilon_N n + \epsilon_T$	$\Delta l = \Delta l_N n + \Delta l_T$
$\epsilon_N = \epsilon_N^{el} + \epsilon_N^{pl}$	$\Delta l_N = \Delta l_N^{el} + \Delta l_N^{pl}$
$\epsilon_T = \epsilon_T^{el} + \epsilon_T^{pl}$	$\Delta l_T = \Delta l_T^{el} + \Delta l_T^{pl}$
$t_\sigma = \sigma_N n + \sigma_T$	$f = f_N n + f_T$
$\sigma_N = \mathcal{E}_N \epsilon_N^{el}$	$f_N = k_N \Delta l_N^{el}$
$\sigma_T = \mathcal{E}_T \epsilon_T^{el}$	$f_T = k_T \Delta l_T^{el}$
$\sigma = \frac{3}{4\pi} \int [t_\sigma \otimes n]^{sym} \, d\Omega$	$\sigma = \frac{Nr}{2V\pi} \int [f \otimes n]^{sym} \, d\Omega$
$\Phi = \|\sigma_T\| - \tan\varphi \, \sigma_N - Y \leq 0$	$\Phi = \|f_T\| - \tan\varphi \, f_N - Y \leq 0$
$\dot{\epsilon}_N^{pl} = \dot{\gamma} \, \mu_N$	$\Delta l_N^{pl} = \dot{\gamma} \, \mu_N$
$\dot{\epsilon}_T^{pl} = \dot{\gamma} \, \mu_T$	$\Delta l_T^{pl} = \dot{\gamma} \, \mu_T$
$\dot{\gamma} = 1/h \, [\nu_N \mathcal{E}_N N + \nu_T \mathcal{E}_T T] : \dot{\epsilon}$	$\dot{\gamma} = \|l\|/h \, [\nu_N k_N N + \nu_T k_T T] : \dot{\epsilon}$

particles. Moreover, their contacts are assumed to be distributed uniformly in space. However, this idealization seems to be too restrictive when compared to reality. For more complex studies, a particle model consisting of particles of different size and shape could be applied to determine micromechanical quantities, for example discrete values of a contact distribution function, which could be used as input parameter for a microplane–based finite element simulation. Since the discrete element method itself is usually too expensive to model complex structures, a finite element simulation based on the additional information provided by microstructural analysis can be considered an appropriate alternative.

Acknowledgement

The authors are grateful for the financial support of the German Science Foundation (DFG) within the research group *Modellierung kohäsiver Reibungsmaterialien* under grant no.VE 163/4-1.4.

References

1. R. J. Bathurst, L. Rothenburg: J. Appl. Mech. **55**, 17 (1988)
2. Z. P. Bažant, P. G. Gambarova: J. Struct. Eng. **110**, 2015 (1984)
3. Z. P. Bažant and B. H. Oh: ZAMM **66**, 37 (1986)
4. Z. P. Bažant, P. Prat: J. Eng. Mech. **114**, 1672 (1988)
5. B. Cambou, P. Dubujet, F. Emeriault, F. Sidoroff: Eur. J. Mech. A / Solids **14**, 255 (1995)
6. I. Carol, Z. P. Bažant, P. Prat: Int. J. Solids & Structures **29**, 1173 (1992)
7. C. S. Chang: 'Dislocation and plasticity of granular materials with frictional contacts'. In: *Powders & Grains 93*, ed. Thornton, pp. 105–110, (1993)
8. C. S. Chang: 'Numerical and analytical modelling of granulates'. In: *Computer Methods and Advances in Geomechanics*, ed. by Yuan, pp. 105–114, (1997)
9. J. Christoffersen, M. M. Mehrabadi, S. Nemat-Nasser: J. Appl. Mech. **48**, 339 (1981)
10. P. A. Cundall, O. D. L. Strack: Géotechnique **29**, 47 (1979)
11. A. Drescher, G. de Josselin de Jong: J. Mech. Phys. Solids **20**, 337 (1972)
12. F. Emeriault, B. Cambou: Int. J. Solids & Structures **33**, 2591 (1996)
13. H. Hertz: J. für die reine u. angew. Math. **92**, 156 (1881)
14. K. -I. Kanatani: Int. J. Eng. Sci. **22**, 149 (1984)
15. N. P. Kruyt, L. Rothenburg: J. Appl. Mech. **118**, 706 (1996)
16. E. Kuhl, E. Ramm: Mech. Coh. Fric. Mat. **3**, 343 (1998)
17. E. Kuhl, G. A. D'Addetta, H. J. Herrmann, E. Ramm: Granular Matter **2**, 113 (2000)
18. M. Lätzel, S. Luding, H. J. Herrmann: Granular Matter **2**, 123 (2000)
19. C. -L. Liao, T. -P. Chang, D. -H. Young and C. S. Chang: Int. J. Solids & Structures **34**, 4087 (1997)
20. V. A. Lubarda, D. Krajcinovic: Int. J. Solids & Structures **30**, 2859 (1993)
21. O. Mohr: Zeitschrift des Vereins Deutscher Ingenieure, **46**, pp. 1524–1530, 1572–1577, (1900)
22. L. Rothenburg, A. P. S. Selvadurai: 'A micromechanical definition of the Cauchy stress tensor for particulate media'. In: *Proceedings of the International Symposium on Mechanical Behavior of Structured Media*, ed. by Selvadurai, pp. 469–486, (1981)

46   E. Kuhl *et al.*

23. A. Suiker, H. Askes, L. J. Sluys: 'Micro-mechanically based 1-d gradient damage models'. To appear in: *Proceedings of the ECCOMAS 2000*
24. K. Walton: J. Mech. Phys. Solids **35**, 213 (1987)
25. O. C. Zienkiewicz, G. N. Pande: Int. J. Num. Anal. Meth. Geom. **1**, 219 (1977)

Short-term creep of shotcrete – thermochemoplastic material modelling and nonlinear analysis of a laboratory test and of a NATM excavation by the Finite Element Method

M. Lechner,[1] Ch. Hellmich,[1] H. A. Mang[2]

Institute for Strength of Materials, Vienna University of Technology, Karlsplatz 13, Vienna, Austria

Abstract. Embedded in a thermochemoplastic material law set up in the framework of thermodynamics, the focus of the work is on the creep characteristics of shotcrete. Short-term creep, with a characteristic duration of several days, turns out to be a fundamental feature for realistic modelling of the structural behaviour of tunnels driven according to the New Austrian Tunnelling Method (NATM). Its origin is a stress-induced water movement within the capillary pores of concrete. This process is related to the accumulation of hydrates, which are initially free of micro-stress. Hence, an incremental formulation for aging viscoelasticity turns out to be a proper tool for modelling this kind of creep. The usefulness of this formulation is tested by re-analyzing a relaxation test with non-constant prescribed strains, showing quantitatively correct results for concrete and qualitatively correct results for shotcrete. The latter results indicate the necessity of classical creep tests for shotcrete.

1 Introduction and motivation for the investigation of creep in shotcrete

When driving a tunnel according to the New Austrian Tunnelling Method (NATM), after the excavation of the tunnel cross section, shotcrete is applied onto the tunnel walls. In this way, a compound structure between the incoming rock and the hydrating shotcrete shell is established. As a consequence, shotcrete is loaded during hydration, i.e., during its chemical 'genesis'. This is the motivation for choosing a material law describing couplings of thermal, chemical and mechanical properties.

According to the pertinent literature on the NATM, the success of this method is strongly based on the extensive creep properties and the rapid hardening characteristics of shotcrete as well as on the in-situ realization by experienced site engineers. In the construction strategy, measurements play an important role as regards monitoring and making decisions for the continuation of the excavation. Especially in case of difficult geological conditions, where the great potential of the NATM can be demonstrated convincingly, there is a great demand for quantitative determination of deformations and stresses.

The latter can be attained by means of numerical simulations, incorporating *in-situ* measurements. This helps to increase the economy and safety standards and to raise the competitiveness of the NATM. To be able to deliver technically relevant statements on the stress state in the tunnel shell, the creep properties of shotcrete must be investigated. Furthermore, the obtained information must be implemented into the thermochemoplastic material model.

2   Thermochemoplastic material model for shotcrete

The thermochemoplastic material model is formulated within the framework of thermodynamics of reactive porous media. It is based on a *macroscopic* description of phenomena on the *microlevel* of the material by means of *state variables* and energetically conjugated *thermodynamic forces*. A general treatment of thermochemomechanical couplings for concrete at early age can be found in [23]. This theory is formulated in terms of small strains.

2.1   State variables

Complicated chemical and physical phenomena on the microlevel of the material are described macroscopically by state variables. It is customary to distinguish between external and internal state variables. In the material model according to [7], [12], and [21], two external (measurable) variables (ε and T) and six internal variables (ξ, ε^p, χ, ε^v, ε^f and γ) are used.

- ε denotes the strain tensor.
- T stands for the absolute temperature.
- The degree of hydration ξ describes the state of hydration. Hydration is the chemical reaction between cement and free water. The reaction products are termed hydrates. ξ is the ratio between the current specific mass of water bound in hydrates and the respective mass at the end of the hydration process.
- ε^p and χ denote the tensor of plastic strains and the vector of hardening variables, respectively. They represent deformations and microstructural changes because of microcracking in shotcrete, respectively.
- According to Ruetz [19], the reason for short-term creep are stress-induced movements of water in the capillary pores of concrete. Their diameter is in the range of micrometers. The resulting viscous strains are denoted by ε^v.
- According to Wittmann [25], long-term or flow creep follows from dislocation-like processes in the nanopores of cement gel. Thus, the observation scale of this phenomena is 1000 times smaller than the one concerning short-term creep. The corresponding macroscopic flow strains are denoted by ε^f. The internal variable γ, called viscous slip [22], respresents microstructural changes resulting from the dislocation-like phenomena.

2.2   Field equations

Field equations are containing spatial derivatives. The physical space is described by means of two field equations. The first one is the equilibrium equation,

$$\mathrm{div}\boldsymbol{\sigma} + \mathbf{k} = \mathbf{0}, \tag{1}$$

with $\mathbf{k}$ denoting the vector of the volume-force density and $\boldsymbol{\sigma}$ standing for the (macroscopic) stress tensor. The second field equation is the first law of thermodynamics. Accounting only for non-negligible terms, this law reads as [23]

$$C\dot{T} - f\dot{\xi} = -\mathrm{div}\mathbf{q}, \tag{2}$$

with C as the specific heat capacity, f as the latent heat of hydration per unit volume of concrete, and $\mathbf{q}$ as the heat flow vector (positive for an efflux). Eqn. (2) states that the change of internal energy, $C\dot{T} - f\dot{\xi}$, equals the external heat supply, $-\mathrm{div}\mathbf{q}$.

2.3   Heat conduction law

The heat flux $\mathbf{q}$ is related to T via Fourier's linear (isotropic) heat conduction law,

$$\mathbf{q} = -k\,\mathrm{grad}T, \tag{3}$$

with k as the thermal conductivity.

2.4   Constitutive equations

State equations State equations are relations between thermodynamic forces and state variables. The (incremental) state equation for the stresses is given as [21],

$$d\boldsymbol{\sigma} = \mathbf{C}(\xi) : [d\boldsymbol{\varepsilon} - d\boldsymbol{\varepsilon}^p - \mathbf{1}d\varepsilon^s(\xi) - \mathbf{1}\alpha_T dT - d\boldsymbol{\varepsilon}^v - d\boldsymbol{\varepsilon}^f], \tag{4}$$

with $\mathbf{C}(\xi)$ as the (aging) isotropic elasticity tensor, depending on Young's modulus $E(\xi)$ and on (constant) Poisson's ratio ν, and $\mathbf{1}\varepsilon^s(\xi)$ as the strains caused by chemical shrinkage. $\mathbf{1}$ is the second-order volumetric unity tensor. α_T is the coefficient of thermal dilatation, which is assumed to be constant. The incremental formulation (4) accounts for the quasi-instantaneous deposition of new hydrates in a state free of microstress [2]. This is illustrated by the 1D rheological model depicted in Fig. 1 (a): Each hydrate is exclusively loaded by microstress resulting from macrostress applied *after* the formation of the respective hydrate.

Evolution equations The set of governing equations is completed by the evolution equations for the internal state variables.

As for ε^p and χ, the concept of multisurface chemoplasticity is applied [12] [24]: Plastically admissible stress states are defined by N loading functions f_α,

$$\boldsymbol{\sigma} \in C_E \Leftrightarrow f_\alpha = f_\alpha(\boldsymbol{\sigma}, \zeta(\chi, \xi)) \leq 0\ \forall \alpha \in [1, 2, ..., N]. \tag{5}$$

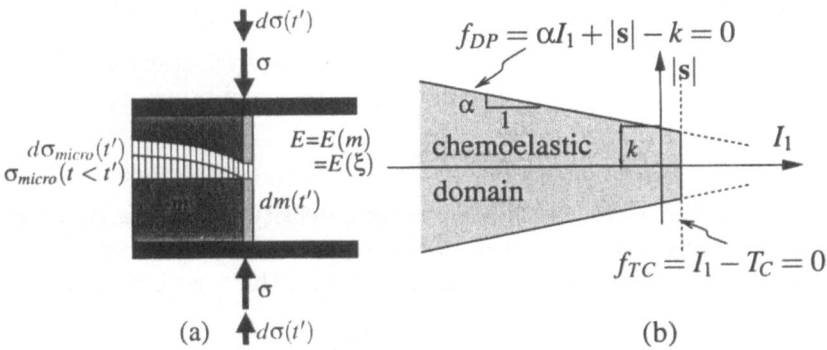

Fig. 1. (a) 1D rheological model illustrating elastic properties of hydrating concrete, with m denoting the mass of formed hydrates, σ_{micro} standing for the microstress in the hydrates, σ for the macroscopic stress used in the material model and E for (aging) Young's Modulus, (b) Drucker-Prager yield surface with tencion-cut-off

In other words, the material can only bear stress states within the chemoelastic domain and on its surface, see Fig. 1 (b). It is seen that the hardening force ζ, which is associated with the size of the yield surface in the stress space, is not only dependent on the plastic hardening variable χ, but also on the degree of hydration ξ (chemoplastic hardening). Assuming, for the sake of simplicity, associative hardening plasticity, the flow and the hardening rules read [14]

$$d\varepsilon^p = \sum_{\alpha \in J_{act}} d\lambda_\alpha \frac{\partial f_\alpha}{\partial \sigma}, \quad d\chi = \sum_{\alpha \in J_{act}} d\lambda_\alpha \frac{\partial f_\alpha}{\partial \zeta}, \tag{6}$$

where $d\lambda_\alpha$ are the plastic multipliers; J_{act} denotes the set of n active yield surfaces, $n \leq N$. It is defined by $J_{act} := \{\alpha \in 1, 2, ..., N | f_\alpha(\sigma, \zeta) = 0\}$. All yield surfaces obey the Kuhn-Tucker loading/unloading (complementarity) conditions,

$$f_\alpha \leq 0, \quad d\lambda_\alpha \geq 0, \quad f_\alpha d\lambda_\alpha = 0, \tag{7}$$

along with the consistency requirement

$$d\lambda_\alpha df_\alpha = 0. \tag{8}$$

For the description of microcracking, by means of (5) to (8), only two yield surfaces are considered. One of them is a Drucker-Prager loading surface, $f_{DP} = 0$, describing the behavior of concrete in (mainly biaxial) compression, which is a typical stress state in a tunnel, see, e.g. [15]. The other one is a tension-cut-off, $f_{TC} = 0$, for the description of the tensile behavior. These loading surfaces are depicted in Fig. 1(b) in the I_1-$|\mathbf{s}|$-stress space. I_1 stands for the first invariant of the stresses, whereas $|\mathbf{s}| = \sqrt{s_{ij}s_{ij}}$ is the norm of the stress deviator. α, k, and T_C are material parameters, which are related to the hardening force ζ, and hence, to the strengths of the material; for details, see [12] [7].

As for ξ, an evolution law of the Arrhenius type reflects the thermally acti-vated nature of the chemical reaction ([3],[24]),

$$\dot{\xi} = \tilde{A}(\xi) \exp\left(-\frac{E_a}{RT}\right), \tag{9}$$

E_a is the activation energy and R is the universal constant for ideal gas. $\tilde{A}(\xi)$ stands for the chemical affinity, which is the driving force of the chemical reac-tion.

The flow strains obey [4]

$$\dot{\boldsymbol{\varepsilon}}^f = \frac{1}{\eta_f} \mathbf{G} : \boldsymbol{\sigma}, \tag{10}$$

with η_f standing for the viscosity governing the kinetics of the aforementioned dislocation-like processes. It can be written as $1/\eta_f = cS\exp[-2U/\bar{k}(1/T - 1/\bar{T})]$, where $U/\bar{k} \approx 2700$ K reflects the thermally activated nature of long-term creep [3]. $\bar{T}$ is a reference temperature, $c=1$ MPa^{-2}s^{-1} is a dimensional constant. The rate of the microprestress force S [4] is linked to the rate of viscous slip γ via [21]

$$\dot{S} = -H\dot{\gamma}, \tag{11}$$

with a constant material parameter H [21]. The evolution law for the viscous slip γ is given as [22]

$$\dot{\gamma} = cS^2 \exp\left[-\frac{U}{\bar{k}}\left(\frac{1}{T} - \frac{1}{\bar{T}}\right)\right]. \tag{12}$$

Like ageing elasticity for hydrating concrete, short-term creep is related to the accumulation of hydrates, which are initially free of microstress (see section 2.4). Hence, an incremental formulation for short-term creep [17], [8] is used for the description of this process. The evolution equation for the viscous strains $\boldsymbol{\varepsilon}^v$ contains the integral over the stress increments $d\boldsymbol{\sigma}(t')$ having occurred until the current time instant t,

$$\dot{\boldsymbol{\varepsilon}}^v(t) = \frac{1}{\tau_w(\xi(t))}\left[\int_{t'=0}^{t} J_\infty^v(\xi(t'))\mathbf{G} : d\boldsymbol{\sigma}(t') - \boldsymbol{\varepsilon}^v(t)\right]. \tag{13}$$

This integral is a so-called Steltjes integral, this means that $d\boldsymbol{\sigma}(t')$ may be an infinitesimal or even a finite value $\Delta\boldsymbol{\sigma}(t')$. In (13), τ_w denotes the characteristic time of the short-term creep process and $\mathbf{G} = E(\xi)\mathbf{C}(\xi)^{-1}$ denotes the normal-ized compliance tensor. According to [21], $\tau_w = \tau_{w,\infty}\xi$. For shotcrete investigated at Lafarge CTEC Mannersdorf [9], $\tau_{w,\infty} \approx 24$ h [7]. For the same shotcrete, the viscous compliance $J_\infty^v(\xi)$ is depicted in Fig. 9(c), [17]. Experimental evidence [16] and the investigations of [21] have shown that in a classical 1D creep test (Fig. 2) the asymptotic creep strain ε_∞^v depends on the time instant of loading, t_0, and hence, on the degree of hydration at this time instant, $\xi(t_0)$. The function

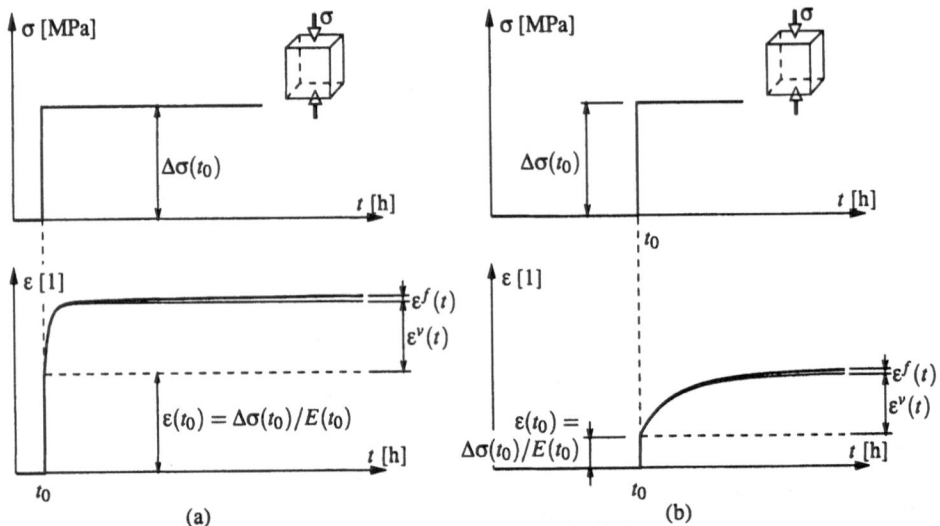

Fig. 2. Classical creep test: Stress history and strain evolution for (a) early and (b) late loading

$J_\infty^v(\xi(t_0)) = \varepsilon_\infty^v(\xi(t_0))/\sigma$ agrees with these qualitative experimental results, see Fig. 2 and 9(c). The evolution law for ε^v given in [21] can be interpreted as an approximation of (13) with a constant $\bar{J}_\infty^v$, see Fig. 9(c),

$$\dot{\varepsilon}^v(t) = \frac{1}{\tau_w(\xi(t))}[\bar{J}_\infty^v \mathbf{G} : \boldsymbol{\sigma}(t) - \boldsymbol{\varepsilon}^v(t)]. \qquad (14)$$

3   Algorithmic treatment of the incremental formulation for short-term creep

The algorithmic translation of the thermochemoplastic material model is performed in the framework of nonlinear finite element method based on a global Newton-Raphson iteration scheme. Return-Mapping-Algorithms together with consistent tangent operators for thermochemoplasticity are used for this purpose [12], [11], [7], [21]. Modifications with regards to [21], resulting from the introduction of the incremental formulation for short-term creep (13), are summarized in the following [17].

3.1   Discretization of the evolution law for short-term creep

The evolution law for the viscous strains ε^v, (13), is discretized by a backward Euler scheme (implicit integration),

$$\frac{\Delta\varepsilon_{n+1}^v}{\Delta t_{n+1}} = \frac{1}{\tau_w(\xi_{n+1})}\left[\sum_{k=1}^{n+1}[J_\infty^v(\xi_k)\mathbf{G} : \Delta\boldsymbol{\sigma}_k] - \boldsymbol{\varepsilon}_{n+1}^v\right], \qquad (15)$$

where n denotes the number of a time increment: $\Delta t_n = t_{n+1} - t_n$. Insertion of

$$\varepsilon_{n+1}^v = \varepsilon_n^v + \Delta\varepsilon_{n+1}^v \tag{16}$$

in (15) and rearrangement of terms yields

$$\Delta\varepsilon_{n+1}^v = \frac{1}{\frac{\tau_w(\xi_{n+1})}{\Delta t_{n+1}} + 1}\left[\sum_{k=1}^{n+1}[J_\infty^v(\xi_k)\mathbf{G}:\Delta\sigma_k] - \varepsilon_n^v\right]. \tag{17}$$

3.2   Discretization of the incremental state equation for the stresses

The discretized form of the incremental formulation for the stresses (4), following from a backward Euler procedure, reads as

$$\Delta\sigma_{n+1} = \mathbf{C}(\xi_{n+1}) : (\Delta\varepsilon_{n+1} - \Delta\varepsilon_{n+1}^p - 1\Delta\varepsilon_{n+1}^s - 1\alpha_T\Delta T_{n+1} - \Delta\varepsilon_{n+1}^v - \Delta\varepsilon_{n+1}^f). \tag{18}$$

Inserting (17) and the discretized form of (10) into (18) as well as accounting for $\sigma_{n+1} = \sum_{k=1}^{n}\Delta\sigma_k + \Delta\sigma_{n+1}$ results in

$$\Delta\sigma_{n+1} = \mathbf{C}(\xi_{n+1}) : \left[\Delta\varepsilon_{n+1} - 1\Delta\varepsilon_{n+1}^s - \Delta\varepsilon_{n+1}^p - 1\Delta\varepsilon_{n+1}^T - \right.$$

$$-\frac{1}{1 + \frac{\tau_w}{\Delta t_{n+1}}}\left[\sum_{k=1}^{n}J_\infty^v(\xi_k)\mathbf{G}:\Delta\sigma_k - \varepsilon_n^v\right] - \frac{\Delta t_{n+1}}{\eta_f(\Gamma_{n+1})}\mathbf{G}:\sigma_n\bigg] -$$

$$-\mathbf{C}(\xi_{n+1}) : \left[\frac{\Delta t_{n+1}}{\eta_f(\Gamma_{n+1})}\mathbf{G}:\Delta\sigma_{n+1} + \frac{1}{1 + \frac{\tau_w}{\Delta t_{n+1}}}J_\infty^v(\xi_{n+1})\mathbf{G}:\Delta\sigma_{n+1}\right]. \tag{19}$$

Exemption of $\Delta\sigma_{n+1}$ gives

$$\Delta\sigma_{n+1} = \frac{1}{K_v}\mathbf{C}(\xi_{n+1}) : \left[\Delta\varepsilon_{n+1} - 1\Delta\varepsilon_{n+1}^s - \Delta\varepsilon_{n+1}^p - 1\Delta\varepsilon_{n+1}^T - \right.$$

$$\frac{1}{1 + \frac{\tau_w}{\Delta t_{n+1}}}\left[\sum_{k=1}^{n}J_\infty^v(\xi_k)\mathbf{G}:\Delta\sigma_k - \varepsilon_n^v\right] - \frac{\Delta t_{n+1}}{\eta_f(\Gamma_{n+1})}\mathbf{G}:\sigma_n\bigg], \tag{20}$$

with the correction coefficient K_v accounting for the influence of creep on the stress-strain relation,

$$K_v = \left[1 + \frac{1}{1 + \frac{\tau_w}{\Delta t_{n+1}}}E(\xi_{n+1})J_\infty^v(\xi_{n+1}) + \Delta t_{n+1}\frac{E(\xi_{n+1})}{\eta_f(\Gamma_{n+1})}\right]. \tag{21}$$

The correction coefficient K_v is identical to the one determined by Sercombe et al. [21].

3.3   Numerical example: creep test with two instants of loading

The most simple way to show the influence of the incremental formulation for short-term creep (13) is a classical creep test extended to two instants of loading (Fig. 3). As shown by the outer curves in Fig. 3, for $\bar{J}_\infty^v$ = const. both for-

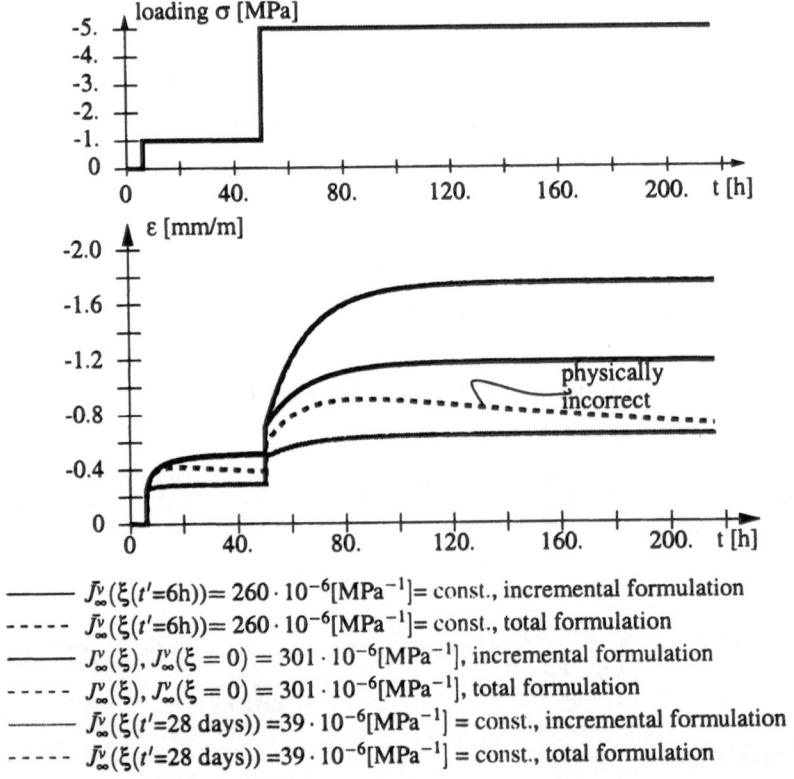

——— $\bar{J}_\infty^v(\xi(t'=6\mathrm{h}))= 260\cdot 10^{-6}[\mathrm{MPa}^{-1}]$= const., incremental formulation
- - - - $\bar{J}_\infty^v(\xi(t'=6\mathrm{h}))= 260\cdot 10^{-6}[\mathrm{MPa}^{-1}]$= const., total formulation
——— $J_\infty^v(\xi), J_\infty^v(\xi = 0) = 301\cdot 10^{-6}[\mathrm{MPa}^{-1}]$, incremental formulation
- - - - $J_\infty^v(\xi), J_\infty^v(\xi = 0) = 301\cdot 10^{-6}[\mathrm{MPa}^{-1}]$, total formulation
——— $\bar{J}_\infty^v(\xi(t'=28\ \mathrm{days})) =39\cdot 10^{-6}[\mathrm{MPa}^{-1}]$ = const., incremental formulation
- - - - $\bar{J}_\infty^v(\xi(t'=28\ \mathrm{days})) =39\cdot 10^{-6}[\mathrm{MPa}^{-1}]$ = const., total formulation

Fig. 3. Influence of the incremental (13) and the total (14) formulation for short-term creep in a stress-controlled experiment

mulations for short-term creep deliver identical results, which are qualitatively correct. The outer curves in Fig. 3 show the range of the system response, depending on the chosen amplitude for $\bar{J}_\infty^v$ = const. This illustrates the necessity to consider J_∞^v as a function of ξ.

The use of a total formulation for short-term creep (14) in combination with variable creep amplitude $J_\infty^v(\xi)$ turns out to deliver physically incorrect results (Fig. 3): When σ is constant and the asymptotic viscous compliance $J_\infty^v(\xi(t'))$ decreases in the course of time, the term $J_\infty^v(\xi(t'))\,\mathbf{G} : \sigma$ decreases, too. After some time, this leads to a negative rate of the compressive viscous strain $\dot\varepsilon^v$, and, consequently to a *decrease* of ε^v, see (14). This is qualitatively **incorrect** from an experimental point of view. Using the incremental formulation for short-term creep (13) instead of the total one (14), the combination of a **decreasing** asymptotic viscous compliance $J^v_\infty(\xi(t'))$ and of stress **increments** $d\boldsymbol{\sigma}(t')$, which are zero except for the two time instants of load change, does not cause a decrease in the term $\int_{t'=0}^{t} J^v_\infty(\xi(t')) \mathbf{G} : d\boldsymbol{\sigma}(t')$ and thus, leads to qualitatively **correct** results.

4 Re-analysis of a laboratory test

Fischnaller [5] performed experiments in order to investigate the relaxation behavior of shotcrete under a loading which is typical for NATM tunnel excavations. Shotcrete was sprayed into special boxes at a tunnel site. Afterwards, the boxes were transported to a laboratory and the shotcrete cylinders were put into test benches for creep (Fig. 4(a)). They were loaded by a strain history (Fig. 4(b)) which is typical for tunnel excavations according the NATM. Additionally, Fischnaller performed creep tests on concrete cast in place. Apart from the accelerator in shotcrete, the composition of concrete is the same as for shotcrete. The laboratory test was re-analyzed in order to check the creep formulation of the thermochemoplastic material model.

4.1 Modelling

For re-analysis of the relaxation test, a 1D stress state was assumed. From an age of 6 and 27 h, respectively (see Fig. 4(b)), on, a 3D Finite Element (Fig. 4(d)) is loaded by prescribed displacements according to the strain history in Fig. 4(b).

4.2 Experimental determination of material properties

All relevant *intrinsic* material functions are extracted from (sligthly extended) standard tests (e.g., [5], [13], [9]) or relations given in the pertinent standardizations for shotcrete/concrete in the open literature (e.g., [6]). The term intrinsic refers to material behaviour, which is independent of field and boundary conditions [24] (e.g., independent of different thermal conditions).

Input For the determination of the intrinsic material functions, information about $E(t)$, $\varepsilon^s(t)$, $T(t)$, the compressive strength $f_c(t)$ and the creep strains $\varepsilon^{creep}(t)$ is required. Fischnaller investigated $E(t)$, $\varepsilon^s(t)$ and $f_c(t)$. The thermal conditions in shotcrete are assumed to be similar to the ones investigated by Huber [13], Fig. 4(c). Concrete was assumed to be under isothermal conditions. As investigated in [17], the influence of non-isothermal conditions can be neglected when re-analyzing the tests of Fischnaller.

Deplorably, in the open literature there is very little information about creep tests performed on shotcrete. This is the reason for using the Model B3 of [1], as

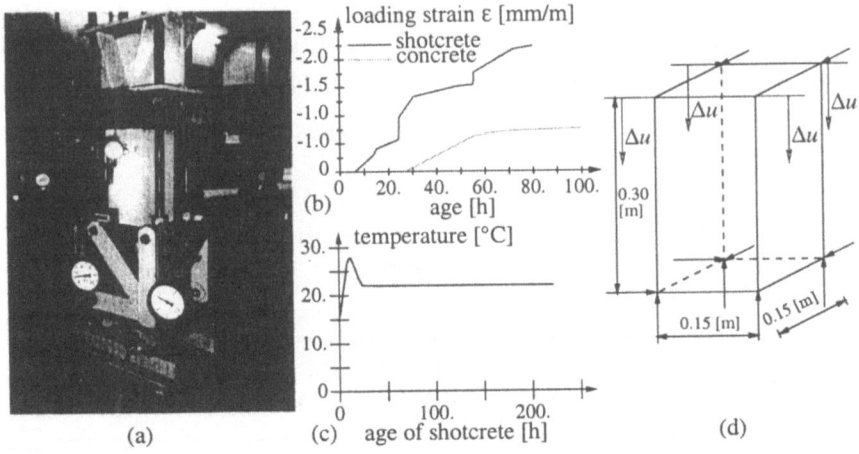

Fig. 4. (a) Experimental setup of creep tests performed by Fischnaller [5], (b) history of prescribed strains, (c) history of temperature in the specimen [13], (d) 3D Finite element loaded by displacements Δu

Table 1. input for Model B3 of [1] for the materials used by [5]

input Model B3		shotcrete	concrete
cement content [kg/m^3]	c	350	350
aggregate/cement ratio [-]	a/c	5.26	5.26
water/cement ratio [-]	w/c	0.55	0.55
compressive strength, 28 d [MPa]	$f_{c,28}$	20	42
age at loading [d]	t_0	.2;.25;.5;1;1.125;3;7;28	1;2;3;5;7;14;21;28

proposed in [21]. It is based on a big quantity of classical creep test for concrete. The input parameters are shown in Table 1.

The output from Model B3 is the compliance function $J(t, t_0) = \varepsilon(t, t_0)/\sigma$ [10^{-6}/MPa], see Fig. 5. Sercombe et al. [21] identified different phenomena, such as short-term creep, long-term creep and 'pseudo'-elasticity by means of different compliance rates, Fig. 5.

The viscous compliance $J^v(t, t_0)$ is characterized by a steep increase after loading followed by a declining compliance rate. For $t \to \infty$, $J^v(t, t_0)$ is limited by its asymptotic value $J^v_\infty(t, t_0)$, the asymptitic viscous compliance. In contrary, the flow compliance $J^f(t, t_0)$ does not have big influence directly after loading, but increases steadily (Fig. 5), thus long-term creep is important for long-term investigations. The shotcrete test of Fischnaller [5] is re-analyzed for 80 h, the concrete test for 100 h, Fig. 4(b). Consequently, the creep behaviour in these experiments is dominated by short-term creep, see Fig. 5 and Fig. 8.

The disadvantage when using the Model B3 is that the admixture of an accelerator cannot be considered.

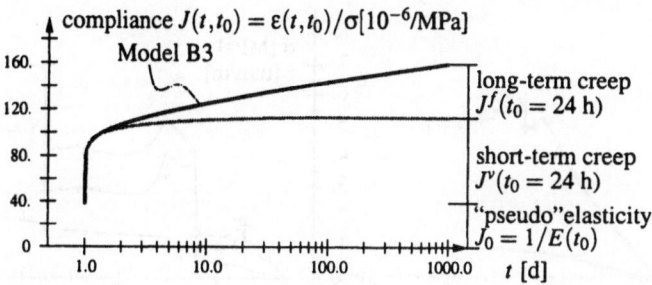

Fig. 5. Compliance $J(t, t_0)$ for concrete of Fischnaller [5] as output of Model B3 and decomposition of $J(t, t_0)$ into the parts stemming from short-term creep, long-term creep and 'pseudo'-elasticity

Output The output of intrinsic material relations (Fig. 6) consists of the strength growth $f_c(\xi)$, the chemical affinity $\tilde{A}(\xi)$, the ageing elasticity modulus $E(\xi)$, the shrinkage strains $\varepsilon^s(\xi)$, all determined according to [12], and the asymptotic viscous compliance $J_\infty^v(\xi)$, the characteristic time for short-term creep $\tau_w(\xi)$, the microprestress force $S(\gamma)$, all determined according to [21].

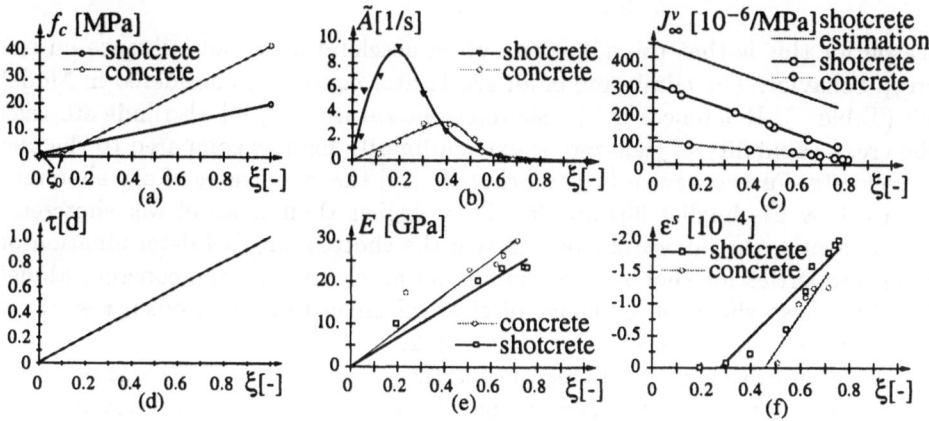

Fig. 6. Intrinsic material functions for the experiments of Fischnaller [5]

4.3 Results

Concrete For concrete the results of the Finite Element simulation approximately coincide with the test results, i.e., they are *quantitatively* correct, see Fig. 7(a).

Shotcrete For shotcrete the stresses attained in the FE simulation are considerably larger than the ones attained in the experiment (Fig. 7(b)). A possible

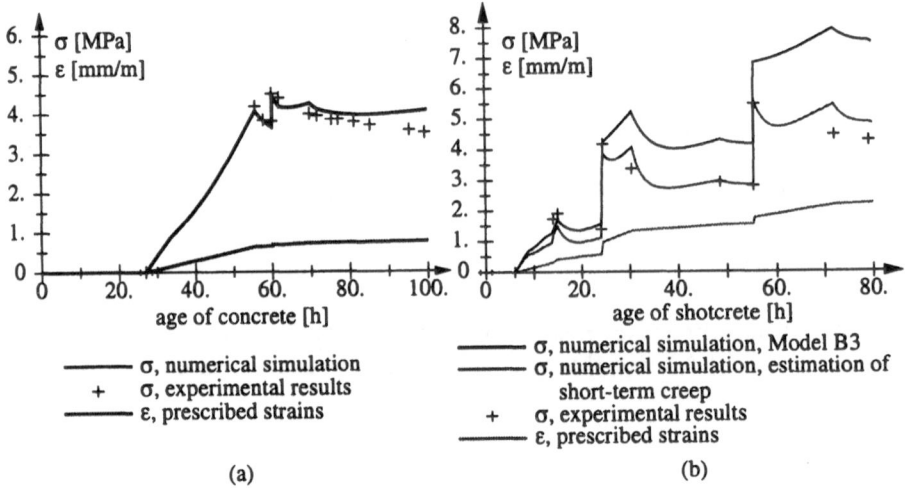

Fig. 7. Simulation of 1D relaxation tests performed by Fischnaller [5], using the incremental formulation for short-term creep (13): (a) results for concrete; (b) results for shotcrete, asymptotic short-term creep compliance based on Model B3 and estimated, respectively (Fig. 6(c))

reason for this is that the admixture of an accelerator has an influence on the creep behavior. The admixture of an accelerator cannot be considered in Model B3 (Table 1). References in the shotcrete literature [18], [5] also indicate that the creep capability of shotcrete is extraordinarily high as compared to the one of concrete. Further, according to Schubert [20] the creep rates of dry shotcrete (as used by Fischnaller [5]) are 25 - 30 % higher than those of wet shotcrete. Consequently, the laboratory tests reveal the shortcomings of determination of creep properties for shotcrete containing an accelerator by the concrete Model B3. They also show the necessity of classical creep tests, i.e., creep tests at a constant stress level, for shotcrete, see Fig. 2.

For an estimation of the creep properties of shotcrete, $J_\infty^v(\xi)$ was enlarged according to Fig. 6(c). The stresses based on this estimation are very close to the experimentally attained stresses (Fig. 7(b)).

Fig. 8 indicates the necessity to use the full thermochemoplastic shotcrete model for re-analysis of the tests. Otherwise the stress level would be by far overestimated. Fig. 8 also shows the numerical results in percentages of the experimental results after 80 h.

5   Simulation of a tunnel driven according to the NATM

3D thermochemomechanical Finite Element analyses were performed for the Siebergtunnel as a real life engineering problem, being part of the high-capacity railway line Vienna-Salzburg. For details of the modelling of the tunnel shell see [7], [10]. For the numerical simulation a 'hybrid method' was used. This method

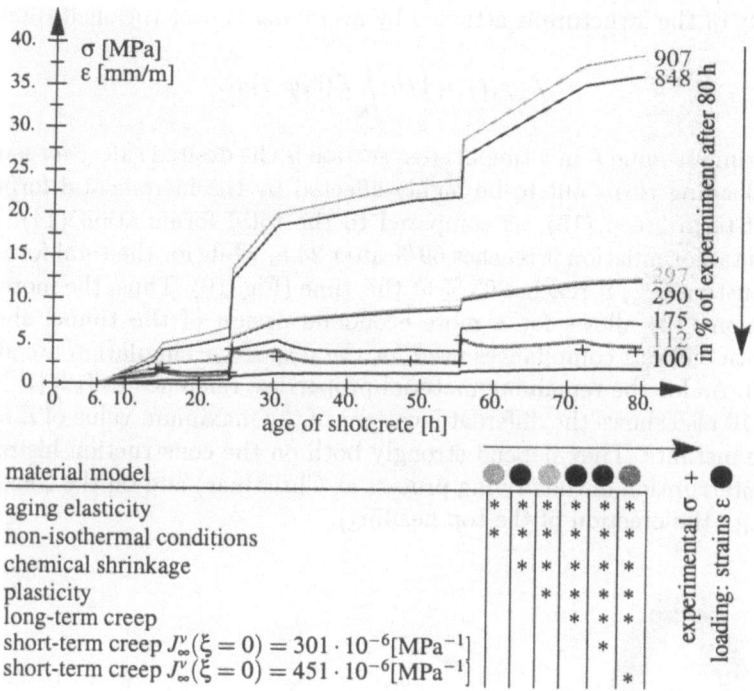

Fig. 8. 1D relaxation test for shotcrete of Fischnaller [5]: influence of micro-cracking (plasticity), chemical shrinkage, long-term creep and short-term creep on the stress state of the material. For all simulations: non-isothermal conditions according to Fig. 4(c), incremental formulation for short-term creep (13)

is a combination of advanced constitutive modelling of the shotcrete shell by the thermochemoplastic material model with loading of the shell by 3D displacement fields. The latter are obtained by suitable temporal and spatial interpolations of 3D displacement measurements between measurement points (MP) (Fig. 9). The method allows quantification of the stress states in the shotcrete tunnel shell as well as an assessment of its safety by defining a level of loading.

The level of loading is based on failure of concrete in the uni- or biaxial compressive domain, described by means of a Drucker-Prager surface ($f_{DP} = 0$) in Fig. 1(b),

$$\mathcal{L} = \frac{\alpha I_1 + |\mathbf{s}|}{k}. \tag{22}$$

For a 1D compression test, (22) becomes $\mathcal{L} = \sigma / f_c$. As investigated by Hellmich [10], the analyzed measurement cross section (km 156.990) is practically not influenced by bending. Thus, if a level of loading of 100 % is reached *locally*, there is no evidence for a failure of the tunnel shell. Hence, a better estimate of the safety of the structure is attained by averaging $\mathcal{L}$ over the shell thickness h,

$$\bar{\mathcal{L}}(\varphi, t) = 1/h \int_h \mathcal{L}(r, \varphi, t) dr. \qquad (23)$$

The maximum value $\bar{\mathcal{L}}$ in a tunnel cross section is the desired safety measure. The level of loading turns out to be highly affected by the incremental formulation for short-term creep (13), as compared to the total formulation (14). For the incremental formulation it reaches 69 % after 24 h, while for the total formulation (with constant $\bar{J}_\infty^v$) it reaches 93 % at this time (Fig. 10). Thus, the more refined formulation (13) allows for a more economic design of the tunnel shell. The asymptotic viscous compliances used for the respective simulation are shown in Fig. 9(c). As for the remaining material properties, reference to [10] [7] is made.

Fig. 10 also shows the different locations of the maximum value of $\bar{\mathcal{L}}$ at different time instants. They depend strongly both on the construction history (e.g., on the interruption of the driving process at Christmas) and on the construction state (e.g., the erection of the top heading).

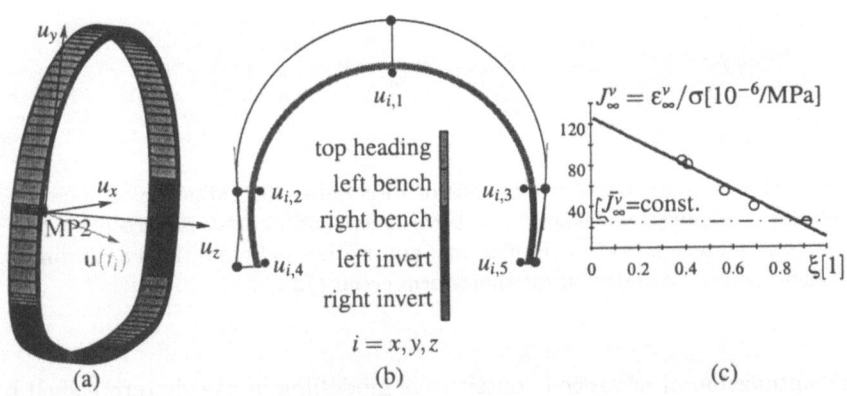

Fig. 9. (a) Hybrid method: Finite Element model and measurement point 2, (b) hybrid method: interpolated displacement fields as boundary conditions, (c) asymptotic compliance function $J_\infty^v(\xi)$ of shotcrete from Lafarge CTEC Mannersdorf [9] and the assumption of constant $\bar{J}_\infty^v(\xi(t' = 28d))$, as used for the total formulation for short term creep [21].

References

1. Z. Bažant and S. Baweja. Short form of creep and shrinkage prediction model B3 for structures of medium sensitivity. *Materials and Structures*, (29):587–593, 1996.
2. Z. P. Bažant. Thermodynamics of solidifying or melting viscoelastic material. *Journal of the Engineering Mechanics Division, ASCE*, 105(6):933–952, 1979.
3. Z. P. Bažant, editor. *Mathematical modelling of creep and shrinkage in concrete*. Wiley, Chichester, England, 1988.

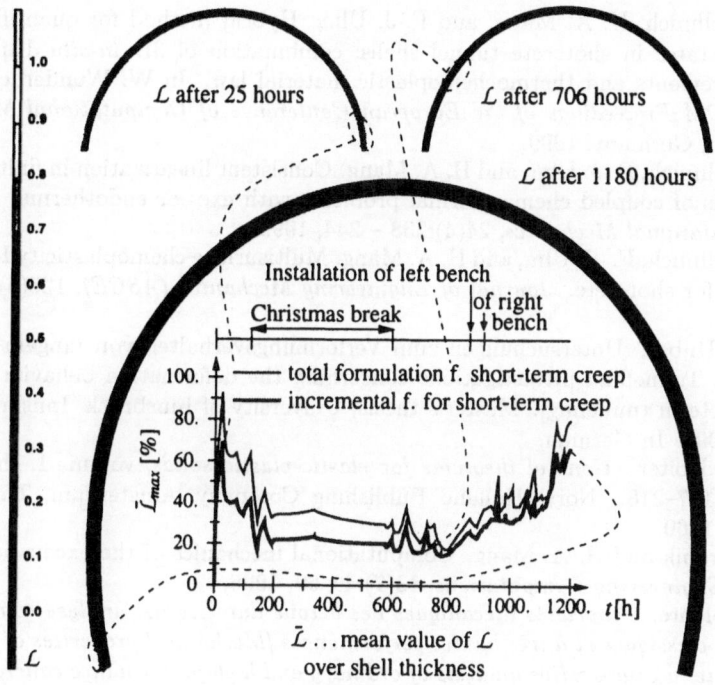

Fig. 10. Level of loading of Sieberg tunnel shell at km 156.990: Influence of incremental formulation for short-term creep (13) versus total formulation for short-term creep (14) and location of the maximum of $\bar{\mathcal{L}}$ in the tunnel shell after 25, 706 and 1180 h

4. Z. P. Bažant, A. B. Hauggard, S. Baweja, and F.-J. Ulm. Microprestress solidification theory for concrete creep, part I: Aging and drying effects. *Journal of Engineering Mechanics, ASCE*, 123(11):1188–1194, 1997.

5. G. Fischnaller. Untersuchungen zum Verformungsverhalten von jungem Spritzbeton im Tunnelbau - Grundlagen und Versuche [Investigation of the deformation behavior of young shotcrete in tunnelling - fundamentals and tests]. Master's thesis, University of Innsbruck, 1992. In German.

6. Guideline. *Richtlinie für Spritzbeton [Guideline for shotcrete]*. Österreichischer Betonverein, Vienna, Austria, 1997. In German.

7. Ch. Hellmich. *Shotcrete as part of the New Austrian Tunneling Method: from thermochemomechanical material modeling to structural analysis and safety assessment of tunnels*. PhD thesis, Vienna University of Technology, Vienna, Austria, 1999.

8. Ch. Hellmich, M. Lechner, R. Lackner, J. Macht, and H. A. Mang. Creep in shotcrete tunnel shells. In S. Murakami and N. Ohno, editors, *Proceedings of the fifth IUTAM Symposium on Creep in Structures*, Nagoya, Japan, 2000. In print.

9. Ch. Hellmich, H. A. Mang, E. Schön, and R. Friedle. Materialmodellierung von Spritzbeton - vom Experiment zum konstitutiven Gesetz [Material modeling of shotcrete - from the experiment to the constitutive law]. In Th. Varga, editor, *Proceedings of the conference held at the 1998 general assembly of the Austrian society for material testing*, Vienna, Austria, 1999. In print. In German.

10. Ch. Hellmich, H. A. Mang, and F.-J. Ulm. Hybrid method for quantification of stress states in shotcrete tunnel shells: combination of 3D *in-situ* displacement measurements and thermochemoplastic material law. In W. Wunderlich, editor, *CD-ROM Proceedings of the European Conference of Computational Mechanics*, Munich, Germany, 1999.

11. Ch. Hellmich, F.-J. Ulm, and H. A. Mang. Consistent linearization in finite element analysis of coupled chemo-thermal problems with exo- or endothermal reactions. *Computational Mechanics*, 24(4):238 – 244, 1999.

12. Ch. Hellmich, F. -J. Ulm, and H. A. Mang. Multisurface chemoplasticity I: Material model for shotcrete. *Journal of Engineering Mechanics (ASCE)*, 125(6):692–701, 1999.

13. H. G. Huber. Untersuchungen zum Verformungsverhalten von jungem Spritzbeton im Tunnelbau [Investigations concerning the deformation behavior of young shotcrete in tunneling]. Master's thesis, University of Innsbruck, Innsbruck, Austria, 1991. In German.

14. W. T. Koiter. *General theorems for elastic-plastic solids*, volume I, chapter IV, pages 167–218. North-Holland Publishing Company, Amsterdam, The Netherlands, 1960.

15. Ch. Kropik and H. A. Mang. Computational mechanics of the excavation of tunnels. *Engineering Computations*, 13(7):49–69, 1996.

16. P. Laplante. *Propriétés mécaniques des bétons durcissants: analyse comparée des bétons classiques et à très hautes performances [Mechanical properties of hardening concrete: a comparative analysis of ordinary and high performance concretes]*. PhD thesis, Ecole Nationale des Ponts et Chaussées, Paris, France, 1993. In French.

17. M. Lechner. Kurzzeitkriechen von Spritzbeton - thermochemoplastische Materialmodellierung und nichtlineare Analysen eines Laborversuchs sowie eines NöT-Tunnelvortriebs mittels Finiten Elementen [Short-term creep of shotcrete - thermochemoplastic material modelling and nonlinear finite element analyses of a laboratory test and of a tunnel driven according to the NATM]. Master's thesis, Vienna University of Technology, Vienna, Austria, 2000. In German.

18. R. Rokahr and K. H. Lux. Einfluß des rheologischen Verhaltens des Spritzbetons auf den Ausbauwiderstand. *Felsbau*, 5:11–18, 1987.

19. W. Ruetz. Das Kriechen des Zementsteins im Beton und seine Beeinflussung durch gleichzeitiges Schwinden [Creep of cement in concrete and the influence on it by simultaneous shrinkage]. *Deutscher Ausschuss Stahlbeton*, Heft 183, 1966. In German.

20. P. Schubert. Beitrag zum rheologischen Verhalten von Spritzbeton [Contribution to the rheological behavior of shotcrete]. *Felsbau*, 6:150–153, 1988.

21. J. Sercombe, Ch. Hellmich, F. -J. Ulm, and H. A. Mang. Modelling of early-age creep of shotcrete. I: Model and model parameters. *Journal of Engineering Mechanics (ASCE)*, 25(3):284–291, 2000.

22. F. -J. Ulm. Couplages thermochémomécaniques dans les bétons : un premier bilan. [Thermochemomechanical couplings in concretes : a first review]. Technical report, Laboratoires des Ponts et Chaussées, Paris, France, 1998. In French.

23. F. -J. Ulm and O. Coussy. Modeling of thermochemomechanical couplings of concrete at early ages. *Journal of Engineering Mechanics (ASCE)*, 121(7):785–794, 1995.

24. F. -J. Ulm and O. Coussy. Strength growth as chemo-plastic hardening in early age concrete. *Journal of Engineering Mechanics (ASCE)*, 122(12):1123–1132, 1996.

25. F. H. Wittmann. *Creep and shrinkage mechanisms*, chapter 6, pages 129–161. Wiley, Chichester, England, 1982.

Thermo-poro-mechanics of rapid fault shearing

I. Vardoulakis

Faculty of Applied Mathematics and Physics, Department of Mechanics, National Technical University of Athens, Greece

Abstract. In this paper the basic mathematical structure of a thermo-poro-mechanical model for faults under rapid shear is discussed. The analysis is 1D in space and concerns the infinitely extended fault. The gauge material is considered as a two-phase material consisting of a thermo-elastic fluid and of a thermo-poro-elasto-viscoplastic skeleton. The governing equations are derived from first principles, expressing mass, energy and momentum balance inside the fault. They are a set of coupled diffusion-generation equations that contain three unknown functions, the pore-pressure, the temperature and the velocity field inside the fault. The original mathematically ill-posed problem is regularized using a viscous-type and a 2nd gradient regularization. Numerical results are presented and discussed.

1 Introduction

According to [4], mechanical energy dissipated in heat inside a slip zone may lead to vaporization of pore-water, creating thus a cushion of zero friction. The idea that a heat generating mechanism might account for the loss of strength of large earth slides due to vaporization has been discussed in the past by [3, 5, 7]. Within 1D-analyses of sliding block-mechanisms (Fig. 1) Anderson [1] first and later Voight and Faust [12] showed, that, even if vaporization does not take place, heat generation may give rise to high pore-water pressures inside the shear-band. Recently Vardoulakis [10] formulated the set of governing equations that account for heat-generated pore-pressures inside rapidly deforming shear-bands, starting from first principles. The gauge material was considered as a two-phase mixture of solids and fluid, and the governing equations were derived from the corresponding balance laws for mass, momentum and energy.

In this paper first we summarize and discuss the equations that govern the phenomenon of heat-generated pore pressures inside a rapidly deforming fault (shear-band). They are a set of coupled diffusion-generation equations that contain three unknown functions, the heat-generated excess pore-fluid pressure, the temperature and the velocity. We remark that these equations constitute together with a set of initial and boundary conditions at the shear-band boundaries a *mathematically ill-posed problem*. This means in turn that in order to solve the corresponding boundary-value problem some kind of mathematical regularization is needed. In our previous papers [10] a "crude" method of regularization was used, namely that of replacing the momentum equation for the velocity field with an ad-hoc solution for the velocity that is correct to the leading, linear-in-z term and is compatible with the boundary conditions.

A more rigorous way to approach this problem is to reconsider the friction law and introduce friction rate hardening or softening. In the last section we discuss critically the viscous regularization and point to its limitations in the light of some experimental results reported recently by Tika and Hutchinson [8], which suggest that for some clay fault materials, frictional rate softening is taking place. In that case a 2nd gradient regularization [9] is proposed and discussed.

2   Formulation

We consider a rapidly deforming and long shear-band of thickness d consisting of water-saturated soil. Inside such a shear-band the pore-pressure $p(z,t)$, the temperature $\theta(z,t)$ and the velocity $v(z,t)$ are assumed to be functions only of the time t and of the position z in normal to the band direction (Fig. 1).

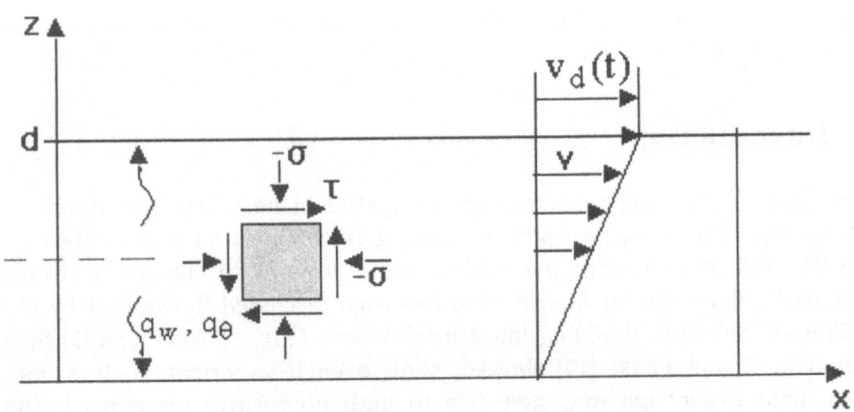

Fig. 1. The shear-band model. In this figure q_θ and q_w denote the heat and fluid flow, respectively

As shown in [10] mass and energy balance equations together with Darcy's and Fourier's laws lead to a set of coupled diffusion-generation equations for the pore-water pressure $p(z,t)$ and the temperature field $\theta(z,t)$ inside the shear band. For easy reference we summarize here these equations and define the pertinent material parameters.

2.1   Mass balance

Mass balance is expressed by the following pore-pressure diffusion-generation equation

$$\frac{\partial p}{\partial t} = \frac{\partial}{\partial z}\left(c_v(\theta)\frac{\partial p}{\partial z}\right) + \lambda_m \frac{\partial \theta}{\partial t} .$$

(1)

In (1) $p(z,t)$ is the excess, heat-generated pore-fluid pressure. We assume that the pore-fluid can drain freely in and out the fault, so that at these boundaries the pore-pressure is always constant. Accordingly we assume homogeneous boundary conditions for the heat-generated pore-pressure

$$p(0,t) = p(d,t) = 0 . \tag{2}$$

In (1) c_v and λ_m are the *consolidation* and the *pore-pressure-temperature* coefficients, respectively. The pore-pressure-temperature coefficient is given in terms of thermal expansion and compressibility coefficients as follows

$$\lambda_m = \left. \frac{\partial p}{\partial \theta} \right|_{V=\text{const undrained}} = \frac{\alpha}{c} . \tag{3}$$

The consolidation coefficient is given in terms of: a) the physical permeability $k, [L^2]$, b) the elastic compressibility of the gauge material skeleton $c, [F^{-1}L^2]$ and c) of the viscosity of the pore-fluid, $\eta_f, [FL^{-2}T]$,

$$c_v = \frac{k}{c\eta_f} . \tag{4}$$

The consolidation coefficient increases with θ, since the viscosity of the pore-fluid is in general a decreasing function of temperature; e.g. for water with $\eta_f = \eta_w$ we have[1]: $\eta_w \approx \eta_{w0}e^{-\theta/\theta_c}$, $\eta_{w0} = 1.47\text{cP}$, and $\theta_c = 55.9°\text{C}$.

2.2   Energy balance

Energy balance results in the following heat diffusion-generation equation,

$$\frac{\partial \theta}{\partial t} = \kappa_\theta \frac{\partial^2 \theta}{\partial z^2} + \frac{\tau}{\varrho_m j C_m} \frac{\partial v}{\partial z} . \tag{5}$$

In (5) κ_m is Kelvin's coefficient of thermal diffusivity of the fluid-solid mixture. The last term in equation (5) accounts approximately[2] for the mechanical work, that is converted to heat due to friction. The parameters ϱ_m and C_m are the density and the specific heat of the mixture, respectively; $j = 4.2\text{J/cal}$, is the mechanical equivalent of heat.

Following Coulomb's friction law and Terzaghi's effective stress principle, the shear stress τ is coupled through the material friction coefficient μ to the pore-pressure inside the shear band as,

$$\tau = (\sigma'_{n0} - p)\mu , \tag{6}$$

where $\mu = \tan\phi$ is the Coulomb friction coefficient and ϕ the corresponding friction angle of the soil in direct shear. $\sigma'_{n0} \approx \text{const}$ is the initial effective stress, acting normal to the shear band.

---

[1] $1\text{cP} = 10^{-9}\text{MPa s}$

[2] This means two assumptions: a) elastic strains are negligible, and b) all plastic work is converted to heat.

From a constitutive point of view the proposed model for thermo-poro-mechanical softening is justified by considering a plasticity model for soil with two-yield surfaces, as is sketched in Fig. 2 (cf. [11]): We assume that the soil inside the shear band is undergoing unloading with respect to the cup, $F_1 = 0$. This is because increasing pore-pressure results in effective stress reduction, and justifies the use of an 'elastic' (unloading) soil compressibility in (3). Accordingly at any instant of the deformation process the effective stress state lies on the Coulomb yield surface, $F_2 = 0$. The corresponding effective stress-path is pointing towards the origin (the stress free or "liquefied" state).

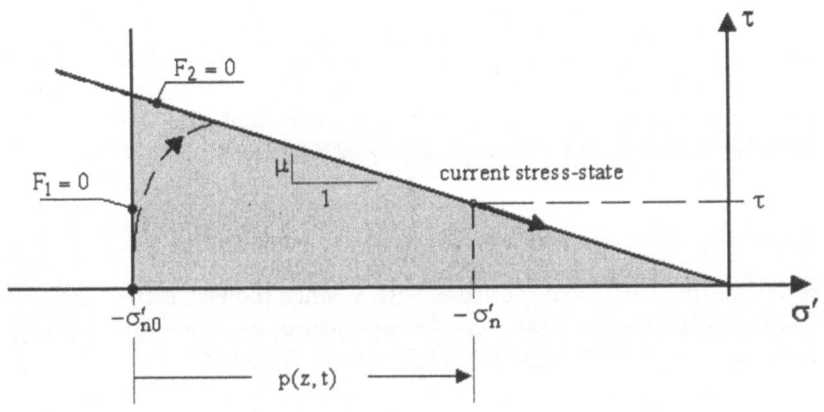

Fig. 2. Assumed soil constitutive model and effective stress-path

From (5) and (6) we obtain finally the following heat diffusion-generation equation,

$$\frac{\partial \theta}{\partial t} = \kappa_m \frac{\partial^2 \theta}{\partial z^2} + \frac{\mu}{\varrho_m j C_m}(\sigma'_{n0} - p)\frac{\partial v}{\partial z}. \tag{7}$$

In the considered shear-band problem we assume that the temperature at the boundaries is kept constant, equal to the ambient temperature

$$\theta(0, t) = \theta(d, t) = \theta_0. \tag{8}$$

2.3 Momentum Balance

Momentum balance inside the shear-band yields

$$\frac{\partial v}{\partial t} = \frac{1}{\varrho_m}\frac{\partial \tau}{\partial z}. \tag{9}$$

This equation together with the assumed friction law for the shear stress, (6), gives

$$\frac{\partial v}{\partial t} = -\frac{1}{\varrho_m}\mu\frac{\partial p}{\partial z}. \tag{10}$$

According to Fig. 1, the boundary conditions for the velocity field are:

$$v(0,t) = 0, \quad v(d,t) = v_d(t) \quad \text{for} \quad t > 0 . \tag{11}$$

First we observe an asymmetry as far as the mathematical structure of the momentum balance equation (10) is concerned in comparison to the other two balance laws, (1) and (7). Due to the Darcy and Fourier gradient-type laws, mass- and energy-balance are expressed by diffusion-type, second order partial differential equations, whereas momentum balance for an ideally plastic frictional material results into a first-order wave-type p.d.e. These equations together with a set of initial and boundary conditions for the fields $p(z,t)$, $\theta(z,t)$ and $v(z,t)$ at the shear-band boundaries, (2), (8) and (11), constitute a *mathematically ill-posed problem*. This is clear from the fact that the velocity appears in the above system of p.d.e. with its first spatial derivative only, and accordingly there is no way to absorb two distinct boundary conditions for it. This means in turn that in order to solve the corresponding boundary-value problem some kind of mathematical regularization is needed.

In our previous paper we used a "crude" method of regularization, namely that of replacing the momentum equation for the velocity field with an ad-hoc solution for the velocity that is correct to the leading, linear-in-z term and is compatible with the boundary conditions. Here we justify this assumption by resorting to a more 'rigorous' approach. Accordingly we consider a viscous-type regularization by assuming that the friction coefficient μ is also function of the shearing velocity gradient,

$$\mu = \hat{\mu}(\dot{\gamma}), \quad \dot{\gamma} = \frac{\partial v}{\partial z} . \tag{12}$$

For example we may assume a 'hyperbolic' law of the form

$$\hat{\mu} = \mu_{\text{dyn}} + (\mu_{\text{stat}} - \mu_{\text{dyn}})\frac{1}{1 + \chi\dot{\gamma}} . \tag{13}$$

In this expression $\mu_{\text{stat}}, \mu_{\text{dyn}}$ are the 'static' and 'dynamic' friction coefficients, respectively, and $\chi > 0$ is a material constant with dimension of time, accounting for rate-sensitive behavior (Fig. 3). If we introduce such a shearing velocity-gradient dependency of the friction coefficient into the friction law for the shear stress, the momentum equation (9) is drastically modified, resulting into a diffusion-generation type p.d.e. for the velocity

$$\frac{\partial v}{\partial t} = \nu_m \frac{\partial^2 v}{\partial z^2} - \mu \frac{\partial p}{\partial z} . \tag{14}$$

The coefficient ν_m in front of the second spatial derivative of the velocity plays the role of a *kinematic viscosity*

$$\nu_m = \frac{1}{\varrho_m}(\sigma'_{n0} - p)H . \tag{15}$$

ν_m is given in terms of the friction rate-sensitivity modulus

$$H = \frac{\mathrm{d}\hat{\mu}}{\mathrm{d}\dot{\gamma}} . \tag{16}$$

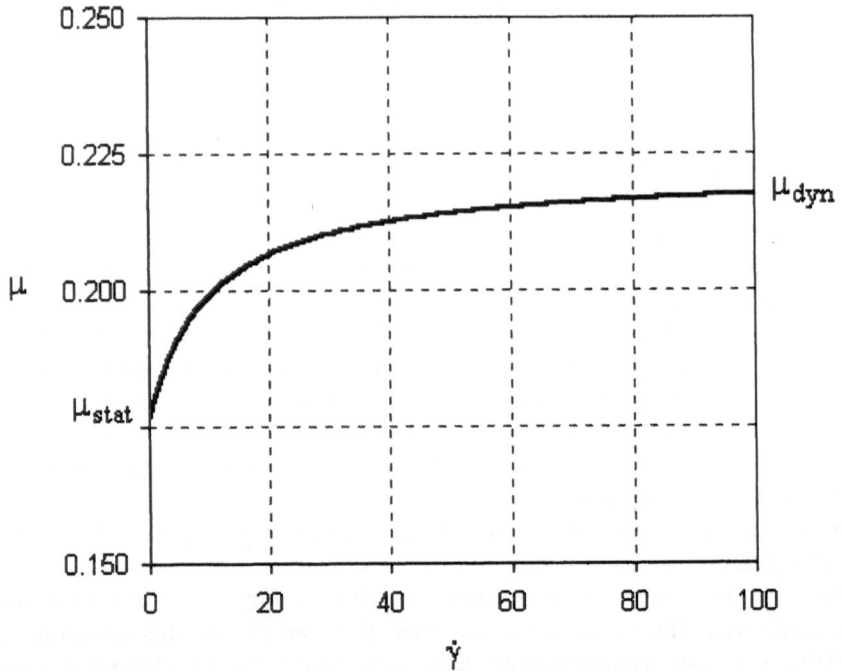

Fig. 3. 'Hyperbolic' friction strain-rate hardening law

3   The Mathematical Model

The problem is re-formulated now in terms of a set of dimensionless variables. For this purpose we select first a set of reference quantities. The geometric scale in the vicinity of the shear-band is set by its thickness d. Thus the shear-band thickness is selected as a reference length

$$d_{\text{ref}} = d \; . \tag{17}$$

The mean initial effective normal stress σ'_{n0} is serving also as a reference pressure or stress,

$$p_{\text{ref}} = \sigma'_{n0} \; . \tag{18}$$

We introduce a 'geostatic' depth

$$h_{\text{ref}} = \frac{\sigma'_{n0}}{\gamma'} \tag{19}$$

where γ' is the effective or buoyant unit weight. For gravity driven phenomena we may use this length scale parameter to define an appropriate reference velocity

$$v_{\text{ref}} = \sqrt{gh_{\text{ref}}} \; , \tag{20}$$

Table 1. System and material parameteres for rapid shearing of a shear band with shearing velocity hardening

ϱ_m	2.435	[g/cm^3]	d_{ref}	0.01	[m]
ν_m	23.887	[kN/m^3]	h_{ref}	169.066	[m]
ν_w	9.81	[kN/m^3]	v_{ref}	40.725	[m/s]
ν	0.18	[-]	t_{ref}	2.46E-04	[s]
ϕ_{stat}	10	[°]	θ_{ref}	12	[°C]
ϕ_{dyn}	15	[°]	σ'_{no}	2.38	[MPa]
χ	0.103	[s]	α	0.2	[-]
c	1.5E-03	[1/MPa]			
kw_{ref}	3.0E-07	[cm/s]			
λ_m	2.0E-02	[MPa/°C]	κ_p	4.9E-05	[-]
$(\varrho_m j C_m)$	2.8E+00	[MPa/°C]	κ_θ	7.4E-07	[-]
c_w	4.9E-04	[1/MPa]	κ_{vo}	2.2E+01	[-]
$c_{v_{\text{ref}}}$	2.0E-05	[m^2/s]	λ	1.0E-01	[-]
κ_m	3.0E-07	[m^2/s]	η_{Fo}	1.3E-02	[-]
v_m	9.0E+00	[m^2/s]	η_{po}	1.0E-01	[-]

and from that we may compute a reference time,

$$t_{\text{ref}} = \frac{d_{\text{ref}}}{v_{\text{ref}}} . \qquad (21)$$

Finally we use the initial ambient temperature θ_0 as reference,

$$\theta_{\text{ref}} = \theta_0 . \qquad (22)$$

With these reference values we define the following set of non-dimensional quantities

$$
\begin{aligned}
z^* &= \frac{z}{d_{\text{ref}}} , \\
t^* &= \frac{t}{t_{\text{ref}}} \geq 0 , \\
v^* &= \frac{v}{v_{\text{ref}}} , \\
p^* &= \frac{p}{p_{\text{ref}}} , \\
\theta^* &= \frac{\theta}{\theta_{\text{ref}}} .
\end{aligned}
\qquad (23)
$$

In addition we define the following dimensionless numbers

$$\kappa_p = \frac{c_v t_{\text{ref}}}{d_{\text{ref}}^2} , \qquad \lambda = \frac{\lambda_m \theta_{\text{ref}}}{\tau_{\text{ref}}} ,$$

$$\kappa_\theta = \frac{\kappa_m t_{ref}}{d_{ref}^2} \ , \qquad \eta_F = \frac{p_{ref}}{(\varrho_m j C_m)\theta_{ref}}\mu \ , \tag{24}$$

$$\kappa_v = \frac{p_{ref} t_{ref}}{\varrho_m d_{ref}^2} H \ , \quad \eta_p = \frac{p_{ref}}{\varrho_m v_{ref}^2}\mu \ .$$

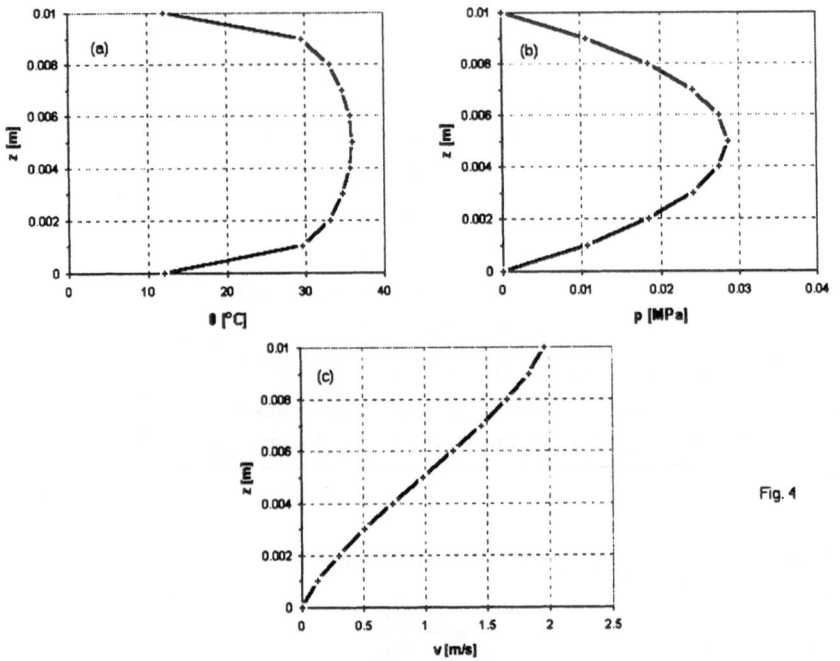

Fig. 4

Fig. 4. Viscous regularization (25): Isochrons for the temperature, heat-generated pore-pressure and velocity at time $t = 1$s (cf. Table 1)

For simplicity in notation, we drop in the resulting set of governing equations the superimposed asterix, and assume that all quantities are dimensionless. The resulting set of the three coupled partial differential equations reads as follows:

$$\frac{\partial p}{\partial t} = \frac{\partial}{\partial z}\left(\kappa_p \frac{\partial p}{\partial z}\right) + \lambda\frac{\partial \theta}{\partial t} \ ,$$

$$\frac{\partial \theta}{\partial t} = \kappa_\theta \frac{\partial^2 \theta}{\partial z^2} + \eta_F(1-p)\frac{\partial v}{\partial z} \ , \tag{25}$$

$$\frac{\partial v}{\partial t} = \kappa_v(1-p)\frac{\partial^2 v}{\partial z^2} - \eta_p\frac{\partial p}{\partial z} \ .$$

The initial conditions for the concerned fields in the domain of definition $0 \leq z \leq 1$ are

$$v(z,0) = 0 \ , \ p(z,0) = 0 \ , \ \theta(z,0) = 1 \ . \tag{26}$$

The boundary conditions are

$$p(0,t) = p(1,t) = 0 \;,$$
$$\theta(0,t) = \theta(1,t) = 1 \;, \tag{27}$$
$$v(0,t) = 0 \;; v(1,t) = v_1(t) \;.$$

As for the boundary velocity is concerned we may assume for example constant acceleration, equal to some fraction $\alpha(0 < \alpha < 1)$ of the gravity acceleration $v_d = \alpha g t$. In Figs. 4(a) to (c) we se the computed isochrons for the temperature and heat-generated pore-pressure and velocity fields. The corresponding material and system parameters are listed in Table 1.

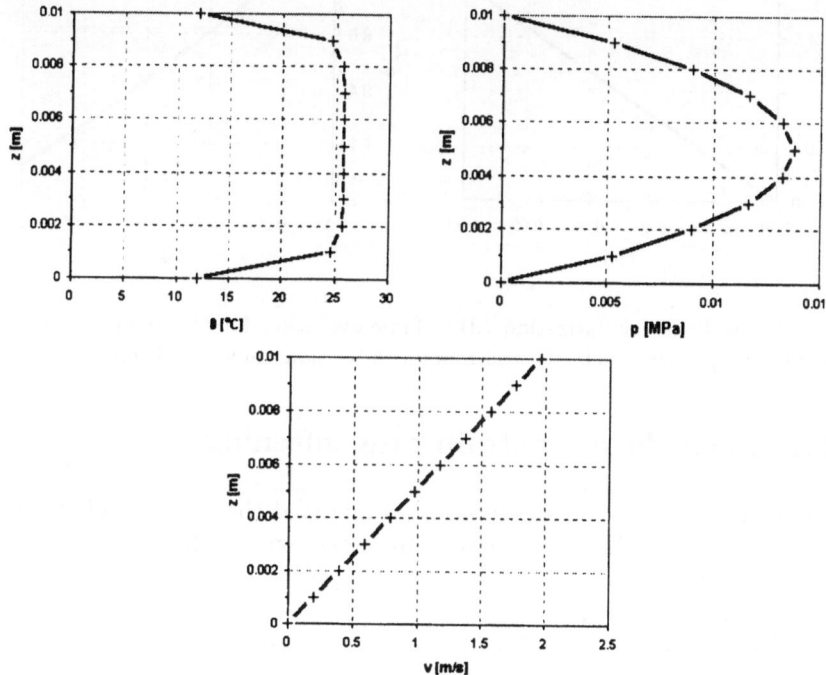

Fig. 5. 2nd gradient regularization (31): Isochrons for the temperature, heat-generated pore-pressure and velocity at time $t = 1$s (cf. Table 2)

One important feature we observe from this typical computation is that, in order to achieve stability, the diffusivity κ_v has to assume relatively high values. If this is not the case, then the velocity profile becomes, after some time, non-linear and results in instability. The value of the diffusivity κ_v is mainly controlled by the hardening modulus H. In the next section we discuss the regularization of the considered problem for positive and negative hardening modulus of relatively small absolute value.

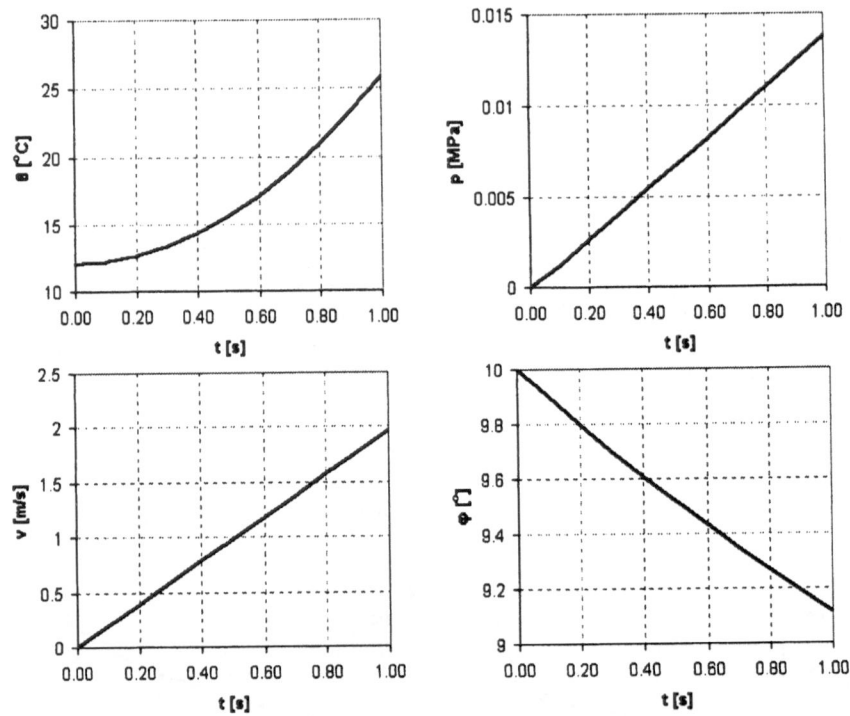

Fig. 6. 2nd gradient regularization (31): Time-evolution for the temperature, heat-generated pore-pressure, velocity and mobilized friction angle (cf. Table 2)

4 Frictional shearing strain-rate softening

In a recent paper Tika and Hutchinson [8] reported the strain- and strain-rate softening of clay soil that was sampled from the exposed failure surface of the Vaiont slide and was tested in the ring shear apparatus of Imperial College. We remark that in a recent paper Di Prisco et. al. ([2]) reported and modeled a similar frictional rate-softening behavior in loose sands, tested in triaxial creep tests.

Notice that in the considered case According to (13) the dynamic friction coefficient is smaller than the static one, and accelerated shear induces additional strength reduction. Friction strain-rate softening results in negative 'kinematic' viscosity, cf. (15). The corresponding momentum balance equation, in terms of dimensionless quantities, has a negative diffusivity coefficient ($\kappa_v \propto H < 0$) and becomes accordingly an 'uphill' diffusion equation

$$\frac{\partial v}{\partial t} = -|\kappa_v|(1-p)\frac{\partial^2 v}{\partial z^2} - \eta_p\frac{\partial p}{\partial z} \tag{28}$$

since for the time running backwards ($t' = -t < 0$), this is an ordinary diffusion-generation equation. Uphill diffusion is known to be mathematically ill posed

Table 2. System and material parameteres for rapid shearing of a shear band with shearing velocity softening

ϱ_m	2.435	[g/cm^3]	d_{ref}	0.01	[m]
ν_m	23.887	[kN/m^3]	h_{ref}	169.066	[m]
ν_w	9.81	[kN/m^3]	v_{ref}	40.725	[m/s]
ν	0.18	[-]	t_{ref}	2.46E-04	[s]
ϕ_{stat}	10	[°]	θ_{ref}	12	[°C]
ϕ_{dyn}	5	[°]	σ'_{n0}	2.38	[MPa]
χ	0.103E-03	[s]	α	0.2	[-]
c	1.5E-03	[1/MPa]			
kw_{ref}	3.0E-09	[cm/s]	ε	0.1	[-]
λ_m	2.0E-02	[MPa/°C]	κ_p	4.9E-05	[-]
$(\varrho_m j C_m)$	2.8E+00	[MPa/°C]	κ_θ	7.4E-07	[-]
c_w	4.93E-04	[1/MPa]	κ_{v0}	2.2E+01	[-]
$c_{v_{ref}}$	2.04E-04	[m^2/s]	λ	1.0E-01	[-]
κ_m	3.0E-07	[m^2/s]	η_{F0}	1.3E-02	[-]
v_m	-8.94E-02	[m^2/s]	η_{p0}	1.0E-01	[-]

(cf. Lattes and Lions [6]), and in the considered setting has no-physical meaning. Following Vardoulakis and Sulem [9] we may regularize such a problem by introducing a 2nd gradient modification of the friction law, by assuming that

$$\mu = \hat{\mu}(\dot{\gamma}) - r\nabla^2\dot{\gamma}, \quad r > 0. \tag{29}$$

Parameter r in (29) has the dimensions: $[r] = TL^2$. If we normalize r by the rate sensitivity parameter χ, with $[\chi] = T$, appearing in the material friction function, (13), we obtain the corresponding 'material length',

$$\ell_c = \sqrt{r/\chi} \rightarrow r = \chi\ell_c^2. \tag{30}$$

This 2nd gradient extension of the friction law yields to the following modification of the momentum equation (28)

$$\frac{\partial v}{\partial t} = -\varepsilon\frac{\partial^4 v}{\partial z^4} - |\kappa_v|(1-p)\frac{\partial^2 v}{\partial z^2} - \eta_p\frac{\partial p}{\partial z} \tag{31}$$

with

$$\varepsilon = r\frac{p_{ref}\,t_{ref}}{\varrho_m\,d_{ref}^4}. \tag{32}$$

According to Lattes and Lions [6] the set of additional boundary conditions that complies with this regularization scheme is a statement for the second spatial derivative of the velocity; e.g.

$$\left(\frac{\partial^2 v}{\partial z^2}\right)_{z=0,1} = 0. \tag{33}$$

Equation (31) is solved together with (25.1) and (25.2), the i.c. (26) and the b.c. (27) and (33), using a Crank-Nicolson integration scheme. Critical for the integration of (31) is the appropriate selection of the singular perturbation parameter ε. The selection of ε depends in turn on the assumed value of the softening modulus

$$H_s = \left| \frac{d\hat{\mu}}{d\hat{\gamma}} \right| . \tag{34}$$

In Fig. 5 we show the computational results for the parameter values listed in Table 2. This analysis suggests that reasonable results were obtained for a value for the perturbation parameter of: $\varepsilon = 0.1$. According to (30) and (32) and for $\chi = 0.1 \cdot 10^{-3}$s, this value corresponds to $\ell_c \approx 0.65$ mm. With a mean particle size for clay material of $d_{50\%} \approx 7\mu$m, this analysis suggests

$$\ell_c \approx 100 \cdot d_{50\%} . \tag{35}$$

Acknowledgement

The Author wants to acknowledge the EU project: *Fault, Fractures and fluids: Golf of Corinth*, in the framework of program Energy (pre-accepted).

References

1. Anderson, D. L. An earthquake induced heat mechanism to explain the loss of strength of large rock and earth slides. *Int. Conf. On Engineering for Protection from natural disasters*, Bangkok, 1980.
2. Di Prisco, C., Imposimato, S. and Vardoulakis, I. Mechanical modeling of drained creep triaxial tests on loose sand. *Geotechnique*, 50:73–82, 2000.
3. Goguel, J. Scale-dependent rockslide mechanisms, with emphasis on the role of pore fluid vaporization. *Developments in Geotechnical Engineering*, Vol. 14a, Rockslides and Avalances, 1, Natural Phenomena. Barry Voight, ed., Elsevier. Ch. 20, pp. 693-705,
4. Habib, P. Sur un mode de glissement des massifs rocheaux. *C. R. Hebd. Seanc. Acad. Sci.*, 264:151–153, Paris, 1967.
5. Habib, P. Production of gaseous pore pressure during rock slides. *Rock Mech.*, 7:193–197, 1975.
6. Lattès, R. and Lions, J. -L. Méthode de Quasi-Réversibilité et Applications, Dunod, Paris, 1967.
7. Romero, S. U. and Molina, R. Kinematic aspects of the Vaiont slide. *Proc. 3rd Congress Int. Soc. Rock Mechanics*, Vol. II, Part B, pages 865–870, Denver, 1974.
8. Tika, Th., E. and Hutchinson, J. N. Ring shear tests on soil from the Vaiont landslide slip surface. *Geotechnique*, 49:59–74, 1999.
9. Vardoulakis, I. and Sulem, J. Bifurcation Analysis in Geomechanics, Chapman & Hall, 1995.
10. Vardoulakis, I. Catastrophic landslides due to frictional heating of the failure plane. *Mech. Coh. Frict. Mat.*, in print.
11. Vermeer, P. A five constant constitutive model unifying well-established concepts. In: *Constitutive Relations for Soils*, G. Gudehus, F. Darve and I. Vardoulakis editors, pages 175–197, Balkema, Rotterdam, 1984.
12. Voight, B. and Faust, C. Frictional heat and strength loss in some rapid landslides. *Geotechnique*, 32:43–54, 1982.

A view on the variational setting of micropolar continua

P. Steinmann

Chair for Applied Mechanics, 67663 Kaiserslautern, University of Kaiserslautern, Germany

Abstract. The objective of this contribution is to elaborate upon the variational setting for micropolar continua with constrained and unconstrained rotations. To this end, several mixed variational principles and their regularizations are considered for both the geometrically linear and nonlinear case. The interrelation between the different formulations are highlighted. The most advantageous result is obtained by translating the insight gained for the geometrically linear case to the geometrically nonlinear case involving large strains and large rotations. It turns out that a particular micropolar description involves standard constitutive models for the symmetric stress part together with a nonsymmetric penalty stress thus circumventing the cumbersome need to describe the constitutive law in terms of a nonsymmetric strain measure.

1 Introduction

Generalized continuum descriptions involving *independent* rotations were introduced by the brothers Cosserat [5] at the beginning of this century. Several decades later further emphasis was directed to this and related theories by a community of researchers among whom we find such prominent members as Günther [9], Grioli [10], Koiter [11], Mindlin [14], Toupin [23] [24], Neuber [17], Schaefer [18], Eringen [7], Lippmann [12] and Besdo [1]. These researchers were obviously attracted by the theoretical challenges and beauties of nonconventional continuum theories. Large scale numerical computations have, obviously, not been in the scope of those years.

Renewed interest for micropolar continua arose recently within the context of inelastic localization computations where boundary layer effects prevail. Thereby, is well understood that the numerical modelling of materials exhibiting softening or nonsymmetric material operators due to e. g. nonassociated flow leads to a pathological mesh dependence of the post peak response within the *classical* continuum description when the deformation pattern obeys a highly localized zone.

The micropolar approach was rediscovered mainly by researchers looking for a remedy for this deficiency of numerical computations within the *classical* continuum theory since it turned out that rotations are an essential ingredient in failure bands where shear failure mechanisms play a dominant role. This new approach to regularize the mesh sensitivity of localization computations was persued e. g. by Mühlhaus & Vardoulakis [16], Mühlhaus [15], de Borst [2][3], Steinmann & Willam [19], Dietsche, Steinmann & Willam [6], Steinmann & Stein [20] and Volk [25] among others.

Nevertheless, it appears that there is a considerable scatter in opinions concerning the appropriate formulation of a *micropolar* continuum theory. For example Fleck & Hutchinson [8] advocate a *constrained* formulation taking into account the *continuum* rotations in order to incorporate higher displacement gradient effects. On the other hand the schools cited in the previous paragraph prefer to formulate a *micropolar* continuum theory in terms of an *independent* rotation field. If there is confusion regarding the more simple geometrically linear case this is even more true in the geometrically nonlinear case involving large strains and rotations. Here the insight into the variational structure of *micropolar* continua is totally blurred from the todays point of view. Therefore, it seems to be in order to clarify the roots of the different formulations and their lateral interrelations. Moreover, it appears to be a pressing task to develop a geometrically nonlinear *micropolar* continuum description extending the *classical* continuum in the most simple way.

Following Steinmann & Stein [22] and to make our derivations as transparent as possible we restrict ourselves to isotropic elastic materials to ensure the existence of a potential energy. Starting from appropriate definitions of strain and curvature measures we will derive the different types of micropolar continua from the corresponding Dirichlet principles. To this end, a combination of strain and curvature measures is selected and the governing differential equations are identified as the Euler-Lagrange equations of the principle of minimum potential energy. For conservative static problems the variation of the total potential energy renders the weak form of the balance equations together with the Neumann boundary conditions. Thereby, the variations of the primary kinematical fields have to satisfy the homogeneous Dirichlet boundary conditions. The weak formulation is equivalent with the principle of virtual displacements (and rotations) and is also valid in the case of nonconservative static problems. The total potential energy is given with W the stored energy function and V^{ext} the external potential energy as

$$V(\bullet) = \int_{\mathcal{B}} W(\bullet) \, dv + V^{ext} \quad \rightarrow \quad \text{stat.} \tag{1}$$

Depending on the arguments of W different sets of differential equations and boundary conditions are obtained. In the sequel we will highlight the relation between the distinct formulations via the introduction of mixed variational principles and regularized mixed principles. First, we will consider the geometrically linear formulations in depth. Guided by these results we will subsequently translate the most important formulations to the geometrically nonlinear case.

2   Geometrically linear micropolar continua

Let $\mathcal{B} \subset \mathbb{R}^3$ denote the domain and $\partial\mathcal{B}$ the boundary occupied by a material body with particles labeled by x. Deformations are described by the standard displacement field $u : \mathcal{B} \to \mathbb{R}^3$ with *continuum* rotation $\Omega = \text{spn}\,\omega = \nabla_x^{skw} u \in$ so(3). In addition an *independent* rotation field with rotation $\bar{\Omega} = \text{spn}\,\bar\omega \in$ so(3) is introduced. Then we define the symmetric *continuum* and the nonsymmetric *micropolar* strain measures

$$\epsilon = \nabla_x u - \Omega \quad \text{and} \quad \bar{\epsilon} = \nabla_x u - \bar{\Omega}. \tag{2}$$

Additionally, the following *continuum* and *micropolar* curvature measures are considered

$$\mathrm{spn}\,\kappa = \nabla_x \Omega \quad \text{and} \quad \mathrm{spn}\,\bar{\kappa} = \nabla_x \bar{\Omega}. \tag{3}$$

Starting from these definitions of strain and curvature measures we will in the sequel highlight the relation between the different types of micropolar continua.

2.1 Gradient type micropolar continuum

First we consider the case of a *constrained* micropolar continuum which corresponds to a continuum incorporating *higher* displacement gradients. This kind of continuum is a special case of a non simple continuum of grade 2 and is connected to the notion of the so called *trièdres cachés*, see Grioli [10], Mindlin & Tiersten [13] or Toupin [23]. The stored energy is assumed to depend on the symmetric strain measure ϵ and the *continuum* curvature measure κ, i. e. the potential energy depends exclusively on the first and second gradients of the displacement field u

$$V^G(u) = \int_B W(\epsilon, \kappa)\, dv + V^{ext}. \tag{4}$$

Thereby, neglecting body forces and body couples the external potential energy V^{ext} is specified as

$$V^{ext} = -\int_{\partial B^\sigma} u \cdot t^p\, da - \int_{\partial B^m} \omega \cdot t_m^p\, da. \tag{5}$$

Variation renders the stationarity condition

$$\delta V^G = \int_B \left[\nabla_x^{sym} \delta u : \sigma(\epsilon) + \nabla_x \delta\omega : m^t(\kappa) \right] dv - G^{ext} := 0 \tag{6}$$

In terms of the symmetric stress and the nonsymmetric couple stress tensor

$$\sigma = \partial_\epsilon W \quad \text{and} \quad m^t = \partial_\kappa W. \tag{7}$$

As usual G^{ext} denotes the contributions to the external virtual work

$$G^{ext} = \int_{\partial B^\sigma} \delta u \cdot t^p\, da + \int_{\partial B^m} \delta\omega \cdot t_m^p\, da. \tag{8}$$

Partial integration together with application of the Gauss divergence theorem renders

$$\delta V^G = -\int_B \left[\delta u \cdot \mathrm{div}\sigma(\epsilon) + \delta\omega \cdot \mathrm{div}m^t(\kappa) \right] dv \tag{9}$$

$$+ \int_{\partial B} \delta u \cdot \sigma(\epsilon) \cdot n\, da + \int_{\partial B} \delta\omega \cdot m^t(\kappa) \cdot n\, da - G^{ext} := 0.$$

Using the identity $\delta\boldsymbol{\omega} \cdot \mathrm{div}\boldsymbol{m}^t(\boldsymbol{\kappa}) = \delta\boldsymbol{\Omega} : \mathrm{spn}\,(\tfrac{1}{2}\mathrm{div}\boldsymbol{m}^t(\boldsymbol{\kappa}))$, introducing the abbreviation

$$\boldsymbol{\xi}(\boldsymbol{\kappa}) := \mathrm{spn}\,(\tfrac{1}{2}\mathrm{div}\boldsymbol{m}^t(\boldsymbol{\kappa})) \quad \rightarrow \quad \delta\boldsymbol{\omega} \cdot \mathrm{div}\boldsymbol{m}^t(\boldsymbol{\kappa}) = -\nabla_x\delta\boldsymbol{u} : \boldsymbol{\xi}^t(\boldsymbol{\kappa}) \qquad (10)$$

and performing again partial integration together with application of the Gauss divergence theorem renders the Euler-Lagrange equations

$$\delta V^G = -\int_B \delta\boldsymbol{u} \cdot \mathrm{div}\,(\boldsymbol{\sigma}(\boldsymbol{\epsilon}) + \boldsymbol{\xi}^t(\boldsymbol{\kappa}))\, dv \qquad (11)$$
$$+ \int_{\partial B^\sigma} \delta\boldsymbol{u} \cdot [\boldsymbol{\sigma}(\boldsymbol{\epsilon}) + \boldsymbol{\xi}^t(\boldsymbol{\kappa})] \cdot \boldsymbol{n}\, da + \int_{\partial B^m} \delta\boldsymbol{\omega} \cdot \boldsymbol{m}^t(\boldsymbol{\kappa}) \cdot \boldsymbol{n}\, da - G^{ext} := 0.$$

Here the homogeneous Dirichlet boundary conditions for $\boldsymbol{u}$ and $\boldsymbol{\omega}$ have been taken into account for the variation of the primary kinematical fields.

Then the system of partial differential equations describing a boundary value problem of the mechanical part of the *constrained* gradient type micropolar continuum theory in $B \subset \mathbb{R}^3$ consists of the balance equations of linear and angular momentum, the kinematic and constitutive relations and a set of Dirichlet and Neumann boundary conditions, see Table 1

Table 1. *Constrained* Gradient Type Case

$$\mathrm{div}\,(\boldsymbol{\sigma}(\boldsymbol{\epsilon}) + \boldsymbol{\xi}^t(\boldsymbol{\kappa})) = 0 \quad \text{in} \quad B$$
$$\mathrm{div}\boldsymbol{m}^t(\boldsymbol{\kappa}) + \overset{3}{\boldsymbol{e}} : \boldsymbol{\xi}(\boldsymbol{\kappa}) := 0 \quad \text{in} \quad B$$

Remark In combining the strong form of the balance of linear momentum and the balance of angular momentum, which comes into play in the form of a definition, the highest displacement gradient in the resulting partial differential equation is of the order four

$$\mathrm{div}\left(\boldsymbol{\sigma}(\nabla_x^{sym}\boldsymbol{u}) - \mathrm{spn}\,(\tfrac{1}{2}\mathrm{div}\boldsymbol{m}^t(\nabla_x\mathrm{axl}(\nabla_x^{skw}\boldsymbol{u})))\right) = 0.$$

Therefore, this kind of continuum compares with Bernoulli beams and Kirchhoff-Love plates and shells.

Remark Observe that the constitutive stress $\boldsymbol{\sigma}(\boldsymbol{\epsilon})$ is symmetric. Therefore, standard constitutive laws calibrated for homogeneous deformations can be used for the stress part. The curvature part then describes the influence of nonhomogeneous deformations.

Remark Note that the balance equation of angular momentum comes into play as a definition for the skewsymmetric stresslike tensor field $\boldsymbol{\xi}$.

Remark Mindlin & Tiersten [13] pointed out that the couple stress Neumann boundary condition may be split up into a normal and a tangential contribution

$$\int_{\partial B^m} \delta\boldsymbol{\omega}\cdot\boldsymbol{m}^t\cdot\boldsymbol{n}\,\mathrm{d}a = \int_{\partial B^m} \delta\boldsymbol{\omega}\cdot[\boldsymbol{I}-\boldsymbol{n}\otimes\boldsymbol{n}]\cdot\boldsymbol{m}^t\cdot\boldsymbol{n}\,\mathrm{d}a + \int_{\partial B^m} \delta\boldsymbol{\omega}\cdot[\boldsymbol{n}\otimes\boldsymbol{n}]\cdot\boldsymbol{m}^t\cdot\boldsymbol{n}\,\mathrm{d}a.$$

With $m_{nn} = \boldsymbol{n}\cdot\boldsymbol{m}^t\cdot\boldsymbol{n}$ the last term is then transformed via partial integration together with application of Stokes theorem into

$$\int_{\partial B^m} m_{nn}\delta\boldsymbol{\omega}\cdot\boldsymbol{n}\,\mathrm{d}a = \frac{1}{2}\oint_C m_{nn}\delta\boldsymbol{u}\cdot\mathrm{d}\boldsymbol{x} - \frac{1}{2}\int_{\partial B^m} \delta\boldsymbol{u}\cdot\mathrm{spn}\,\nabla_x m_{nn}\cdot\boldsymbol{n}\,\mathrm{d}a.$$

Therefore, if ∂B^m is smooth and for m_{nn} uniformly distributed the boundary conditions may alternatively be specified in terms of the *effective* stress vector and the *tangential* couple stress vector as

$$\left[\boldsymbol{\sigma}+\boldsymbol{\xi}^t-\frac{1}{2}\mathrm{spn}\,\nabla_x m_{nn}\right]\cdot\boldsymbol{n}-\bar{\boldsymbol{t}}^p = 0 \quad\text{and}\quad [\boldsymbol{I}-\boldsymbol{n}\otimes\boldsymbol{n}]\cdot\boldsymbol{m}^t\cdot\boldsymbol{n}-\bar{\boldsymbol{t}}^p_m = 0.$$

2.2   Cosserat type micropolar continuum

Next we consider the case of an *unconstrained* micropolar continuum which corresponds to a continuum incorporating an *independent* rotation field, see e. g. the brothers Cosserat [5], Günther [9] or Schaefer [18]. These kinematics are the essential ingredients for a Cosserat continuum and are connected to the notion of the so called *trièdres mobiles*. The stored energy is assumed to depend on the nonsymmetric *micropolar* strain measure $\bar{\boldsymbol{\epsilon}}$ and the *micropolar* curvature measure $\bar{\boldsymbol{\kappa}}$, i. e. the potential energy depends on both, the displacement field $\boldsymbol{u}$ and the independent rotation field $\bar{\boldsymbol{\omega}}$

$$V^C(\boldsymbol{u},\bar{\boldsymbol{\omega}}) = \int_B W(\bar{\boldsymbol{\epsilon}},\bar{\boldsymbol{\kappa}})\,\mathrm{d}v + V^{ext} \tag{12}$$

where V^{ext} represents the external potential energy

$$V^{ext} = -\int_{\partial B^\sigma} \boldsymbol{u}\cdot\boldsymbol{t}^p\,\mathrm{d}a - \int_{\partial B^m} \bar{\boldsymbol{\omega}}\cdot\boldsymbol{t}^p_m\,\mathrm{d}a. \tag{13}$$

Variation renders the stationarity condition

$$\delta V^C = \int_B \left[\nabla_x\delta\boldsymbol{u}:\boldsymbol{\sigma}^t(\bar{\boldsymbol{\epsilon}}) + \nabla_x\delta\bar{\boldsymbol{\omega}}:\boldsymbol{m}^t(\bar{\boldsymbol{\kappa}}) - \delta\bar{\boldsymbol{\Omega}}:\boldsymbol{\sigma}^t(\bar{\boldsymbol{\epsilon}})\right]\mathrm{d}v - G^{ext} := 0. \tag{14}$$

Here, we defined the nonsymmetric stress and couple stress tensors as

$$\boldsymbol{\sigma}^t = \partial_{\bar{\boldsymbol{\epsilon}}}W \quad\text{and}\quad \boldsymbol{m}^t = \partial_{\bar{\boldsymbol{\kappa}}}W \tag{15}$$

As usual G^{ext} denotes the contributions to the external virtual work

$$G^{ext} = \int_{\partial B^\sigma} \delta \boldsymbol{u} \cdot \boldsymbol{t}^p \, da + \int_{\partial B^m} \delta \bar{\boldsymbol{\omega}} \cdot \boldsymbol{t}^p_m \, da. \qquad (16)$$

Partial integration together with application of the Gauss divergence theorem renders the Euler-Lagrange equations

$$\delta V^C = - \int_B \left[\delta \boldsymbol{u} \cdot \mathrm{div} \boldsymbol{\sigma}^t(\bar{\boldsymbol{\epsilon}}) + \delta \bar{\boldsymbol{\omega}} \cdot \left[\mathrm{div} \boldsymbol{m}^t(\bar{\boldsymbol{\kappa}}) + \overset{3}{\boldsymbol{e}} \colon \boldsymbol{\sigma}(\bar{\boldsymbol{\epsilon}}) \right] \right] dv \qquad (17)$$

$$+ \int_{\partial B^\sigma} \delta \boldsymbol{u} \cdot \boldsymbol{\sigma}^t(\bar{\boldsymbol{\epsilon}}) \cdot \boldsymbol{n} \, da + \int_{\partial B^m} \delta \bar{\boldsymbol{\omega}} \cdot \boldsymbol{m}^t(\bar{\boldsymbol{\kappa}}) \cdot \boldsymbol{n} \, da - G^{ext} := 0.$$

Here the homogeneous Dirichlet boundary conditions for $\boldsymbol{u}$ and $\bar{\boldsymbol{\omega}}$ have been taken into account for the variation of the primary kinematical fields.

Then the system of partial differential equations describing a boundary value problem of the mechanical part of the *unconstrained* Cosserat micropolar continuum theory in $B \subset \mathbb{R}^3$ consists of the balance equations of linear and angular momentum, the kinematic and constitutive relations and a set of Dirichlet and Neumann boundary conditions, see Table 2

Table 2. *Unconstrained* Cosserat Type Case

$$
\begin{aligned}
\mathrm{div} \boldsymbol{\sigma}^t(\bar{\boldsymbol{\epsilon}}) &= 0 \quad \text{in} \quad B \\
\mathrm{div} \boldsymbol{m}^t(\bar{\boldsymbol{\kappa}}) + \overset{3}{\boldsymbol{e}} \colon \boldsymbol{\sigma}(\bar{\boldsymbol{\epsilon}}) &= 0 \quad \text{in} \quad B
\end{aligned}
$$

Remark In combining the strong form of the balance of linear momentum and the balance of angular momentum the highest displacement gradient in the resulting partial differential equation is of the order two and the highest gradient of the *independent* rotation field is of the order three

$$\mathrm{div} \left(\boldsymbol{\sigma}^{sym}(\nabla_x \boldsymbol{u} - \bar{\boldsymbol{\varOmega}}) - \mathrm{spn} \left(\frac{1}{2} \mathrm{div} \boldsymbol{m}^t (\nabla_x \bar{\boldsymbol{\omega}}) \right) \right) = 0.$$

Therefore, the Cosserat continuum compares with Timoshenko beams and Mindlin-Reissner plates and shells.

Remark Observe that the constitutive stress $\boldsymbol{\sigma}(\bar{\boldsymbol{\epsilon}})$ is nonsymmetric. Therefore, standard constitutive laws calibrated for homogeneous deformations can only be used for the symmetric stress part. The skewsymmetric stress part together with the curvature part describes the influence of nonhomogeneous deformations.

2.3   Mixed formulation gradient type case

The Dirichlet principle for the *constrained* micropolar continuum incorporates second order gradients of the displacement field $\boldsymbol{u}$. As a consequence, the displacement expansions of e. g. a numerical Galerkin-Bubnov solution scheme have to satisfy C^1 continuity, e. g. by Hermite expansions. In order to circumvent these severe continuity requirements and to allow for C^0 low order displacement expansions we adhere to a mixed regularized variational principle. To this end, an *auxiliary* rotation field with spin tensor $\tilde{\Omega} \in so(3)$ and *auxiliary* curvature $\tilde{\kappa} = \nabla_x \tilde{\omega}$ is introduced under the subsidary condition $[\nabla_x u - \tilde{\Omega}]^{skw} = 0$ which has to be enforced by a skewsymmetric stresslike Lagrange multiplier $\ell \in so(3)$. The stored energy is assumed to depend on the symmetric strain measure ϵ and the *auxiliary* curvature measure $\tilde{\kappa}$, i. e. the potential energy depends on both, the displacement field u and the *auxiliary* rotation field $\tilde{\omega}$. The resulting mixed functional is then given as

$$V_\ell^G(u, \tilde{\omega}, \ell) = \int_B \left[W(\epsilon, \tilde{\kappa}) + \ell^t : [\nabla_x u - \tilde{\Omega}]^{skw} \right] dv + V^{ext}. \qquad (18)$$

The associated Euler-Lagrange equations summarized in the following Table 3 coincide with those of the *constrained* gradient type micropolar continuum.

Table 3. Mixed Formulation *Constrained* Gradient Type Case

$$\mathrm{div}\left(\sigma(\epsilon) + \ell^t\right) = 0 \quad \mathrm{in} \quad B$$

$$\mathrm{div} m^t(\tilde{\kappa}) + \overset{3}{e} : \ell = 0 \quad \mathrm{in} \quad B$$

$$[\nabla_x u - \tilde{\Omega}]^{skw} = 0 \quad \mathrm{in} \quad B$$

2.4 Regularized mixed formulation gradient type case

From a numerical point of view it is often convenient to regularize the variational principle in the Lagrange multiplier. Therefore, we consider the mixed regularized functional with $2\mu_c \in \mathbb{R}_+$ the regularization parameter

$$V_\mu^G(u, \tilde{\omega}) = \int_B \left[W(\epsilon, \tilde{\kappa}) + \mu_c [\nabla_x u - \tilde{\Omega}]^{skw} : [\nabla_x u - \tilde{\Omega}]^{skw} \right] dv + V^{ext}. \quad (19)$$

Clearly, the penalty parameter μ_c has to approach infinity $\mu_c \to \infty$ in order to retrofit the original problem. Variation then renders the stationarity condition

$$\int_B \left[\nabla_x \delta u : [\sigma(\epsilon) + p^t] + \nabla_x \delta \tilde{\omega} : m^t(\tilde{\kappa}) - \delta \tilde{\Omega} : p^t \right] dv - G^{ext} := 0. \qquad (20)$$

Here, we identified $2\mu_c [\nabla_x u - \tilde{\Omega}]^{skw}$ with the skewsymmetric stress p^t of the *unconstrained* micropolar continuum, i. e. $2\mu_c [\nabla_x u - \tilde{\Omega}]^{skw} := p^t$, thus highlighting the connection between the regularized version of the *constrained* gradient type micropolar continuum and the *unconstrained* Cosserat formulation.

We therefore conclude that the Euler-Lagrange equations of the regularized *con-strained* gradient type micropolar variational principle with relaxed subsidary condition $[\nabla_x u - \Omega]^{skw} = 0$ in Table 4 coincide with those of the linear elastic *unconstrained* Cosserat variational principle. The Euler-Lagrange equations of the original *constrained* micropolar continuum are retrofitted as $\mu_c \to \infty$.

Table 4. Regularized Mixed Formulation *Constrained* Gradient Type Case

$$\operatorname{div}\left(\sigma(\epsilon) + p^t\right) \qquad = 0 \quad \text{in} \quad \mathcal{B}$$
$$\operatorname{div} m^t(\tilde{\kappa}) + \overset{3}{e} : p \qquad = 0 \quad \text{in} \quad \mathcal{B}$$
$$p^t - 2\mu_c[\nabla_x u - \tilde{\Omega}]^{skw} = 0 \quad \text{in} \quad \mathcal{B}$$

Remark Within a computational setting penalty procedures become dramatically ill-conditioned as the penalty parameter is increased. These calamities may be avoided by the means of an *augmented* Lagrange multiplier method. Thereby, the *augmented* Lagrange multiplier method allows to satisfy the subsidary condition $[\nabla_x u - \tilde{\Omega}]^{skw} = 0$ with arbitrary accuracy.

3   Geometrically nonlinear micropolar continua

Let $\mathcal{B}_0 \subset \mathbb{R}^3$ denote the domain and $\partial\mathcal{B}_0$ the boundary occupied by a material body with particles labeled by X in the reference configuration. Deformations are described by the standard nonlinear map $\varphi(X) : \mathcal{B}_0 \to \mathcal{B}$ taking particles labeled by their position X in the reference configuration $\mathcal{B}_0$ to their placement $x = \varphi(X)$ in the spatial configuration $\mathcal{B}$. Then the associated linear tangent map follows as the deformation gradient $F = \nabla_X \varphi : T\mathcal{B}_0 \to T\mathcal{B}$ with *continuum* rotation $R \in \mathrm{SO}(3)$. In addition an *independent* rotation field with rotation tensor $\bar{R} \in \mathrm{SO}(3)$ is introduced. Variation leads to the corresponding spatial spin tensors, i. e.

$$\Omega = \delta R \cdot R^t = \operatorname{spn}\omega \in \mathrm{so}(3) \quad \text{and} \quad \bar{\Omega} = \delta\bar{R} \cdot \bar{R}^t = \operatorname{spn}\bar{\omega} \in \mathrm{so}(3), \qquad (21)$$

respectively. Then we define the symmetric *continuum* stretch and the nonsymmetric *micropolar* stretch as strain measures

$$U = R^t \cdot \nabla_X \varphi \quad \text{and} \quad \bar{U} = \bar{R}^t \cdot \nabla_X \varphi. \qquad (22)$$

Additionally, the following second order *continuum* and *micropolar* curvature measures are considered

$$\operatorname{spn} K = R^t \cdot \nabla_X R \quad \text{and} \quad \operatorname{spn}\bar{K} = \bar{R}^t \cdot \nabla_X \bar{R}. \qquad (23)$$

Taking directional derivatives the following relations are established, see Steinmann [21]

$$\delta \bar{U} = \bar{R}^t [\nabla_x \delta x - \bar{\Omega}] \cdot F \quad \text{and} \quad \delta \bar{K} = \bar{R}^t \cdot \nabla_X \bar{\omega} \quad \text{and} \quad \delta K = R^t \cdot \nabla_X \omega. \quad (24)$$

Based on these definitions we will extend the results obtained for the geometrically linear case to the geometrically nonlinear case involving large strains and large rotations.

3.1   Mixed formulation gradient type case

First we consider the case of the geometrically nonlinear version of a *constrained* micropolar continuum which corresponds to a continuum incorporating *higher* deformation gradients. In accordance with the requirements of objectivity the stored energy is assumed to depend on the symmetric right stretch tensor U and the *continuum* right curvature tensor K. Since the right stretch tensor U is symmetric it proofs convenient to substitute it by the right Cauchy-Green tensor $C = U^2$. Therefore, the potential energy depends exclusively on the first and second gradients of the deformation map $x = \varphi(X)$

$$V^G(x) = \int_{\mathcal{B}_0} W(C, K) \, dV + V^{ext}. \quad (25)$$

Guided by the results obtained for the geometrically linear theory and in view of the difficulties encountered when operating directly in terms of the *continuum* rotation R we prefer to start from the onset with a mixed formulation where we introduce the *auxiliary* rotation $\tilde{R}$ with corresponding spatial spin tensor $\tilde{\Omega} = \delta \tilde{R} \cdot \tilde{R}^t = \text{spn}\,\tilde{\omega} \in \text{so}(3)$ and *auxiliary* curvature measure $\text{spn}\,\tilde{K} = \tilde{R}^t \cdot \nabla_X \tilde{R}$. Thereby, the skewsymmetric Biot stresslike Lagrange parameter $\Lambda \in \text{so}(3)$ enforces the subsidary condition $[\tilde{R}^t \cdot F]^{skw} = \tilde{U}^{skw} = 0$. The resulting mixed functional is then given as

$$V_\ell^G(x, \tilde{R}, \Lambda) = \int_{\mathcal{B}_0} \left[W(C, \tilde{K}) + \Lambda^t : [\tilde{R}^t \cdot F]^{skw} \right] dV + V^{ext}. \quad (26)$$

Variation of the restricted functional renders the stationarity condition

$$\delta V_\ell^G = \int_{\mathcal{B}_0} \left[\nabla_X \delta x : [F \cdot S(C) + \tilde{R} \cdot \Lambda^t] + \delta \Lambda^t : [\tilde{R}^t \cdot F]^{skw} \right] dV \quad (27)$$

$$+ \int_{\mathcal{B}_0} \left[\nabla_X \tilde{\omega} : [\tilde{R} \cdot M^t(K)] - \tilde{\Omega} : [\tilde{R} \cdot \Lambda^t \cdot F^t] \right] dV - G^{ext} := 0.$$

Thereby, the constitutive law for the symmetric 2. Piola-Kirchhoff stress and the nonsymmetric Biot couple stress was introduced as

$$S = 2\partial_C W \quad \text{and} \quad M^t = \partial_{\tilde{K}} W. \quad (28)$$

Invoking spatial stress and couple stress quantities via *push-forward* as

$$J\sigma = F \cdot S \cdot F^t \quad \text{and} \quad Jm^t = \tilde{R} \cdot M^t \cdot F^t \quad \text{and} \quad J\ell^t = \tilde{R} \cdot \Lambda^t \cdot F^t \quad (29)$$

we arrive at the spatial formulation in terms of Cauchy stress and couple stress measures

$$\int_{\mathcal{B}_0} \left[\nabla_x \delta x : [\sigma + \ell^t] + \nabla_x \tilde{\omega} : m^t - \tilde{\Omega} : \ell^t + \delta \Lambda^t : [\tilde{R}^t \cdot F]^{skw} \right] dV - G^{ext}. \quad (30)$$

Additionally, neclecting body forces and body couples the external virtual work G^{ext} is specified as

$$G^{ext} = \int_{\partial \mathcal{B}_0^\sigma} \delta x \cdot T^p \, dA + \int_{\partial \mathcal{B}_0^m} \tilde{\omega} \cdot T_m^p \, dA. \quad (31)$$

The remaining steps to obtain the Euler-Lagrange equations are standard and therefore omitted.

Then the system of partial differential equations describing a boundary value problem of the mechanical part of the *constrained* micropolar continuum theory in the spatial configuration $\mathcal{B} \subset \mathbb{R}^3$ consists of the balance equations of linear and angular momentum, the kinematic and constitutive relations and a set of Dirichlet and Neumann boundary conditions, see Table 5

Table 5. Nonlinear Mixed *Constrained* Gradient Type Case

$$\mathrm{div}\left(\sigma + \ell^t\right) = 0 \quad \text{in} \quad \mathcal{B}$$

$$\mathrm{div} m^t + \overset{3}{e} : \ell = 0 \quad \text{in} \quad \mathcal{B}$$

$$\left[\tilde{R}^t \cdot F\right]^{skw} = 0 \quad \text{in} \quad \mathcal{B}_0$$

Remark Observe that here the constitutive Cauchy stress σ is symmetric. Therefore we could invoke standard constitutive laws calibrated with homogeneous deformations for the stress part. Again the curvature part is only activated in the case of nonhomogeneous deformations.

3.2   Cosserat type micropolar continuum

Next we consider the geometrically nonlinear case of an *unconstrained* Cosserat micropolar continuum incorporating an *independent* rotation field. Here the stored energy is assumed to depend on the nonsymmetric micropolar stretch tensor $\bar{U}$ and the right curvature tensor $\bar{K}$, i. e. the potential energy depends on both, the nonlinear deformation map $x = \varphi(X)$ and the independent *micropolar rotation* $\bar{R}$

$$V^C(x, \bar{R}) = \int_{\mathcal{B}_0} W(\bar{U}, \bar{K}) \, dV + V^{ext} \quad (32)$$

Variation renders the stationarity condition expressed in terms of Biot stress and couple stress measures

$$\int_{B_0} \left[\nabla_X \delta x : [\bar{R} \cdot \Sigma^t] + \nabla_X \bar{\omega} : [\bar{R} \cdot M^t] - \bar{\Omega} : [\bar{R} \cdot \Sigma^t \cdot F^t] \right] dV - G^{ext}.$$

Thereby, the constitutive law for the nonsymmetric Biot stress and couple stress measures was introduced as

$$\Sigma^t = \partial_{\bar{U}} W \quad \text{and} \quad M^t = \partial_K W. \tag{33}$$

By *push-forward*, i. e. invoking spatial Cauchy stress and couple stress measures

$$J\sigma^t = \bar{R} \cdot \Sigma^t \cdot F^t \quad \text{and} \quad Jm^t = \bar{R} \cdot M^t \cdot F^t \tag{34}$$

the stationarity condition is expressed as

$$\delta V^C = \int_B \left[\nabla_x \delta x : \sigma^t + \nabla_x \bar{\omega} : m^t - \bar{\Omega} : \sigma^t \right] dV - G^{ext} := 0. \tag{35}$$

As usual G^{ext} denotes the contributions to the external virtual work

$$G^{ext} = \int_{\partial B_0 \sigma} \delta x \cdot t^p \, dA + \int_{\partial B_0 m} \bar{\omega} \cdot t_m^p \, dA. \tag{36}$$

Partial integration together with application of the Gauss divergence theorem renders the Euler-Lagrange equations. The system of partial differential equations describing a boundary value problem of the mechanical part of the *unconstrained* micropolar continuum theory in the spatial configuration $B \subset \mathbb{R}^3$ consists of the balance equations of linear and angular momentum, the kinematic and constitutive relations and a set of Dirichlet and Neumann boundary conditions, see Table 6

Table 6. Nonlinear *Unconstrained* Cosserat Type Case

$$\text{div}\sigma^t \qquad = 0 \quad \text{in} \quad B$$
$$\text{div}m^t + \overset{3}{e} : \sigma = 0 \quad \text{in} \quad B$$

Remark Observe that the constitutive Cauchy stress σ is in general nonsymmetric thus neccesitating nonstandard constitutive laws for the stress part.

3.3   Regularized formulation gradient type case

A regularized functional for the mixed formulation of the *constraint* gradient type micropolar continuum with $\mu_c \in \mathbb{R}_+$ the regularization parameter is given by

$$V_{\mu}^{G}(\boldsymbol{x}, \tilde{\boldsymbol{R}}) = \int_{\mathcal{B}_0} \left[W(\boldsymbol{C}, \tilde{\boldsymbol{K}}) + \mu_c [\tilde{\boldsymbol{R}}^t \cdot \boldsymbol{F}]^{skw} : [\tilde{\boldsymbol{R}}^t \cdot \boldsymbol{F}]^{skw} \right] \mathrm{d}V + V^{ext}. \quad (37)$$

Recall, that the penalty parameter μ_c has to approach infinity $\mu_c \to \infty$ in order to retrofit the original problem. Variation renders the stationarity condition

$$\int_{\mathcal{B}_0} \left[\nabla_X \delta \boldsymbol{x} : [\boldsymbol{F} \cdot \boldsymbol{S} + \tilde{\boldsymbol{R}} \cdot \boldsymbol{\Pi}^t] + \nabla_X \tilde{\boldsymbol{\omega}} : [\tilde{\boldsymbol{R}} \cdot \boldsymbol{M}^t] - \tilde{\boldsymbol{\Omega}} : [\tilde{\boldsymbol{R}} \cdot \boldsymbol{\Pi}^t \cdot \boldsymbol{F}^t] \right] \mathrm{d}V - G^{ext}.$$

Here, we reiterate the constitutive law for the 2. Piola-Kirchhoff stress and the Biot couple stress

$$\boldsymbol{S} = 2 \partial_{\boldsymbol{C}} W \quad \text{and} \quad \boldsymbol{M}^t = \partial_{\tilde{\boldsymbol{K}}} W \quad \text{and} \quad \boldsymbol{\Pi}^t = 2 \mu_c [\tilde{\boldsymbol{R}}^t \cdot \boldsymbol{F}]^{skw}. \quad (38)$$

Additionally, we identified $2\mu_c [\tilde{\boldsymbol{R}}^t \cdot \boldsymbol{F}]^{skw}$ with the skewsymmetric Biot stress $\boldsymbol{\Pi}^t$. Obviously, we arrived at a special kind of an *unconstrained* micropolar continuum with $\boldsymbol{\Pi}^t = 2\mu_c \tilde{\boldsymbol{U}}^{skw}$ being the skewsymmetric Biot stress of the *unconstrained* micropolar continuum with linear relation between the skewsymmetric parts of the micropolar Biot stress $\boldsymbol{\Sigma}^t$ and the micropolar right stretch tensor $\tilde{\boldsymbol{U}}$. Thus the connection between the regularized version of the *constrained* micropolar continuum and the *unconstrained* Cosserat formulation is established. The Euler-Lagrange equations of the original *constrained* micropolar continuum are retrofitted as $\mu_c \to \infty$. Spatial stress and couple stress measures follow by *push-forward*

$$J\boldsymbol{\sigma} = \boldsymbol{F} \cdot \boldsymbol{S} \cdot \boldsymbol{F}^t \quad \text{and} \quad J\boldsymbol{m}^t = \tilde{\boldsymbol{R}} \cdot \boldsymbol{M}^t \cdot \boldsymbol{F}^t \quad \text{and} \quad J\boldsymbol{p}^t = \tilde{\boldsymbol{R}} \cdot \boldsymbol{\Pi}^t \cdot \boldsymbol{F}^t. \quad (39)$$

Then the stationarity condition is expressed in spatial Cauchy stress and couple stress measures

$$\delta V_{\mu}^{G} = \int_{\mathcal{B}} \left[\nabla_x \delta \boldsymbol{x} : [\boldsymbol{\sigma} + \boldsymbol{p}^t] + \nabla_x \tilde{\boldsymbol{\omega}} : \boldsymbol{m}^t - \tilde{\boldsymbol{\Omega}} : \boldsymbol{p}^t \right] \mathrm{d}v - G^{ext}. \quad (40)$$

The system of partial differential equations describing a boundary value problem of the mechanical part of the regularized *constrained* micropolar continuum theory in the spatial configuration is summarized in the following Table 7

Table 7. Nonlinear Regularized Mixed *Constrained* Gradient Type Case

$$\begin{aligned}
\operatorname{div}(\boldsymbol{\sigma} + \boldsymbol{p}^t) &= 0 \quad \text{in} \quad \mathcal{B} \\
\operatorname{div}\boldsymbol{m}^t + \overset{3}{\boldsymbol{e}} : \boldsymbol{p} &= 0 \quad \text{in} \quad \mathcal{B} \\
2\mu_c [\tilde{\boldsymbol{R}} \cdot \boldsymbol{F}]^{skw} - \boldsymbol{\Pi}^t &= 0 \quad \text{in} \quad \mathcal{B}_0
\end{aligned}$$

Remark Observe that the constitutive Cauchy stress σ is symmetric while the penalty Cauchy stress p is in general nonsymmetric.

4 Conclusion

It has been the aim of this work to highlight the variational interrelation between two prominent formulations of micropolar continua : the gradient type continuum with *constrained* rotations and the Cosserat type continuum with *unconstrained* rotations. In the first part of this paper mixed and regularized mixed variational principles for the geometrically linear case have been discussed in detail. The main result is the coincidence of the regularized mixed *constrained* gradient type and the *unconstrained* Cosserat type formulation. Thereby, the former principle invokes symmetric strains and stresses with a skewsymmetric penalty stress which is in contrast to the later principle. The benefit of the regularized formulation is apparant since standard constitutive laws may be invoked for the stress part.

On the other hand the *unconstrained* rotations of the Cosserat continuum may be restricted to coincide with the *continuum* rotations in order to end up with the *constrained* gradient type continuum. Thereby, the cumbersome C^1 continuity requirements of the original *constrained* formulation are circumvented.

The results obtained for the geometrically linear case have been compiled to the geometrically nonlinear case in the second part of the paper. Besides the formulation of nonlinear *constrained* gradient type and *unconstrained* Cosserat type continua the essential result is the interpretation of the regularized mixed *constrained* gradient formulation as a particular case of an *unconstrained* Cosserat continuum. Standard constitutive models for the stress part relying on the symmetric Cauchy-Green strain may be invoked, thus alleviating the difficulties encountered if one whishes to formulate the stress response in terms of the nonsymmetric micropolar Biot strain. Here the nonsymmetric stress contributions are incorporated merely due to the penalty formulation.

It is believed that the results achieved in this work represent a point of departure for an elegant and efficient implementation of the geometrically nonlinear micropolar continuum into a numerical solution scheme.

References

1. BESDO D., [1974], "Ein Beitrag zur nichtlinearen Theorie des Cosserat-Kontinuums", *Acta Mech.*, Vol. 20, pp. 105-131.
2. BORST R. DE, [1991], "Simulation of Strain Localization : A Reappraisal of the Cosserat Continuum", *Engr. Comp.*, Vol. 8, pp. 317-332.
3. BORST R. DE, [1993], "A Generalisation of J_2-Flow Theory for Polar Continua", *Comp. Meth. Appl. Mech. Engr.*, Vol. 103, pp. 347-362.
4. BORST R. DE & H. B. MÜHLHAUS, [1992], "Finite Deformation Analysis of Inelastic Materials with Micro-Structure", in *Proc. Iutam Symp. on Finite Inelastic Deformations, Hannover 1991*, Eds. D. Besdo & E. Stein, Springer-Verlag, Berlin etc.

5. COSSERAT E. & F. COSSERAT, [1909], *Théorie des Corps Déformables*, Librairie Scientifique A. Hermann et Fils, Paris.
6. DIETSCHE A., P. STEINMANN & K. WILLAM, [1992], "Micropolar Elasto-Plasticity and its Role in Localization Analysis", *Int. J. Plast.*, Vol. 9, pp. 813-831.
7. ERINGEN A. C., [1968], "Theory of Micropolar Elasticity", in *Fracture, An Advanced Treatise*, Ed. H. Liebowitz, Academic Press, New York etc.
8. FLECK N. A. & J. W. HUTCHINSON, [1993], "A Phenomenological Theory for Strain Gradient Effects in Plasticity", *J. Mech. Phys. Solids*, Vol. 41, pp. 1825-1857.
9. GÜNTHER W., [1958], "Zur Statik und Kinematik des Cosseratschen Kontinuums", *Abh. Braunschweig. Wiss. Ges.*, Vol. 10, pp. 195-213.
10. GRIOLI G., [1960], "Elasticità Asimmetrica", *Ann. di Mat. Pura ed Appl.*, Ser. IV 50, pp. 389-417.
11. KOITER W. T., [1964], "Couple-Stresses in the Theory of Elasticity, I & II", in *Proc. Roy. Netherlands Acad. Sci., B67*, pp. 17-44.
12. LIPPMANN H., [1969], "Eine Cosserat-Theorie des plastischen Fließens", *Acta Mech.*, Vol. 8, pp. 255-284.
13. MINDLIN R. D. & H. F. TIERSTEN, [1962], "Effects of Couple-Stresses in Linear Elasticity", *Arch. Rational. Mech. Anal.*, Vol. 11, p. 415-448.
14. MINDLIN R. D., [1964], "Microstructure in Linear Elasticity", *Arch. Rational. Mech. Anal.*, Vol. 16, pp. 51-78.
15. MÜHLHAUS H. B., [1989], "Application of Cosserat Theory in Numerical Solutions of Limit Load Problems", *Ing. Arch.*, Vol. 59, pp. 124-137.
16. MÜHLHAUS H.B. & I. VARDOULAKIS, [1987], "The Thickness of Shear Bands in Granular Materials", *Géotechnique*, Vol. 37, pp. 271-283.
17. NEUBER H., [1966], "Über Probleme der Spannungskonzentration im Cosseratkörper", *Acta Mech.*, Vol. 2, pp. 48-69.
18. SCHAEFER H., [1967], "Das Cosserat-Kontinuum", *Z. Angew. Math. Mech.*, Vol. 47, pp. 485-498.
19. STEINMANN P. & K. WILLAM, [1991], "Localization within the Framework of Micropolar Elasto-Plasticity", in *Advances in Continuum Mechanics*, Eds. O. Brüller, V. Mannl & J. Najar, Springer-Verlag, Berlin etc.
20. STEINMANN P. & E. STEIN, [1994], "Finite Element Localization Analysis of Micropolar Strength Degrading Materials", in *Proc. EUROC'94, Innsbruck, Austria*, Eds. H. Mang, N. Bicanic & R. de Borst.
21. STEINMANN P., [1994], "A Micropolar Theory of Finite Deformation and Finite Rotation Multiplicative Elastoplasticity", *Int. J. Solids Structures*, Vol 31, pp. 1063-1084.
22. STEINMANN P. & E. STEIN, [1997], "A Unifying Treatise of Variational Principles for two Types of Micropolar Continua", *Acta Mechanica*, Vol 121, pp. 215-232.
23. TOUPIN R. A., [1962], "Elastic Materials with Couple-Stresses", *Arch. Rational Mech. Anal.*, Vol. 11, pp. 385-414.
24. TOUPIN R. A., [1964], "Theory of Elasticity with Couple-Stress", *Arch. Rational Mech. Anal.*, Vol. 17, pp. 85-112.
25. VOLK W., [1999], *Untersuchung des Lokalisierungsverhaltens mikropolarer poröser Medien mit Hilfe der Cosserat Theorie*, Diss, Uni Stuttgart.

Macromodelling of softening
in non-cohesive soils

T. Marcher, P. A. Vermeer

Institute of Geotechnical Engineering, University of Stuttgart, Germany

Abstract. Considering non-cohesive material, the void ratio is used to control softening on the friction angle. In order to obtain well-posed boundary-value problems, the softening parameter is defined as a nonlocal state parameter, so that computational results remain independent of the FEM-discretization. The model is evaluated through a comparison of numerical analysis and experimental data on Hostun sand. For a calibration of the constitutive model triaxial tests and oedometer tests with near-homogeneous deformations were used. Biaxial tests with strong shear-banding for dense sand were used to study the evolution of strain localization. The resulting shear band thickness, the role of the internal length and the input of combining both material and geometric softening are discussed.

1 Introduction

In non-cohesive sand the degradation of the friction angle evolves from dilatant plastic shearing, being a consequence of particle motion and particle rearrangement. Especially densely packed sands become looser during plastic flow until reaching the so-called critical-(steady) state. During this process the peak friction angle, φ, decreases down to the critical state friction angle, φ_{cs}. Due to the fact that the number of particle contacts is hardly changed during dilatant shearing, the unloading stiffness remains insensitive to this type of material softening. As a consequence, constitutive models have to be formulated within the theory of rate-independent plasticity [1] rather than the damage theory of fracture mechanics [2]. The promising continuum models, as proposed in current mechanics literature, are Cosserat models [3] and nonlocal models [4], including gradient plasticity [5, 6]. In this paper the simplest framework of modelling is chosen, i.e. nonlocal plasticity as originally proposed by Eringen [7, 8] and extended by Bazant [9] a.o..

In contrast to concrete, very little quantitative comparisons of numerical and experimental data have been published for geological materials. To improve this situation, attention is focused on the well-known test results of Hostun 'RF' sand [10]. A large database on Hostun sand containing the results of axisymmetric and plane strain tests under drained conditions is used. The main part of the tests were performed at Grenoble [11–14].

The paper begins by formulating friction softening as an extension of the simple Drucker-Prager model. It is shown that the local form leads to considerable mesh-dependency of the post-peak load-displacement curve. On the other hand, the nonlocal form, which involves the input of an internal length, shows

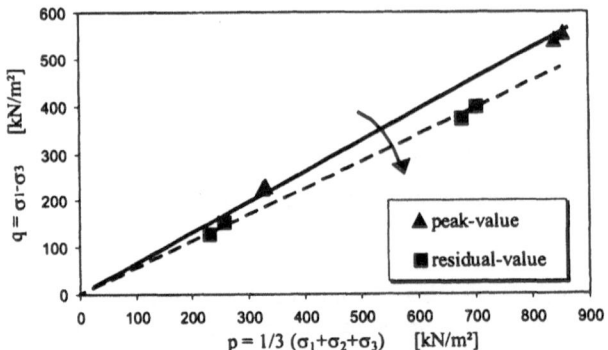

Fig. 1. Rotation of the strength envelope from triaxial test data on 'dense' Hostun sand

fully regularized curves and a qualitative agreement with experimental data. For obtaining quantitative agreement the simple Drucker-Prager model is replaced by a more advanced Hardening Soil model which includes deviatoric hardening/softening, a MC-yield criterion and addionally a cap [15–17]. First of all, it is shown that this model is well able to predict triaxial test data. Then full attention is paid to shear-banding in biaxial tests. For a quantitative back-analysis, both the use of an advanced nonlocal model and its embedment in an updated Lagrangian FE-formulation appears to be essential.

2 Approach to friction softening

Softening behaviour of dense sand is always accompanied with large changes of the void ratio in the zone of localization and involves a reduction of the friction angle. Using the triaxial test results for the 'dense' Hostun sand, one can observe in the stress-plane of Fig. 1 that the strength envelope rotates from a peak value to the residual critical state. It appears that the degradation of the friction angle can be specified by the void ratio e as a softening parameter and a softening modulus h_φ. A simple linear relation is indicated in Eq. (1), where a superimposed dot is used to denote time rates:

$$\dot{\varphi} = -h_\varphi\,\dot{e} \quad \text{for} \quad \varphi > \varphi_{cs} \ . \tag{1}$$

Evidence for this relationship is a.o. put forward by Teferra [18] and Schultze [19]. Both authors compared a lot of different equations with experimental data and advocated the use of a simple linear relation, i.e. a constant softening modulus until the critical state angle φ_{cs} is reached.

For the present study, the final decision to use only a constant softening modulus h_φ is based on the data for Hostun sand, an empirical relationship as presented by Bolton [20] and the Danish Code of Practice [21], as illustrated

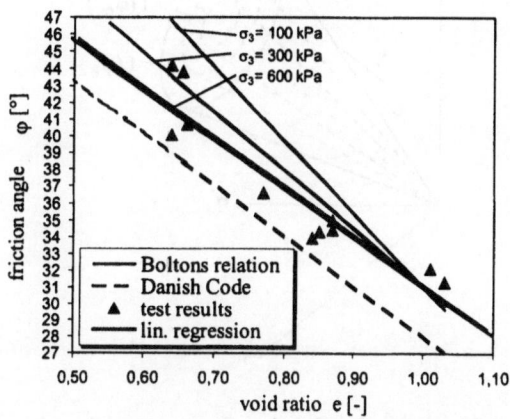

Fig. 2. Relation between friction angle φ and void ratio e for Hostun sand

in Fig. 2. Using drained triaxial tests on Hostun sand with different confining pressures and for varying initial void ratios, a more or less linear relationship between the peak friction angle φ and the void ratio e can be observed in Fig. 2. The points (marked by triangles) indicate test results and the solid line represents the linear regression giving $\varphi = 60.9° - 30° e$. It is assumed that this relation will hold also in localized zones.

The empirical relationship for triaxial conditions of stress and strain from Bolton [20] is described by Eq. (2). I_R is the so-called relative dilatancy index according to Eq. (3), where I_D indicates the relative density and p' the mean effective stress, using kN/m^2 as unit:

$$\varphi = \varphi_{cs} + 3 I_R \quad , \tag{2}$$

$$I_R = I_D(10 - \ln(p')) - 1; \quad I_D = \frac{e_{max} - e}{e_{max} - e_{min}} \quad . \tag{3}$$

The Danish Code of Practice [21] indicates for triaxial compression conditions

$$\varphi = 30° + \left(14 - \frac{4}{U}\right) I_D - \frac{3}{U} - K \quad , \tag{4}$$

where U is the so-called coefficient of uniformity (d_{60}/d_{10}) and K is a reduction factor, which depends on grain shape and silt content.

The void ratio is related to the volumetric strain $\dot{\varepsilon}_v$ according to the formula

$$\dot{\varepsilon}_v = \frac{\dot{e}}{1 + e} \approx \frac{\dot{e}}{1 + e_0} \quad . \tag{5}$$

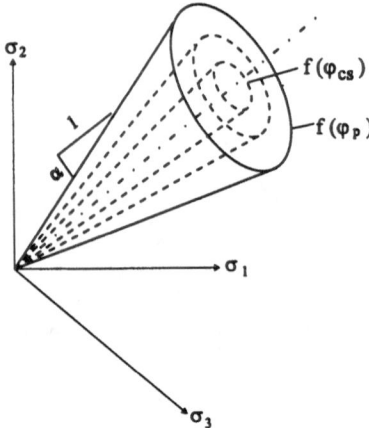

Fig. 3. Yield surface for Drucker-Prager model with softening

Applying this equation, the degradation of the friction angle can be controlled by the volumetric strain rate instead of the void ratio. Eq. (5) is linearised by using the initial void ratio e_0. Eqs. (1) and (5) can now be combined to give

$$\dot{\varphi} = -h_\varphi \left(1 + e_0\right) \dot{\varepsilon}_v \quad .$$ (6)

3   Drucker-Prager model with local softening

Only the most important equations of the well-known DP-model will be described here. For cohesionless material, the yield function can be formulated in terms of the isotropic stress p and the deviatoric stress q by the equations

$$f = q - \alpha p \quad \text{with} \quad \alpha = \alpha_0 - h_\alpha \, \varepsilon_v^p \geq \alpha_{cs} \quad .$$ (7)

The yield condition $f = 0$ describes a cone in the principle stress space, where the coefficient α is related to the friction angle φ and indicates the steepness of the cone (also see Fig. 3). The softening expression for α agrees with Eq. (6), but for the use of plastic volumetric strain ε_v^p instead of total volumetric strain ε_v. For soils elastic strains are relatively small, so that difference between ε_v^p and ε_v tends to vanish for monotonic deformation. Obviously the well-known circular yield cone of the DP-model presents a fairly crude option to describe soil failure. In this study it is only a simple introduction to friction softening. In the particular case of plane strain, a realistic relation is $\alpha \approx \sqrt{3} \sin\varphi$ [22], at least for small angles of dilatancy.

A non-associated flow rule of the form

$$\dot{\varepsilon}^p = \dot{\lambda} \frac{\partial g}{\partial \sigma} \quad \text{with} \quad g = q - \beta p$$ (8)

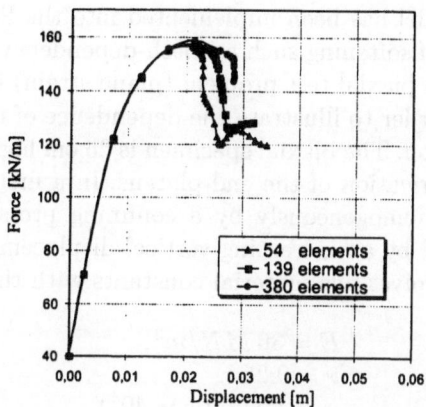

Fig. 4. Computed biaxial curves for the local DP-model show post-peak mesh-dependency

is introduced, where the coefficient β describes soil dilatancy. Here we use the vector notation for stress and strain, i.e. $\varepsilon^T = (\varepsilon_{xx}, \varepsilon_{yy}, \varepsilon_{zz}, \varepsilon_{xy}, \varepsilon_{yz}, \varepsilon_{zx})$ and $\sigma^T = (\sigma_{xx}, \sigma_{yy}, \sigma_{zz}, \sigma_{xy}, \sigma_{yz}, \sigma_{zx})$ and D^e for the elastic stiffness matrix. The plastic multiplier λ in Eq. (8) has to be found by using the consistency condition

$$\dot{f} = \left(\frac{\partial f}{\partial \sigma}\right)^T \dot{\sigma} + \frac{\partial f}{\partial \alpha}\dot{\alpha} = 0 \quad . \tag{9}$$

Together with $\dot{\sigma} = D^e \dot{\varepsilon}^e = D^e (\dot{\varepsilon} - \dot{\varepsilon}^p)$ and Eq. (8) one obtains

$$\left(\frac{\partial f}{\partial \sigma}\right)^T D^e \left(\dot{\varepsilon} - \lambda\frac{\partial g}{\partial \sigma}\right) + \frac{\partial f}{\partial \alpha}\dot{\alpha} = 0 \quad . \tag{10}$$

For computing λ one has to consider the softening rule for $\dot{\alpha}$. The softening rule is defined analogous to Eq. (6) and fully specified by Eq. (7). On differentiating the equation for α one obtains

$$\dot{\alpha} = -h_\alpha (1 + e_0) \dot{\varepsilon}_v^p = -h_\alpha (1 + e_0) \delta^T \dot{\varepsilon}^p = -h_\alpha (1 + e_0) \delta^T \lambda \frac{\partial g}{\partial \sigma} \quad , \tag{11}$$

where $\delta^T = (1, 1, 1, 0, 0, 0)$. Together with Eq. (10) this yields:

$$\lambda = \frac{1}{d}\left(\frac{\partial f}{\partial \sigma}\right)^T D^e \dot{\varepsilon} \quad \text{with}$$

$$d = \left(\left(\frac{\partial f}{\partial \sigma}\right)^T D^e + \frac{\partial f}{\partial \alpha} h_\alpha (1 + e_0) \delta^T\right)\frac{\partial g}{\partial \sigma} \quad . \tag{12}$$

4   Necessity of regularization

The softening DP-model has been implemented into the PLAXIS FE-Code [17] to investigate effects of softening such as mesh-dependency of load-displacement curves. In this study a biaxial test problem (plane strain) is calculated with different FE-meshes in order to illustrate the dependence of the strain localization on the FE-discretization. The biaxial specimen is 25 cm high and 10 cm wide and is tested with full lubrication of the end-platens. In a first phase, the specimen will be consolidated homogeneously by a confining pressure. Thereafter axial compression is applied by an increasing vertical displacement at the top side.

The DP-model involves five material constants with the following values:

stiffness modulus	$E = 30\,MN/m^2$,
Poisson ratio	$\nu = 0.25$,
friction constant	$\alpha = 1.11 \quad (\varphi = 40°)$,
dilatancy constant	$\beta = 0.3 \quad (\psi = 10°)$,
softening modulus	$h_\alpha = 0.7 \quad (h_\varphi = 30)$.

The above set of parameters represents typical values for a dense sand. In order to trigger a shear band, a reduced friction angle $\varphi = \varphi_{cs} = 34°$ is assigned to one Gaussian integration point of the specimen. All analysis were carried out using six-noded triangular elements. The load-displacements curves resulting from three calculations all with a confining pressure of $400\,kN/m^2$, but varying FE-dicretizations (54/139/380 elements) are presented in Fig. 4. Up to the peak, the three curves are equal, thereafter the softening behaviour differs strongly. The calculation results show an unstable failure mechanism and indicate the well-known mesh-dependent softening behavior of local softening models.

Due to the missing physical length scale, i.e. a distinct thickness of a shear band, the applications of a softening rule in local plasticity formulations results in a severe mesh-dependence. In order to avoid such FE-mesh-influenced shear bands, it is most important to extend the present model and to include a so-called internal length that regularizes the thickness of the shear band.

5   Nonlocal DP-model

The regularization applied in the present paper is the so-called nonlocal theory. This theory was established by Eringen [7,8] a.o. and later extended by Bažant [9]. A full nonlocal method involves a relation between average stresses and average strains, i.e.:

$$\sigma(x_n)^* = \frac{1}{A}\iiint w(x'_n)\,\sigma(x_n + x'_n)\,dx'_1 dx'_2 dx'_3\,, \tag{13}$$

$$\varepsilon(x_n)^* = \frac{1}{A}\iiint w(x'_n)\,\varepsilon(x_n + x'_n)\,dx'_1\,dx'_2\,dx'_3\,. \tag{14}$$

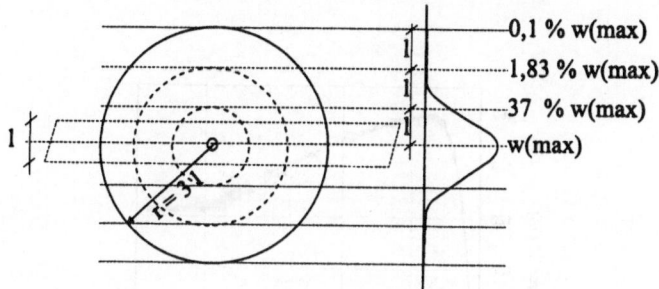

0,1 % w(max)

1,83 % w(max)

37 % w(max)

w(max)

Fig. 5. Spherical weighting function which restricts zone of spatial averaging

where a superimposed star is used to denote the nonlocal method. x_n is a global, x'_n is a local coordinate with $n = 1, 2, 3$. $w(x'_n)$ is a weighting function, as illustrated in Fig. 5, and A indicates a weighted volume

$$A = \iiint w(x'_n) \, dx'_1 \, dx'_2 \, dx'_3 \ . \tag{15}$$

In homogeneous bodies the function $w(r)$ will have spherical symmetry and the above equations can simply be written as

$$\sigma^* = \frac{1}{A} \int_V w(r) \, \sigma \, dV; \quad \varepsilon^* = \frac{1}{A} \int_V w(r) \, \varepsilon \, dV \ , \tag{16}$$

where r is the distance from the material point considered to the various points of integration. Usually the following error function (visualized in Fig. 5) is used as a weighting function for the nonlocal theory:

$$w(r) = \frac{1}{l \sqrt{\pi}} e^{\left(-\frac{r}{l}\right)^2} \ . \tag{17}$$

In this function l is the internal length parameter as visualized in Fig. 5. At a distance of a few times the internal length the function is practically equal zero. The averaging is thus restricted to a small representative area around the material point considered. In the present paper, the use of a fully nonlocal theory with σ^* and ε^* is avoided. Instead the use of nonlocal variables is restricted to the softening parameter employed:

$$\dot{\varepsilon_v}^* = \frac{1}{A} \int_V w(r) \, \dot{\varepsilon_v}^P \, dV \ . \tag{18}$$

The nonlocal DP-model is obtained by using ε_v^* instead of ε_v^p within the yield condition in Eq. (7). In order to illustrate the effectiveness of the nonlocal method according to Eq. (18) the biaxial test problem, as described in Section 4, will now be analyzed again. In addition to the parameters presented in Section 4 the internal length parameter l has to be determined. In the current example the internal length is selected arbitrarily with a value of $4\,cm$. In fact, the assumed value is much higher than required for a realistic shear band thickness, but

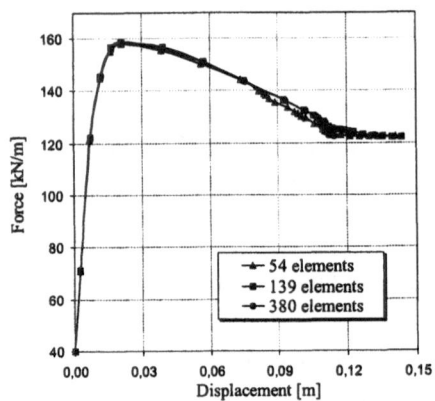

Fig. 6. Computed biaxial curves for nonlocal DP-model show full mesh-independency.

it is adequate for demonstrating the effectiveness of the regularization-method applied.

Fig. 6 presents the load-displacement curves from three calculations with the same confining pressure of $400\,kN/m^2$, but varying FE-discretizations (54/139/380 elements). Compared with the 'local' curves of Fig. 4, the present curves match sufficiently well. It is remarkable that the iteration-process within the 'nonlocal' calculations show fast convergence of equilibrium iterations, whereas the 'local' model showed a very slow convergence. Moreover, the calculation results show a stable failure mechanism when using the nonlocal theory.

6   Internal length and numerical shear band thickness

In the computations for Fig. 6 a large value of 4 cm was chosen for the internal length and resulting shear bands were found to have a thickness of about 8 cm. For determining the shear band thickness t_s as a function of the internal length l, a large series of FE-analysis were carried out for different values of l in the range between 2.5 and 30 mm. For all computations sufficient finess of the mesh was guaranteed by satisfying the criterion

$$l > \mu \; L_{el} \quad \text{with} \quad \mu \approx 0.8 \quad, \tag{19}$$

where L_{el} is the length of the elements being used. For 6-noded triangular elements with 3 Gaussian integration points, the value of μ was found to be somewhat below 0.8. This means that the averaging procedure, as indicated in Fig. 5, should at least comprise a single element between the points of contraflexure of the Gaussian weighting function.

Fig. 7 shows computed biaxial force-displacement curves for different values of the internal lengths. All computations were carried out with the DP-model and input data as presented in Section 4. As in all previous calculations the confining pressure was taken to be $400kN/m^2$. The sample had a height of 25 cm and a with of 10 cm. The end platens were taken to be rough and the top platen was allowed to slide horizontally. On choosing these boundary condition, no perturbation for triggering shear banding appeared to be necessary.

Fig. 8 shows resulting shear banding for two different values of the internal length. A relatively thin band is observed for the small internal length and a much thicker band results for the larger internal length. For a precise determination of the thickness one has to consider the velocity distributions across the bands, as also illustrated in Fig. 8. Within the shear band one observes a near linear velocity distribution and a more or less sharp transition to the rest of the biaxial sample. This enables a straight forward determination of the shear band thickness t_s.

Data from 16 FE-computations with various different values of the internal length has been assembled in Fig. 9. For saving computer run time, large values of l were combined with coarse meshes and the effect of small internal lengths were analysed using fine meshes. A total of 6 different meshes was used as also indicated in Fig. 9. In all cases the criterion $l > 0.8\ L_{el}$ according Eq. (19) was satisfied to guarantee nonlocal regularization. From Fig. 9 one observes a linear growth of the shear band thickness with the internal length. For the 6-noded elements it follows that $t_s = 2\ l$. It would seem that a thickness of $2\ l$ is more or less precisely found when the mesh is relatively fine with respect to the internal length. Small deviations from the line $t_s = 2\ l$ may also be due to our procedure of determining t_s on the basis of velocity distributions as indicated in Fig. 9.

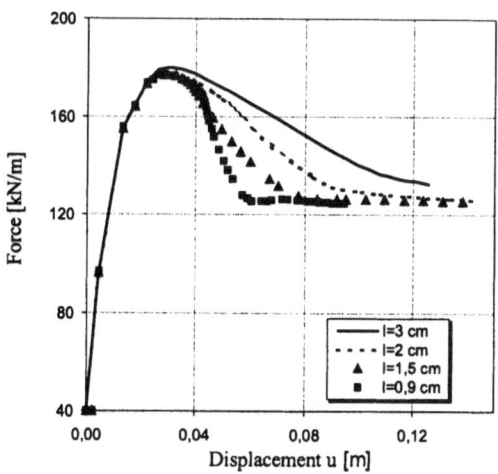

Fig. 7. Load-displacement curves for different values of the internal length l.

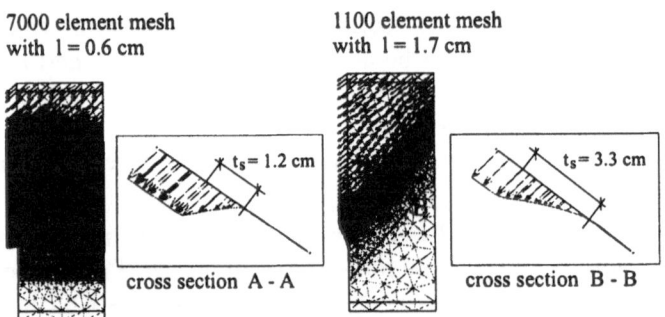

Fig. 8. Resulting shear banding for two different internal length parameters.

7 Empirical shear band thicknesses

Desrues et al. [14] presented one of the first publications with measurements on shear band thicknesses. Using stereophotogrammetry they measured t_s-values for Hostun 'RF' sand to be in between 15 and 20 times the mean grain size d_{50}. In 1994 Yoshida et al. [23] presented data for 7 different sands. Again stereophotogrammetric measurements were taken and depending on the sand considered shear band thicknesses were found to be in the relatively wide range between 10 and 20 times the mean grain size.

A new technique for measuring shear band thicknesses was recently introduced by Oda et al. [24]. With this procedure the deformed sand sample is impregnated with a special resin to fix the particles in place. After hardening, thin slices of the impregnated sand are cut to study the dilatant shear band.

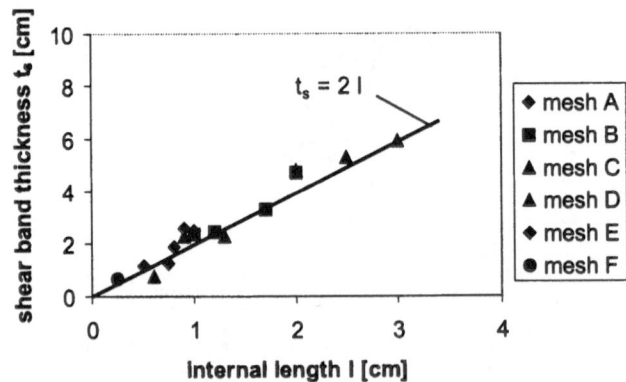

Fig. 9. Relationship between the internal length parameter l and the shear band thickness t_s for different FE-meshes.

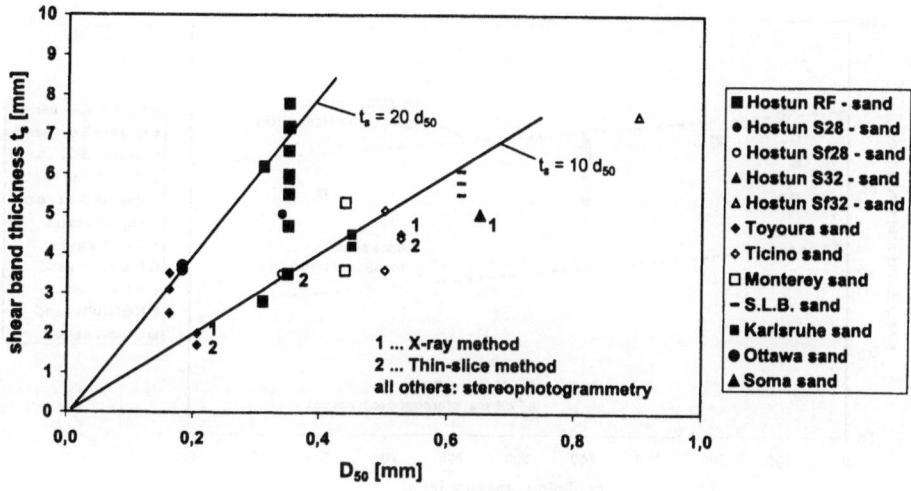

Fig. 10. Shear band thickness as a function of the medium grain size for different sands.

The authors have tried this method to find astonishing clear fixed in place shear bands. Moreover sharp transitions were observed that separated the band from the rest of the material, enabling a precise measurement of the shear band thickness. Yust like Oda et al. [24] we measured thicknesses of about 10 times the mean grain size.

Assembling data from the above authors as well as Viggiani et al. [25], we produced Fig. 10. Most data clearly suggest the correlation $t_s = 10\ d_{50}$. Moreover this correlation is fully supported by the thin-slice method, and the X-ray method, being more reliable than the stereophotogrammetric method. Indeed the latter method involves optical measurements through a rubber membrane that might cause some deviations [24]. The authors thus consider the correlation $t_s = 10\ d_{50}$ as most likely; with $t_s = 2\ l$ we have a solid correlation $l \approx 5\ d_{50}$.

Another interesting observation of biaxial test concerns the inclination of the shear band. Considering data of various different authors on dense sands with relative densities in the range between $I_D = 0.7$ and $I_D = 1.0$, inclination angles were found to range from 58 to 70°. Fig. 11 shows observed inclination angles as a function of the applied confining pressure. In addition the Coulomb inclination ($\theta = 45 + 1/2\ \varphi$) and the Roscoe inclination ($\theta = 45 + 1/2\ \psi$) are indicated using data for friction angles and dilatancy angles as published by [14] for dense Hostun 'RF' sand. It would seem that these data are also a best guess for the other sands. Indeed, all data comes from dense subrounded poorly graded quartz sands, which suggests compareable data on φ and ψ. Fig. 11 illustrates that the

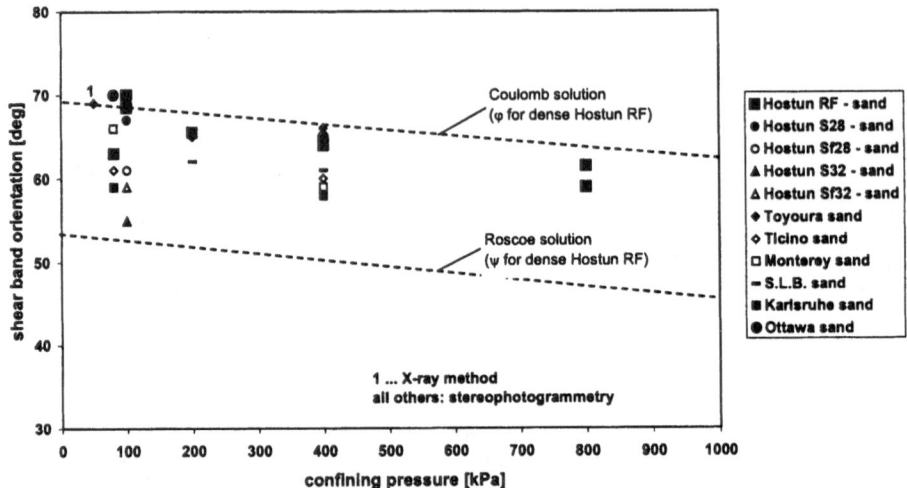

Fig. 11. Shear band inclination (to the horizontal) as a function of the applied confining pressure for different sands.

inclination angles are in between the classical ones with a tendency towards the steep Coulomb band.

8   Softening scaling on h and l

Adopting the correlation $t_s = 10\ d_{50}$, it follows that shear bands in a coarse sand with a mean particle size of 1 mm will have a thickness of 10 mm. On the other hand, a shear band in a fine sand with a mean particle size of 0.2 mm will have a thickness of only 2 mm. It thus follows from the relationship $t_s = 2\ l$ that fine sands should be modeled with an internal length of only 1 mm. Considering the constraint $l > 0.8\ L_{el}$ (see Eq. 19), it follows that very fine meshes with element size below 2 mm would be needed. For coarse sands, the situation is slightly better, as here we arrive at a minimum element size of approximately 6 mm.

For the numerical analysis of practical boundary value problems, the use of element sizes below 6 mm is not feasible. Indeed, one might use adaptive remeshing techniques [26] which restrict mesh refinement to the localized zones, but even this would require tremendously large computer capacities. A feasible solution to this problem is the softening scaling as proposed by [27]. Instead of striving for modeling the real shear band thickness, a much larger non-physical thickness is adopted, which is the minimum thickness that can be resolved by a given practical finite element mesh. Hence L_{el} is given and one accepts relatively thick shear bands of $t_s \approx 1.6\ L_{el}$ (according $l > 0.8\ L_{el}$ and $t_s = 2\ l$). This leads to two different internal lengths:

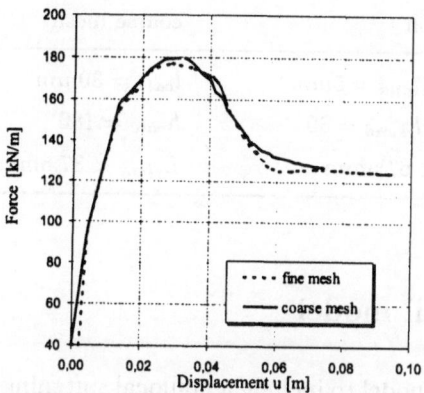

Fig. 12. Load displacement curves for a fine and a coarse FE-mesh.

$$l_{calc} = 0.5 \ t_{s,calc} > 0.8 \ L_{el} \quad , \tag{20}$$

$$l_{sand} = 0.5 \ t_{s,sand} \approx 5 \ d_{50} \quad . \tag{21}$$

Here l_{calc} is an internal length that suits the adopted mesh and l_{sand} is the real one of the sand considered. The global softening behaviour depends linearly on the ratio of h/t_s where h is the softening modulus and t_s the shear band thickness. Large values of h/t_s lead to steep softening curves and small values yield small degrees of global softening . Scaling is thus possible by following the rule

$$\frac{h_{sand}}{t_{s,sand}} = \frac{h_{calc}}{t_{s,calc}} \quad , \tag{22}$$

where h_{calc} is the scaled softening modulus that suits the mesh, i.e.:

$$h_{calc} = h_{sand} \frac{t_{s,calc}}{t_{s,sand}} = h_{sand} \frac{l_{calc}}{l_{sand}} \quad . \tag{23}$$

For demonstrating the softening scaling, the biaxial test of Section 4 is reconsidered. Two different analysis were performed according to table 1:

the fine mesh is used for a straight forward analysis using the sand data of Section 4. The coarse mesh is combined with scaling in order to match the sand data indirectly. Fig. 12 shows that both analysis lead to nearly indentical load-displacement curves.

Table 1. Softening Scaling

fine mesh	coarse mesh
$l_{calc} = l_{sand} = 5mm$	$l_{calc} = 30mm$
$h_{calc} = h_{sand} = 30$	$h_{calc} = 180$
$L_{elem} < 6.25mm$	$L_{elem} < 37.5mm$

9   Hardening soil model

On extending the DP-model to include a nonlocal softening parameter, the back-analysis of biaxial tests with strong shear-banding is shown to be feasible. Obviously the DP-model has to be replaced by a more advanced constitutive model in order to arrive at proper results. For this purpose the so-called Hardening Soil model [15] is briefly presented. In fact, this is an improved version of the so-called Double Hardening model as described by [28]. Only the main characteristics of this elastoplastic model are presented in the present paper.

The essential ideas of the HS-model are the simulation of the hardening as observed in triaxial tests and the confining stress dependence of the stiffness, which is also reflected in the oedometer findings. For a deviatoric load under triaxial conditions the following hyperbolic stress-strain relation in terms of effective principal stresses was chosen according to [29]:

$$\varepsilon_1 = \frac{q_a}{2E_{50}} \frac{(\sigma_1 - \sigma_3)}{q_a - (\sigma_1 - \sigma_3)} \ . \tag{24}$$

The asymptotic stress q_a and the secant module E_{50} are defined as

$$q_a = \frac{q_f}{R_f} = \frac{6 \sin \varphi}{R_f \, (3 - \sin \varphi)} \, (p + c \cot \varphi) \ , \tag{25}$$

$$E_{50} = E_{50}^{ref} \left(\frac{\sigma_3 + c \cot \varphi}{p^{ref} + c \cot \varphi} \right)^m \ . \tag{26}$$

R_f stands for the relation between the maximum stress q_f and the asymtotic stress q_a. As a standard setting there is $R_f = 0.9$. With an increasing load, the yield surface approaches a Mohr Coulomb boundary surface with q_f in accordance with Eq. (25). This equation involves the soil cohesion c and the peak friction angle φ. E_{50}^{ref} is determined from a standard triaxial test with $\sigma_3 = p^{ref}$ as indicated in Fig. 13. The power m can either be determined by experiments or estimated using empirical values. For sand, m lies somewhere between 0.35 and 0.65.

For triaxial conditions of stress and strain the shear yield function f^s reads

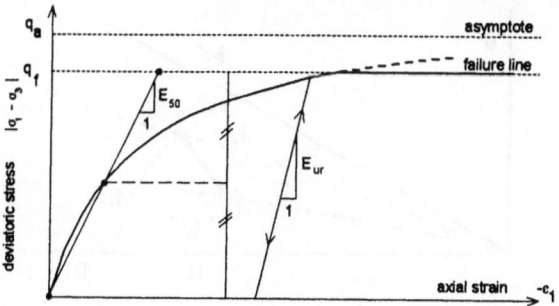

Fig. 13. Hyperbolic stress-strain relation in primary loading for a standard drained triaxial test.

$$f^s = \bar{f} - \gamma^p \quad \text{with} \quad \bar{f} = \frac{1}{E_{50}} \frac{q}{\left(1 - \frac{q}{q_a}\right)} - \frac{2q}{E_{ur}} \; , \tag{27}$$

where E_{ur} is Young's modulus for elastic unloading and reloading and γ^p is a plastic-shear-strain related hardening parameter ($\gamma^p = \varepsilon_1^p - \varepsilon_2^p - \varepsilon_3^p$).

As for all plasticity models, the Hardening Soil model involves a relationship between rates of plastic strain, i.e. a relationship between $\dot{\varepsilon}_v^p$ and $\dot{\gamma}^p$:

$$\dot{\varepsilon}_v^p = \sin \psi_m \, \dot{\gamma}^p \; . \tag{28}$$

The plastic potential does not agree with the shear yield function in Eq. (27) and so this is the case of a non-associated flow rule $g^s \neq f^s$. The mobilized dilatancy angle ψ_m is defined by the equation

$$\sin \psi_m = \frac{\sin \varphi_m - \sin \varphi_{cs}}{1 - \sin \varphi_m \sin \varphi_{cs}} \; , \tag{29}$$

where φ_{cs} is the friction angle at the critical state. The mobilized friction angle φ_m is defined as

$$\sin \varphi_m = \frac{\sigma_1 - \sigma_3}{\sigma_1 + \sigma_3 - 2c \cot \varphi} \; . \tag{30}$$

In the p-q-plane, the shear yield surface is closed by means of a yield cap. This cap moves independently from the shear yield surface. Fig. 14 shows the form of the yield surfaces as well as the limiting conditions according to Mohr-Coulomb in the p-q-plane. The plastic area is marked by the two yield functions f^s and f^c, the plastic strain rates are defined with the help of the plastic potentials g^s and g^c. The shape of the cap is defined by means of the cap yield function

$$f^c = \frac{q^2}{\alpha^2} - p^2 - p_p^2 \, , \tag{31}$$

where α is an auxiliary model parameter that relates to K_0^{nc}. In fact α is chosen in such a way that $K_0^{nc} = 1 - \sin \varphi$. The position and size of the cap in the stress

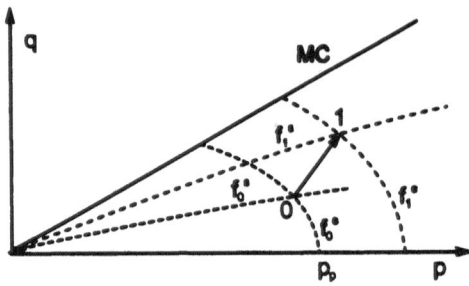

Fig. 14. Yield functions f^s and f^c.

space is controlled via p_p, the so-called preconsolidation pressure in isotropic loading. The hardening law describes the relation between the preconsolidation pressure p_p and the volumetric strains of the cap ε_v^{pc}:

$$p_p = p^{ref} \left(\frac{1-m}{\beta} \varepsilon_v^{pc} \right)^{\frac{1}{1-m}} \tag{32}$$

with

$$\varepsilon_v^{pc} = \int \dot{\varepsilon}_v^{pc} dt \quad \text{and} \quad \dot{\varepsilon}_v^{pc} = \dot{\lambda}^c \, \delta^T \frac{\partial g^c}{\partial \sigma} \ . \tag{33}$$

The constant β is not a direct input parameter. Instead we prefer to use a (tangent) oedometer modulus as an input parameter. The yield cap is combined with an associated flow rule $g^c = f^c$ to define plastic straining due to a shift of the yield cap.

The total strain rates are composed of an elastic part, a plastic part resulting from the shear hardening $\dot{\varepsilon}^{ps}$ and a part resulting from the cap hardening $\dot{\varepsilon}^{pc}$.

$$\dot{\varepsilon} = \dot{\varepsilon}^e + \dot{\varepsilon}^{ps} + \dot{\varepsilon}^{pc} =$$
$$= D^{e(-1)} \dot{\sigma} + \dot{\lambda}^s \frac{\partial g^s}{\partial \sigma} + \dot{\lambda}^c \frac{\partial g^c}{\partial \sigma} \ . \tag{34}$$

The factors λ^s and λ^c are determined by means of the consistency conditions $\dot{f}^s = 0$ and $\dot{f}^c = 0$. For more details regarding the formulation of the HS-model the reader is referred to the recent papers [16, 30].

10   HS-model with nonlocal softening

The original HS-model is based on a constant friction angle independent of Lode's angle and soil density. However density dependence can be simply introduced by

$$\varphi = \varphi_{cs} + h_\varphi \left(e_{cs} - e^* \right) \ , \tag{35}$$

where e^* is a nonlocal void ratio in the sense of Eq. (18). It now follows that

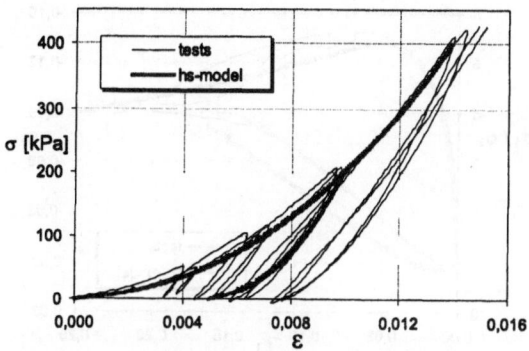

Fig. 15. Oedometer test - calibration of HS-model using test data on 'dense' Hostun sand.

$$\dot{\varphi} = -h_\varphi \, \dot{e}^* = -h_\varphi \, (1 + e_0) \, \dot{\varepsilon}_v^* \; , \tag{36}$$

where ε_v^* is the nonlocal volumetric strain as defined by Eq. (18).

The density dependency of the friction angle allows for both softening as well as hardening. The softening is obviously related to shear dilatancy (see Eq. 28). Within the HS-model ψ_m is a mobilized dilatancy angle as expressed by Eq. (29). Positive angles lead to softening. Material densification occurs both for negative values of ψ_m and obviously because of cap yielding in compression. For clay-type soils such a densification is known to be significant, but for sandy soils it is relatively small. However, since now upgrading of the friction angle due to densification has not been studied. Instead of rigorously applying Eq. (36), we have assumed $\dot{\varepsilon}_v \approx \sin \psi_m \, \dot{\gamma}^p$ to obtain

$$\begin{aligned} \dot{\varphi} &= -h_\varphi \, (1 + e_0) \sin \psi_m \, \dot{\gamma}^* \quad \text{for} \quad \psi_m \geq 0 \; , \\ \dot{\varphi} &= 0 \quad \text{for} \quad \psi_m < 0 \; , \end{aligned} \tag{37}$$

where γ^* is a nonlocal plastic shear strain in the sense of Eq. (18). Hence upgrading of the friction angle due to densification has been excluded.

The HS-model with nonlocal softening was calibrated by calculating triaxial compression tests and oedometer tests using the database on Hostun sand, as indicated in Figs. 15 and 16 (calibration was described in full detail in [32]). A total of 8 material parameters are given below for the so-called 'dense' Hostun sand:

critical-state angle	$\varphi_{cs} = 34°$
secant stiffness	$E_{50}^{ref} = 30 \, MN/m^2$
tangent oedometer stiffness	$E_{oed}^{ref} = 30 \, MN/m^2$

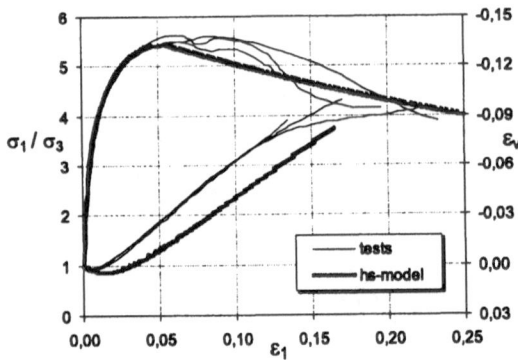

Fig. 16. Triaxial compression test ($\sigma_3 = 100\,kN/m^2$) - calibration of the HS-model using test data on 'dense' Hostun sand.

power in stiffness law	$m = 0.55$
un/reloading stiffness	$E_{ur}^{ref} = 90\,MN/m^2$
un/reloading Poisson ratio	$\nu_{ur} = 0.25$
softening modulus	$h_\varphi = 30$
internal length	$l = 3\,cm$
	$(\sigma^{ref} = 100\,kN/m^2)$

11   Geometrical Nonlinearity

Large deformations induce significant changes of the initial geometry of slender structures such as beams, plates and shells. In general soil bodies are far from being slender and consequently, FE-analysis of soil bodies often tacitly disregard changes in geometry. On the other hand, the biaxial sample considered is relatively slender and all previous analysis have indicated significant changes of geometry. Therefore, material nonlinearity is now combined with geometrical nonlinearity, where the latter is encountered by using an updated Lagrangian technique, as described a.o. by [33]. Constitutive equations as given by the HS-model relate measures of strain to objective measures of stress. In the present case we used the co-rotational rate of Kirchhoff stress [33], being referred to as the Biezeno-Hencky stress rate by Breinlinger [34].

Attention is again focussed on the influence of the internal length, l, on the width of the shear band, whilst the biaxial sample is modelled by a relatively coarse mesh of 110 elements. Different internal lengths (3, 4 and 5 cm) have been selected for the investigation. Resulting stress-strain curves are shown in Fig. 17 for $\sigma_3 = 400\,kN/m^2$. In fact, the particular values being selected have an influence on the post-peak curve, but differences are modest.

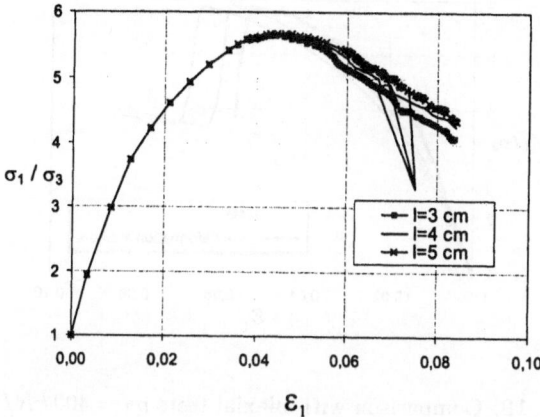

Fig. 17. Computed biaxial curves for the nonlocal HS-model including geometrical nonlinearity.

Finally, numerical data can be compared with experimental biaxial data as presented in [12]. On considering this data for dense Hostun sand, one observes a sharp stress drop, as indicated by the thin lines in Fig.18, followed by a gradual increase of strength again. The increase of the apparent strength after the first minimum is most probably linked to a perturbative effect caused by very large strains inside the shear band, which induces local membrane reactions [11].

The numerical data in Fig. 18 stems from a computation with coarse mesh (110 elements) as also used for the analysis of Fig. 17. For the final computation, we selected $l_{calc} = 40$ mm, whereas $l_{sand} \approx 2$ mm, at least for the Hostun sand considered. To compensate for the high ratio of l_{calc}/l_{sand}, the same ratio was used for h_{calc}/h_{sand}. Having $h_{sand} = 30$ as given in Section 2, we thus used $h_{calc} = 600$ (Eq. 23). The result is a very steep softening curve, which fits the experimental data extremely well. On the other hand, the hardening part of the numerical curve is somewhat too stiff with respect to the experimental data. No doubt, it would be easy to adjust the input parameters of the HS-model to match biaxial experiments most precisely. In such a case, however, we would not completely match the input data of oedometer tests (Fig. 15) and triaxial tests (Fig. 16).

12   Conclusions

The present study was based on a strict separation between input data from triaxial and oedometer tests and output data for biaxial tests. In fact, the calibration of the model for obtaining input data was done more than a year ago and published by Marcher et al. [32]. More recently the HS-model has been extended

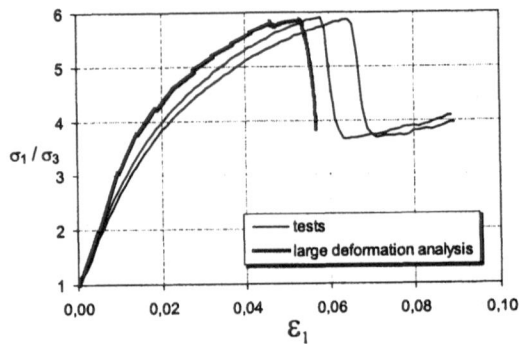

Fig. 18. Comparison with biaxial tests $\sigma_3 = 400 \, kN/m^2$.

to include nonlocal softening, which yielded both the internal length and the softening modulus as additional parameters [31]. On combining the original parameters of the HS-model with objective data on the softening modulus and the internal length one obtains a reasonable prediction of hardening and softening in biaxial tests, as illustrated in Fig. 18.

The computational effort for analyzing softening problems would seem to be acceptable when using a nonlocal-type model. This is first of all achieved by restricting the nonlocal formulation to a simple parameter, i.e. the hardening-softening parameter. The weighted averaging is thus restricted to a single parameter. An additional measure for restricting the computational effort is to apply softening scaling. In this procedure both the softening modulus and the internal length (l_{calc}) are proportionally, so that l_{calc} exceeds well the length of individual finite elements. Rather than choosing elements below the size of the internal length, it is thus suggested to select an internal length that suits the (relatively coarse) finite element mesh.

Acknowledgments

The authors gratefully acknowledge the financial support provided by the German Research Council (DFG). Dr. J. Desrues and Dr. E. Flavigny are acknowledged for providing valuable test results.

References

1. Hill, R.: The mathematical theory of plasticity, Oxford University Press, 1950
2. Pijaudier-Cabot, G., Bažant, Z. P.: Nonlocal damage theory, J. Eng. Mech. ASCE, 113, *1512-1533*, 1987
3. Ehlers, W., Volk, W.: On shear band localization phenomena of liquid-saturated granular elastoplastic porous solid materials accounting for fluid viscosity and micropolar solid rotations, Mech. Cohesive-frictional Mater., 2, *301-320*, 1997

4. De Borst, R., Sluys, L. J., Mühlhaus, H. -B., Pamin, J.: Fundamental issues in finite element analysis of localization of deformation, Eng. Comput., 10, *99-121*, 1993

5. Mühlhaus, H. -B., Aifantis, E. C.: A variational principle for gradient plasticity, Int. J. Solids Structures, 28, *845-857*, 1991

6. Vardoulakis, I.: 2nd gradient constitutive models, in: Kolymbas (Ed.), Constitutive modelling of granular materials, *225-248*, 2000

7. Eringen, A. C.: Nonlocal polar elastic continua, Int. J. Eng. Sci., 10, *1-16*, 1972

8. Eringen, A. C.: On nonlocal plasticitiy, Int. J. Eng. Sci., 19, *1461-1474*, 1981

9. Bažant, P., Gambarova, B.: Shear crack in concrete: Crack band microplane model, J. Struct. Eng. ASCE, 110, *2015-2036*, 1984

10. Flavigny, E., Desrues, J., Palayer, B.: Le sable d'Hostun RF, Rev. Franc. Géotechnique, 53, *67-70*, 1990

11. Desrues, J.: La localisation de la déformation dans les matériaux granulaires, Thèse de Docteur des Sciences, Institut de Mécanique de Grenoble, 1984

12. Hammad, W. I.: Modélisation non linéaire et étude expérimetale des bandes de cisaillement dans les sables, Thèse de Docteur des Sciences, Institut de Mécanique de Grenoble, 1991

13. Mokni, M.: Relations entre déformations en masse et déformations localisées dans les matériaux granulaires, Thèse de Docteur des Sciences, Institut de Mécanique de Grenoble, 1992

14. Desrues, J., Hammad, W.: Shear banding dependency on mean stress level in sand, International Workshop: Numerical methods for localization and bifurcation of granular bodies, Gdansk-Sobieszewo, *57-67*, 1989

15. Schanz, T.: Zur Modellierung des Mechanischen Verhaltens von Reibungsmaterialien, Mitteilung 45 des Instituts für Geotechnik der Universität Stuttgart, 1998

16. Schanz, T., Vermeer, P. A., Bonnier, P. G.: The hardening soil model - formulation and verification, Plaxis-Symposium: Beyond 2000 in computational geotechnics, Amsterdam, *281-296*, 1999

17. Vermeer, P. A., Brinkgreve, R. B. J.: Plaxis; Finite element code for soil and rock analyses. Version 7.1, Balkema, Rotterdam, 1998

18. Teferra, A.: Beziehungen zwischen Reibungswinkel, Lagerungsdichte und Sondierwiderständen nichtbindiger Böden mit verschiedener Kornverteilung, Heft 1, Aachen 1975

19. Schultze, E.: Lockere und dichte Böden, Mitt. Inst. Verkehrswasserbau, Grundbau und Bodenmechanik, TH Aachen, 1968

20. Bolton, M. D.: The strength and dilatancy of sands, Géotechnique 36, No. 1, *65-78*, 1986

21. Danish Code of Practice: Danish Geotechnical Institute, DGI-Bulletin No. 36, Copenhagen, 1985

22. Chen W. -F., Limit analysis and soil plasticity, Elsevier, 1975

23. Yoshida, T., Tatsuoka, F., Siddiquee, M. S. A., Kamegai, Y., Park, C. -S.: Shear banding in sands observed in plane strain compression, in: Chambon, Desrues, Vardoulakis (Eds.), Loacalisation and Bifurcation Theory for Soils and Rocks, *165-179*, 1994

24. Oda, M., Kazama, H.: Microstructure of shear bands and its relation to the mechanisms of dilatancy and failure of dense granular solis, Géotechnique 48, No. 4, *465-481*, 1998

25. Viggiani, G., Küntz, M., Desrues, J.: Does shear banding in sand depend on grain size distibution?, to be published in: Proceedings CDM2000 Stuttgart, International symposium on continuous and discontinuous modelling of cohesive frictional materials, Springer, 2000

26. Zienkiewicz, O. C., Zhu, J. Z.: A simple error estimator and adaptive procedure for practical Engineering analysis, Int. J. Num. Meth. Eng., 24, *337-357*, 1987
27. Pietruszczak S., Mroz, Z.: Finite element analysis of deformation of strain-softening materials, Int. J. Num. Meth. Eng., 17, *327-334*, 1981
28. Vermeer, P. A.: A double hardening model for sand, Géotechnique, 28, *413-433*, 1978
29. Kondner, R. L., Zelasko, J. S.: A hyperbolic stress strain formulation for sands, Proc. 2nd Pan. Am. Int. Conf. Soil Mech. Found. Engrg., Brazil, Vol. 1, *289- 394*, 1963
30. Vogt, C.: Experimentelle und numerische Untersuchungen zum Tragverhalten und zur Bemessung horizontaler Schraubanker, Mitteilung 47 des Instituts für Geotechnik der Universität Stuttgart, 1999
31. Marcher, T., Vermeer, P. A., Bonnier, P. G.: Application of a nonlocal model to the softening behaviour of Hostun sand, to be published in: Proceedings IWBL Perth, International Workshop on Bifurcation and Localisation, Balkema, 2000
32. Marcher, T., Vermeer, P. A., von Wolfferdorff P. -A.: Hypoplastic and elastoplastic modelling - a comparison with test data, in: Kolymbas (Ed.), Constitutive modelling of granular materials, *353-374*, 2000
33. McMeeking, R. M., Rice, J. R.: Finite-element formulations of large elastic-plastic deformation, Int. J. Solids Struct., 11, *601-616*, 1975
34. Breinlinger, F.: Bodenmechanische Stoffgleichungen bei großen Deformationen sowie Be- und Entlastungsvorgängen, Mitteilung 30 des Instituts für Geotechnik der Universität Stuttgart, 1989

An experimental investigation of the relationships between grain size distribution and shear banding in sand

G. Viggiani,[1] M. Küntz,[2] J. Desrues[1]

[1] Laboratoire 3S, UJF-INPG-CNRS, Grenoble, France
[2] Physics Department, UQAM, Montréal, Canada

Abstract. This paper experimentally investigates the possibility to describe the influence of microstructure on shear banding in sand in terms of grain size distribution. Different gradations of the same sand were tested, differing from each other in terms of both the mean grain size and uniformity. *Monodisperse* sands as well as binary mixtures were tested. Water–saturated specimens of the different sands were tested in plane strain under drained conditions, starting from high relative densities. False relief stereophotogrammetry was used to capture the onset of strain localization and for accurately measuring the width and orientation of shear bands. Results from a total of 11 tests are summarized herein. The influence of mean grain size and grain size distribution uniformity on the occurrence of strain localization and shear band thickness and orientation is discussed. While the results confirm the dependence of the band thickness on the mean grain size, they demonstrate that the orientation of a shear band is not related to the mean size of sand particles size, in contradiction with previous experimental findings. Moreover, shear band orientation can neither be simply related to the uniformity of sand grading. The major result of this study is that while grain size distribution can greatly affect shear banding characteristics, there is not such a thing as a direct relationships between characteristics of localization and mean grain size, or degree of uniformity, or other parameters describing grain size distribution in a simple way. Sand microstructure is a key factor yet its influence on shear banding cannot be described in terms of grain size distribution only.

Keywords: strain localization, shear band, granular materials, grain size distribution, plane strain.

1 Introduction

Understanding strain localization in granular soils such as sand and gravel remains a critical issue in many fields of geotechnical engineering, as stability and deformation of soil bodies strongly depend on the occurrence of shear bands. In conjunction with major developments in both theoretical and numerical aspects of modeling, *e.g.* [24], various experimental studies have attempted to analyze shear banding in sand. See for example the recent contributions by Desrues *et al.* [12], Finno *et al.* [13], Saada *et al.* [23], Desrues & Viggiani [11]. These investigations have provided important data concerning shear band features such as orientation and width, as well as patterns of shear band formation and evolution. In particular, they have pointed out the influence on strain localization of such variables as the initial state of sand (effective stress and relative density), as well as the specimen geometry (size and slenderness).

There is also substantial experimental evidence showing that strain localization in granular soils depends on their microstructure. For a given state and loading path, different sands exhibit different responses, both in terms of orientation and width of shear bands as well as the strain level at which shear banding occurs, *e.g.* [25]. Such differences cannot be explained in terms of macroscopic state variables such as void ratio and stress state only. Because granular soils are very much particulate materials, their macroscopic mechanical behavior in fact depends on the nature and evolution of interparticle contacts through which stresses are transmitted, *e.g.* [1, 6, 8, 21]. Many parameters such as shape and angularity of grains, grain size distribution as well as the way grains were deposited contribute to define the microstructure of a granular assembly. However, these factors are usually difficult to study separately in natural sands and their influence on shear banding has still to be established. The question then remains of which are the variables appropriately describing the influence of the microstructure of a granular assembly on strain localization.

This paper specifically investigates the influence of grain size distribution on strain localization in sand. Among the several factors that may influence microstructure, grading is the most easy to control and has therefore received more attention in previous experimental studies on strain localization. For instance, it has been found that the shear band thickness is about 10 to 20 times the mean grain size [20, 22]. Arthur & Dunstan [4] and Mokni [19] also reported experimental data suggesting that the angle of shear band measured with respect to the direction of axial loading as well as the strain level at which shear banding occurs, increase with the mean grain size, although the physical basis for such a relation has not been clarified yet.

For this study, 8 sands were considered, differing from each other in terms of both the mean grain size and uniformity. The different materials come from only two different gradations of the same natural sand, to limit as much as possible the influence of any other parameters than grain size distribution. Among the 8 sands, 4 were obtained by mixing two grain size populations in various proportions. The relation between grain size distribution and microstructure in binary mixtures of regular spheres as well as irregular grains has been extensively investigated in the last two decades. The influence of both the composition and size ratio between the two grain sizes on the density of mixtures has been pointed out *e.g.* in [16, 21]. Oger [21] also reports experimental evidence that stress is preferentially transmitted through the coarse grains in binary assemblies. Binary mixtures are thus good candidates for analyzing the influence of microstructure in relation with the phenomenon of strain localization.

A total of 11 plane strain compression tests were carried out under drained conditions on water-saturated, initially dense specimens of the various sands. False Relief Stereophotogrammetry (FRS) was used to capture the onset of strain localization, and to measure shear band width and orientation The experimental device and testing procedures are briefly described in section 2. The physical characteristics of the different granular soils used in this study are detailed in section 3. The obtained results are reported in section 4. Differences and similarities between the results are examined in terms of both the global stress-strain and volumetric strain response, and thickness and orientation of the shear band. The influence of mean grain size and grain size distribution uniformity on the occurrence of strain localization, shear band thickness and orientation is discussed in the last section.

2   Experimental device and testing procedure

The experiments reported herein were performed in the plane strain compression apparatus originally developed in Grenoble by Desrues [9] and later modified by Hammad [17]. The biaxial apparatus was specifically conceived to allow free shear band formation and progression in a soil specimen. The features of the apparatus have been recently detailed by Desrues & Viggiani [11] and are shortly summarized herein. Within this device, a prismatic sand specimen, surrounded by a latex membrane, is mounted between two rigid walls inducing plane strain conditions (see Fig. 1). The side walls are glass plates which allow photographs to be taken of the in-plane deformation of a specimen during the test. All surfaces in contact with the specimen are enlarged and lubricated to minimize friction. The lower and the upper loading platens house porous stones connected to drainage lines. The top platen is free to rotate and to slide horizontally in the plane of deformation. This allows free lateral displacement of the upper part of the specimen once a shear band forms due to deviatoric loading. A large cell, filled with oil, surrounds the specimen. The cell has two opposite pairs of large Plexiglas windows on its lateral surface. Strain-controlled axial loading is applied through a screw jack which rests atop the device. Vertical displacement of the top platen is measured by a LVDT, and the axial load is measured by an internal load cell at the top. The traditional self-compensating mercury control is used to apply a back pressure to the pore water. Global volume change is monitored with an LVDT measuring the drop in mercury level in the upper cylinder. The output signals of all transducers are conditioned by a process interface unit which is linked to a microcomputer controlling all operations.

Sand specimens, 35 mm thick, 176 mm high and 88 mm wide, were formed in a mold hosting a latex membrane, pre-formed to prismatic shape. In order to obtain initially dense specimens, the method of dry pluviation was used, which is well known to yield very uniform specimens, e.g. [15]. Before transporting the specimen (under a vacuum of 80 kPa) into the biaxial apparatus, a regularly spaced grid was drawn and a paint was sprayed on the larger side of the specimen to assist in the identification of individual points for stereophotogrammetry. After placing the specimen into the apparatus, a confining pressure equivalent to the vacuum was applied as the vacuum was removed. To saturate a specimen, first carbon dioxide and then de-aired water were passed upwards through the sample. Finally, the specimen was back pressure saturated. After isotropic consolidation, specimens were sheared under displacement control at a constant rate of 1.0 mm/min, corresponding to a nominal axial strain rate of approximately 34% per hour for a 176 mm specimen height. All tests were performed under drained conditions. In all the experiments reported herein, the confining pressure was equal to 180 kPa and the back pressure to 80 kPa, *i.e.*, the horizontal in–plane effective stress σ_3' was equal to 100 kPa.

As mentioned above, photographs of the in–plane specimen deformation were taken throughout each test. Use of False Relief Stereophotogrammetry allowed to characterize the evolution of deformation throughout the test. In particular, the stereoscopic view of successive photographs allowed the measurement of the thickness and orientation of a shear band, which is of interest herein, with an accuracy of $\pm 1°$ and ± 2 mm respectively. The principles, details, and accuracy of stereophotogrammetry are described elsewhere [10, 11].

3   Tested sands

The granular materials selected for this study initially consisted of two natural sands, siliceous in nature, coming from the Hostun quarry (Drôme, France) which is operated by Sika Ltd. The two sands, referred to as S28 and S32, are uniform, poorly graded granular materials, with a coefficient of uniformity D_{60}/D_{10} approximately equal to 1.5 and a mean particle diameter (D_{50}) of 0.34 mm and 0.90 mm, respectively (Fig. 2 and Tab. 1). The specific gravity of the quartz grains is $G_s = 2.65$ [14].

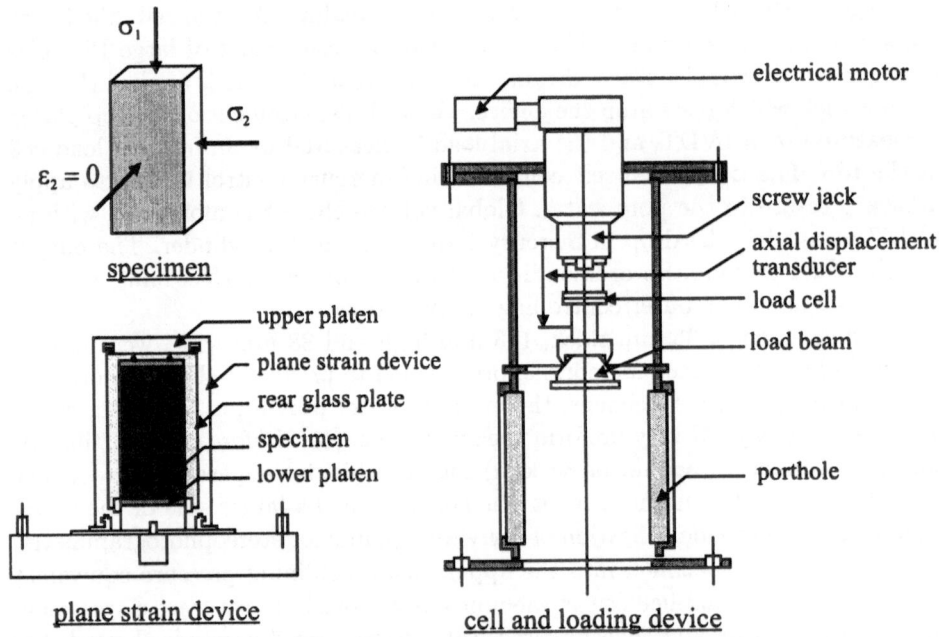

Fig. 1. Schematic view of the biaxial apparatus

Table 1. Grain size distribution of sands S28 and S32. The fractions in boldface correspond to Sf28 and Sg32, respectively

D (mm)	0.13	0.16	0.20	0.25	**0.32**	0.40	0.50	0.63	**0.80**	1.00	1.25	1.60	2.00
S28 (%)	0	2	10	27	**43**	17	1	0					
S32 (%)					0	2	9	20	**40**	20	8	1	0

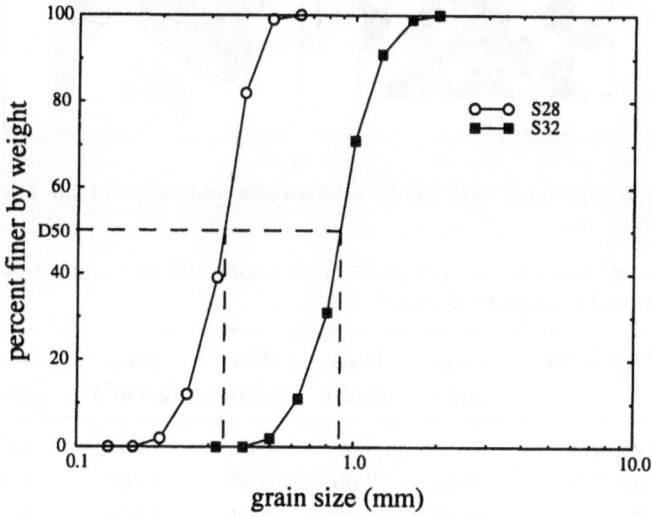

Fig. 2. Grain size distribution of S28 and S32 sands

Two uniform (almost monogranular) sands were obtained from S28 and S32, and will be referred to as Sf28 and Sg32 in the following sections. The Sf28 sand is a narrow gradation of S28 about its mean particle diameter, with 100% of the material passing a 0.40 mm sieve and retained on a 0.32 mm sieve. This fraction represents 43% of the initial S28 sand (Tab. 1). The Sg32 sand was obtained by retaining the principal grain fraction (40%) of S32 between 0.80 and 1.00 mm (Table 1). Sf28 and Sg32 represent very uniform materials having the same D_{50} as S28 and S32, respectively. The two populations of grains appear both subangular in shape in microscopic view, although it must be noted that the grains of the Sg32 fraction are slightly more rounded (Fig. 3), as it was already observed by Mokni [19] and to be expected, as grain angularity is known to decrease with grain size (*e.g.*, [27]). The minimum, maximum, and mean particle diameters of the four sands S28, S32, Sf28, and Sg32, are given in Tab. 2, along with the minimum and maximum values of density (ρ_{min} and ρ_{max}). These last values were determined according to the procedure suggested by the Japanese Geotechnical Society [26]. See [7] for further details. Note that the coarser sands (S32 and Sg32) systematically show slightly higher densities than fine materials (S28 and Sf28), which is likely due to the slight differences in shape between the fine and coarse grains.

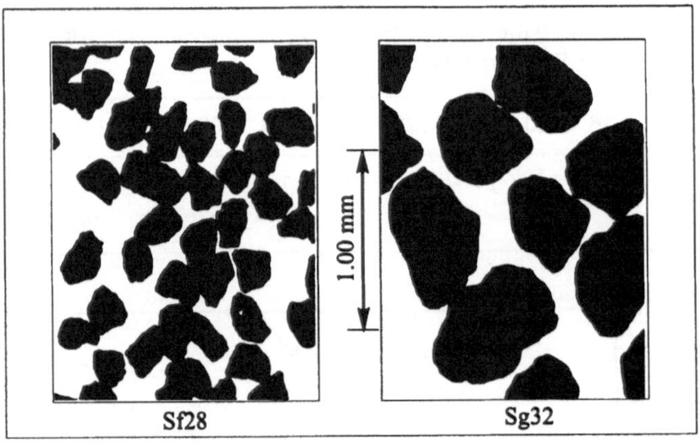

Fig. 3. Sf28 and Sg32 sands: relative size and shape of the grains

Table 2. Monodisperse sands: minimum, maximum, and mean grain particle diameters; minimum and maximum density

sand gradation	D_{min} (mm)	D_{max} (mm)	D_{50} (mm)	ρ_{min} (g/cm^3)	ρ_{max} (g/cm^3)
S28 (%)	0.13	0.50	0.34	1.353	1.630
S32 (%)	0.40	2.00	0.90	1.487	1.711
Sf28 (%)	0.32	0.40	0.34	1.344	1.614
Sg32 (%)	0.80	1.00	0.90	1.451	1.664

A second set of materials was obtained by mixing Sf28 and Sg32 in various proportions. Note that the size ratio between the two fractions of grains equals approximately 2.7. Four such binary mixtures were tested, with the coarse fraction (Sg32) representing respectively 10, 25, 35, and 50% (in weight) of the final granular material. By considering also the tests performed with Sf28 and Sg32, the coarse fraction actually varied from 0 to 100%.

An experimental procedure, originally suggested in [16], was developed to obtain homogeneous mixtures of the two grain sizes. The two fractions, or grain size populations, are pluviated with the help of two electronic vibrating devices. As the specific weight is the same for the two grain sizes, the final proportions of the mixture depend on the ratio of the flow rates of the two sets of grains, which in turn is controlled by the frequency ratio of the two vibrating devices. Sand grains fall through a series of sieves on which they rebound and mix up; the size of the mesh is approximately four times the maximum size of the grains, in order to improve grain mixing. Mixed grains then flow within an oversized tube of pre–determined length attached to a funnel, guiding the flow of sand and setting the drop height. The mold is moved downward as the sand is evenly deposited, keeping the bottom of the tube at a constant distance from the level of the sand in the mold. The flow rate of sand remains constant, as it is controlled by the nozzle diameter of the funnel. The overall procedure minimizes the possible segregation of the two fractions and yields reasonably homogeneous mixtures. Homogeneity of the mixture was checked for each specimen after testing, by cutting the specimen in twelve portions and sieving the relevant material to determine the local grain size proportions. As an example, the measured standard deviation was of 2.5% for a specimen with an average 11.6% coarse proportion (nominal value 10%), and 2.6% for a specimen having average coarse proportion of 50.9% (nominal value 50%).

4 Experimental results

Tab. 3 summarizes the experimental program and lists for each test the values of initial void ratio e_0 (before consolidation), along with the percent of coarse fraction for the binary mixtures. To make the results comparable, all tests were performed at 100 kPa effective confining stress, starting from approximately the same relative initial density (around 80%, see Tab. 3).

Table 3. Summary of experimental program

monodisperse sands				binary mixtures			
test	ρ_0 (g/cm^3)	e_0	D_r (%)	test	ρ_0 (g/cm^3)	e_0	coarse fract. (%)
TS2801	1.577	0.680	81	00BM01	1.541	0.720	0
TS2802	1.563	0.695	76	00BM02	1.577	0.680	0
00BM01	1.541	0.720	73	10BM01	1.619	0.637	11.6
00BM02	1.577	0.680	86	25BM01	1.670	0.587	22.6
100BM01	1.633	0.623	85	35BM01	1.642	0.614	35.1
100BM02	1.625	0.630	82	50BM01	1.651	0.605	51.1
TS3201	1.659	0.597	75	100BM01	1.633	0.623	100
				100BM02	1.625	0.630	100

Strain localization was consistently observed in all tests. It took place after a few percent of axial deformation, slightly before the peak of effective stress ratio, and gave always way to a single, persistent shear band. The global volumetric response was always dilative as the shear band formed, as expected for a dense sand. Results are presented in terms of global stress-strain response, and orientation and width of the shear band. The effective stress ratio σ_1'/σ_3' and volumetric strain $\epsilon_v = -\ln(V/V_0)$ are plotted as a function of the axial strain $\epsilon_1 = -\ln(H/H_0)$, where H_0 [V_0] and H [V] are the initial and current

height [volume] of the specimen, respectively. Note that all strains are positive
in compression. The orientation of the shear band is expressed by θ, the angle
between the direction of axial loading and the shear band. Both width and orientation are based on the first available photographic increment after the band
has formed, as suggested in [18] for instance. As an example, Fig. 4 shows strain
localization as observed through stereo–comparison of the first two photographs
after shear band formation in four different tests. When viewed in stereo, the
zone of concentrated strain appears as a slope connecting two planes of different
elevations, a false relief. Fig. 4 shows the projection of the boundaries of this
slope — which is the shear band — on the plane of the photograph.

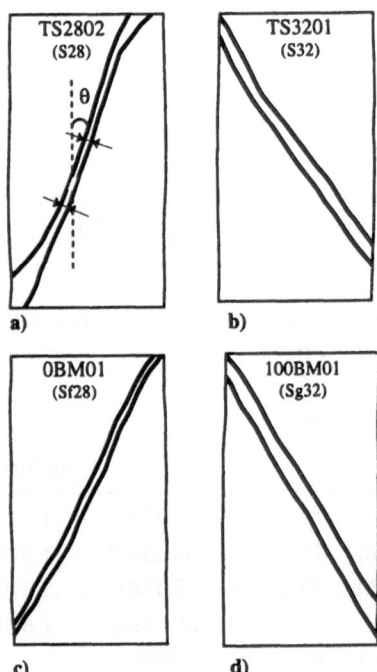

Fig. 4. Illustration of shear bands as observed through the stereoscopic view

For convenience, experimental results will be presented as follows: a) results
obtained on the sands S28, S32, Sf28 and Sg32; and b) results obtained on binary
mixtures. The former four sands will be termed *monodisperse* hereafter, as op-
posed to the binary mixtures. In the following, some comparisons are made focus-
ing on differences and similarities between the results. Tab. 4 lists, for each test,
the peak value of the mobilized friction angle $\phi = \arcsin\left[(\sigma_1' - \sigma_3')/(\sigma_1' + \sigma_3')\right]$
along with the (axial) strain level at which this is attained, and the value of
the dilatancy angle ψ at the peak. In the table are also listed, for each test, the
width of the shear band and its orientation θ as measured through stereophotogrammetry.

4.1   Monodisperse sands

The values of initial void ratio e_0 measured before consolidation as well as relative densities of the specimens are listed in Tab. 3 for each test. It can be observed that initial value of relative density D_r varied from one test to another, with a maximum difference of 12%. Therefore, sand grading was strictly not the only varying parameter from a test to another. The possible effect on strain localization of the differences of initial relative density will be commented later in the paper. However, it can be anticipated that such an effect was not very important, as it can be inferred from the comparison of the results obtained from the two tests on S28 sand (00BM01 and 00BM02). Despite a difference of 12% in the initial relative density, these two tests yielded almost coincident results, both in terms of the stress-strain response and orientation and width of the shear band. More generally, it should be stressed that, whenever a test was duplicated (as for tests 100BM01 and 02, and TS2801 and 02), a quite good repeatability was found, independent of the difference in the initial value of D_r. This is a direct measure of the quality and reliance of all results obtained.

Fig. 5 shows the stress-strain responses obtained in the tests on S28 and S32 sands. Note that stress–strain response from test TS2801 was lost due to troubles in the acquisition system. The response of the two sand is essentially similar in the pre–peak portion of the tests, that is before a complete shear band has formed. The peak effective stress ratio was attained at larger axial strain for S32, with a higher value of the maximum mobilized friction angle ϕ (approximately 51°, as compared to the value of 48°, obtained for S28). Also the volumetric behavior is almost coincident in the two tests, at least in the pre–peak regime. In both tests, the globally measured dilatancy, indicated by the slope of the volumetric strain curve, reaches a maximum around the peak of effective stress ratio and approaches zero at the end of the test. However, after the peak the deformation has concentrated into a narrow zone, and globally–measured dilatancy does not characterize the local behavior any more. In fact, it should be recalled that axial and volumetric strain are computed from boundary measurements of displacements and global volume change of the specimen, respectively. In the presence of localized strains, this data reduction scheme is no longer applicable. Similar considerations apply to the axial stress, which is determined from the axial load cell and the cross–sectional area of the specimen. Once again, in the post–peak portion of a test, i.e. when a shear band has formed, the stress–strain curve should be regarded only as a nominal response, which does not reflect the actual behavior of the sand. It is for this reason that the observed differences between the results are not to be overlooked in the post–peak portion of the test.

As far as the shear band is concerned, major differences were observed between S28 and S32 as for both thickness and orientation. See Fig. 4a,b and Tab. 4. The measured value for the angle θ was equal to 21° and 23° in the tests on S28 sand, whereas a less steep band ($\theta = 35°$) was observed in the test on S32. The thickness of the shear band varied from 5 mm (S28) to 7.5 mm (S32). It is interesting to note that the value of θ is close to the Coulomb value $\theta_C =$

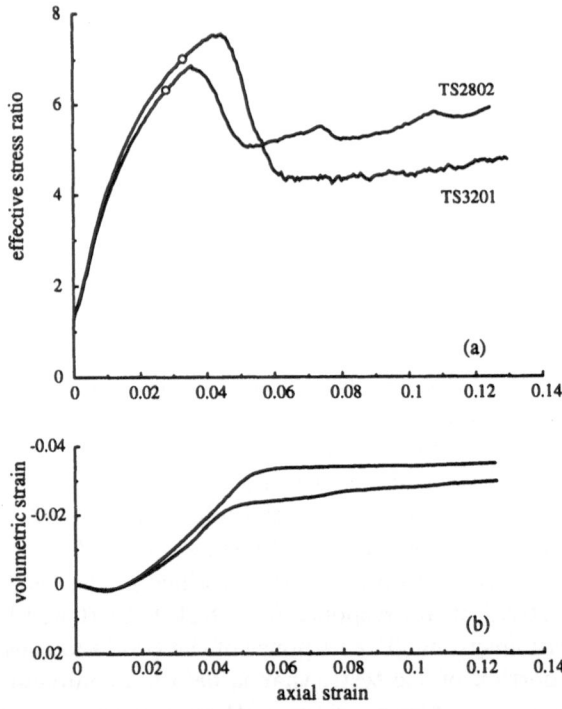

Fig. 5. Global stress–strain responses obtained on S28 and S32: effective stress ratio (a) and volumetric strain (b) as a function of axial strain

$\pi/4 - \phi/2 = 20.9°$ for S28, whereas it is closer to the so–called Roscoe orientation, $\theta_R = \pi/4 - \psi/2 = 36.8°$ for S32. Note also that the steepest band (21°) was obtained for the S28 specimen with the smallest relative density (76%).

Figs. 6 and 7 compare the stress–strain responses obtained in the tests on S32 and Sg32, and S28 and Sf28, respectively. Recall that Sg32 and Sf28 were obtained as almost monogranular fractions of S32 and S28, respectively. Significant differences of global response were observed between the S32 and Sg32 sands, having the same D_{50} but a different degree of uniformity (Fig. 6). Both the strain and effective stress ratio at peak are different, with a difference of more than 4° as for the maximum mobilized friction angle. It should be noted, however, that the volumetric strain curves of the two sands are remarkably coincident. The orientation of the shear band, θ, was equal to 35° for S32 and 30 − 31° for Sg32 (see Tab. 4). This latter value closely corresponds to the value suggested by Arthur *et al.* [3] $\theta_A = \pi/4 - (\phi + \psi)/4 = 29.7°$. Note that the steepest band was obtained in the more uniform sand. As for the band thickness, a value of 7.5 mm was measured for both S32 and Sg32.

Table 4. Main data from the tests

test	ϵ_1 at peak (%)	ϕ at peak (°)	ψ at peak (°)	band width (mm)	θ (°)
TS2801	–	–	–	5.0	23
TS2802	3.55	48.2	14.4	5.0	21
00BM01	3.94	46.7	14.3	3.5	29
00BM02	3.83	46.9	14.6	4.5	28
100BM01	3.90	46.1	–	7.5	31
100BM02	3.92	46.7	14.4	7.5	30
TS3201	4.45	51.2	16.4	7.5	35
10BM01	3.70	46.6	13.4	5.0	30
25BM01	3.71	45.9	14.2	4.5	31
35BM01	3.37	46.1	14.1	5.5	27
50BM01	3.61	46.9	14.6	6.0	29

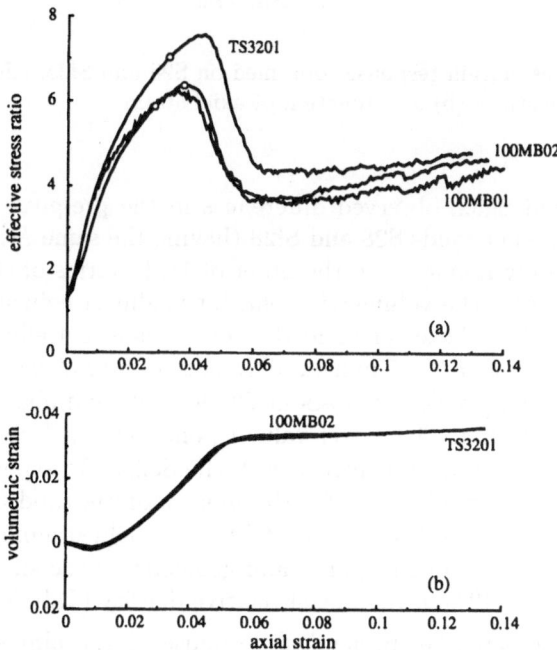

Fig. 6. Global stress–strain responses obtained on S32 and Sg32: effective stress ratio (a) and volumetric strain (b) as a function of axial strain

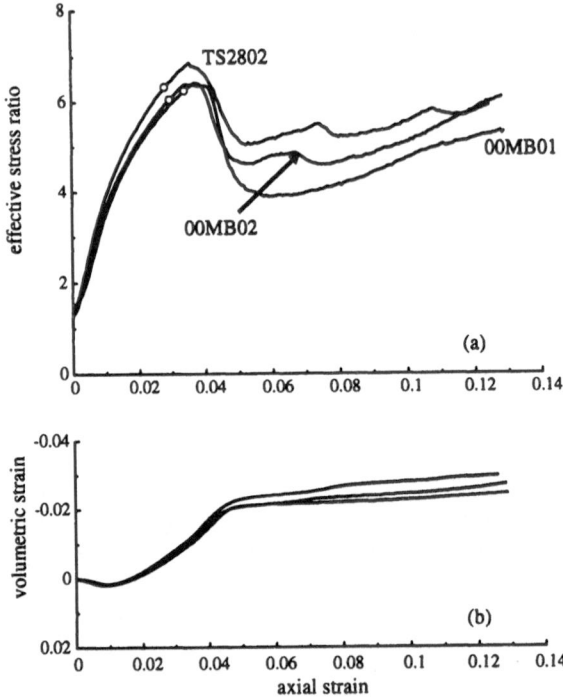

Fig. 7. Global stress–strain responses obtained on S28 and Sf28: effective stress ratio (a) and volumetric strain (b) as a function of axial strain

Unlike S32 and Sg32, observed differences in the pre-peak stress-strain response of the two finer sands S28 and Sf28 (having the same D_{50} but a different gradation) were only minor — on the order of 1°, in terms of the friction angle at peak (Fig. 7). Also the volumetric behavior is almost coincident for the two sands. However, while the shear band thickness was essentially the same in the four tests (in the range $3.5 - 5$ mm), a difference of 5 to 8° was observed in the two sands tested. The angle θ was about 22° for S28, and about 28.5° for Sf28. See Fig. 4a,c and Tab. 4. This latter value is, once again, close to the value θ_A. It must be also noted that in contrast with the S32/Sg32 experiments, the less steep shear band is now obtained for the more uniform sand Sf28. As already pointed out, note that the two tests on Sf28 sand yielded almost coincident results both in terms of global response and geometry of the shear band, despite a difference of about 12% in the initial relative density (Tab. 3).

If we now compare the mechanical response of the almost monogranular fractions Sf28 and Sg32 (which essentially differ from each other only by their mean grain size), the stress-strain responses obtained in the two pairs of tests are coincident — but in the post–peak portion — (Fig. 8) and the shear band had a similar orientation in all four tests ($28 - 29° \leq \theta \leq 30 - 31°$, Tab. 4). The only but significant difference concerned the shear band thickness, approximately equal to 5 mm and 7.5 mm for Sf28 and Sg32 respectively (Figs. 4b and d). Such a difference can be unambiguously related to the mean grain size (D_{50}), as already suggested by several authors. It should be also noted that the volumetric behavior of Sg32 was remarkably more dilatant as compared to Sf28.

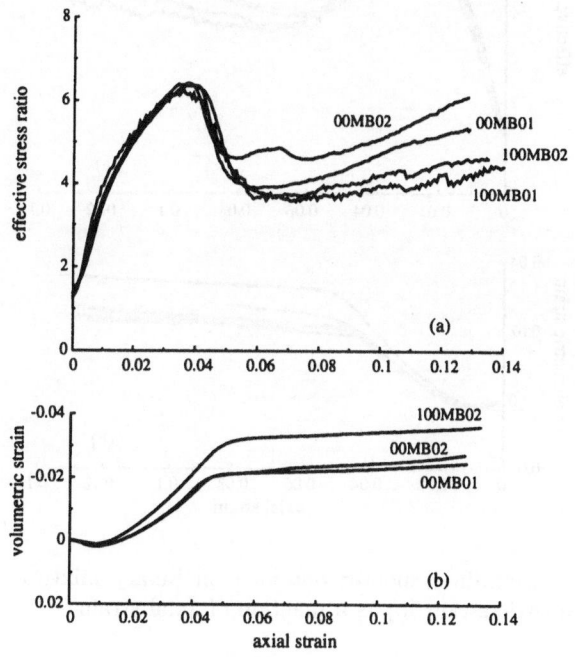

Fig. 8. Global stress–strain responses obtained on Sf28 and Sg32: effective stress ratio (a) and volumetric strain (b) as a function of axial strain

4.2   Binary mixtures

Four binary mixtures of the two sands Sf28 and Sg32 were tested, with the coarse fraction (Sg32) representing respectively 11.6, 22.6, 35.1, and 51.1% (in weight) of the final granular material. In this case, only the initial density r0 and not the relative density of the specimens is given in Tab. 3, as the minimum and maximum densities were not determined for the binary mixtures. Stress-strain responses obtained in the tests performed on binary mixtures are shown in Fig. 9, including data from the tests on Sf28 and Sg32 already presented in Fig. 8. Therefore, six different mixtures are reflected in the plot, with a coarse fraction varying from 0 (Sf28) to 100% (Sg32).

The major observation is that all specimens globally exhibited the same response, whatever the composition of the mixture. All the stress-strain curves in Fig. 9 essentially coincide, with a peak friction angle of $46-47°$ (Tab. 4). The volumetric strain evolution is also quite similar from one test to another. Shear

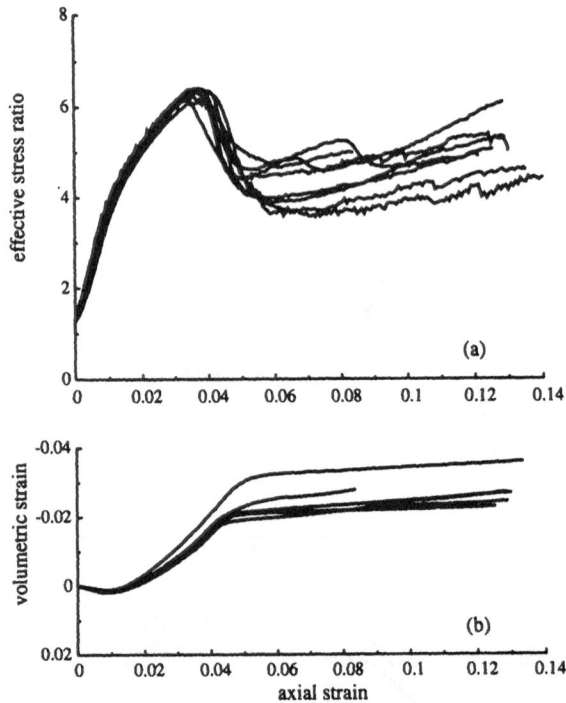

Fig. 9. Global stress–strain responses obtained on binary mixtures: effective stress ratio (a) and volumetric strain (b) as a function of axial strain

band orientation varies only slightly from one test to another, in the range 27 to 30°, that is close to the value θ_A. It must also be pointed out that these slight differences do not appear to be correlated to the composition of the mixtures. The only difference concerned the shear band thickness, which is shown to increase with increasing fraction of coarse grains (Tabs. 3 and 4).

5   Discussion

All the experiments presented above were conducted under the same experimental conditions starting from dense specimens, with relative densities in the range 75 to 85%. It is important to stress that when tests were duplicated, the effect of a different initial relative density on shear banding was found negligible, in that both the stress-strain response and orientation and width of the shear band were almost coincident. Therefore, all differences in the mechanical behavior exhibited by the different sands used in this study can be directly related to the corresponding differences of grain size distribution.

The experimental results clearly indicate that shear band orientation is not related to the mean particle size of granular materials, in contradiction with the previous experimental findings of Arthur & Dunstan [4] and Mokni [19]. This result has been established in two ways, by comparing the response of: i) granular materials with the same mean grain size, but different gradation (S28 *vs.* Sf28 and S32 *vs.* Sg32); and ii) granular materials with a different mean grain size but similar gradation (Sf28 *vs.* Sg32). In the former case, a difference of at least 5° of the shear band orientation was systematically observed. In the latter case, the shear bands which have formed in Sf28 and Sg32 specimens have the same orientation, despite an average size ratio of about 2.5 between the two sands. This last result seems to confirm the assumption that the geometry of monogranular assemblies is independent from the size of grains — as suggested *e.g.* in [2, 5]. Note that the slight difference in angularity between the fine and coarse grains (see Fig. 3) seems to have a negligible influence on strain localization in this case. A possible explanation for the discrepancy between the results of this study and those of Arthur & Dunstan [4] and Mokni [19] is that other factors than mean grain size were also varying in their experiments. Caution should then be used when comparing localization in different sands in term of mean grain size.

The results also indicate that shear band orientation can neither be simply related to the uniformity of the granular materials. Although the shear band inclination was shown to vary with the degree of uniformity, the steepest band was obtained in the more uniform sand in the case of the S32–Sg32 pair of sands, whereas the opposite was observed for S28 and Sf28. It is therefore not possible to establish a definite trend between shear band orientation and the degree of uniformity of sand grading from our experiments. However, since this conclusion is only based on two examples, the influence of sand grading on strain localization should be more systematically investigated, making sure that uniformity is the only varying parameter from a test to another.

As for the influence of mean grain size on shear band thickness, the above results do not add any new finding, that is, they just confirm the well-known dependence of the band thickness on the mean grain size, *e.g.*, [20, 22].

In contrast with *monodisperse* granular materials, all the binary mixtures globally behave in the same way both in terms of stress–strain response and the geometry of the localized shear band, whatever the composition of the specimens (Fig. 9 and Tab. 4). The only correlation that can be established between grain size distribution and shear banding in binary mixtures concerns once again the shear band thickness, which becomes larger and larger as the coarse fraction increases. The variations of the microstructure induced by the variation of the proportions of the two grain size populations are not reflected into corresponding measurable variations of the mechanical properties of sands at the macroscopic level. The fact that different microstructures may give rise to a similar mechanical response — in particular, similar shear band characteristics — is indeed the salient result of the tests on binary mixtures. This might suggest that strain localization in granular materials depends on the collective behavior of the ensemble of grains than upon the details of the microstructure. Although changes in grain size distribution may have a strong influence on shear banding, as shown for monodisperse materials, grain size distribution imperfectly describes the fabric of sands. The mean grain size, the coefficient of uniformity (and more generally any other parameters that may be used to characterize the grain size distribution) are not by themselves significant parameters on which the phenomenon of strain localization depends. While this was the initial (and, *a posteriori*, possibly naive) idea of the authors, the simple and attracting idea to describe the relation between microstructure and shear banding in terms of the grain size distribution only seems inappropriate.

6   Conclusions

An experimental program has been conducted to investigate the possible influence of grain size distribution on the phenomenon of shear banding in sand. All sands tested came from the same natural deposit, which means that while different in terms of grading, they are the same material, as far as the mineralogy and shape of the grains are concerned. Since all the specimens were tested under the same experimental conditions, grain size distribution was the only physical parameter varying from one test to another.

The experimental results show that, while the relation between the band thickness and the mean grain size is confirmed, the orientation of a shear band can neither be related to the mean particle size nor to the degree of uniformity of a sand in a simple way.

More generally, the major result of this study is that while grain size distribution can greatly affect shear banding characteristics, there is not such a thing as a direct relationships between characteristics of localization and mean grain size, or degree of uniformity, or other parameters describing grain size distribution in a simple way. Whether this means that there is no direct relationships between microstructure and localization cannot be clarified by the results. In the authors' opinion, shear banding does indeed depend on microstructure. However, on the basis of the above results it can be concluded that the considered parameters of grain size distribution are not pertinent ways of describing microstructure in relation to strain localization phenomena. In other words, microstructure is a key factor yet its influence on shear banding cannot be described in terms of grain size distribution only.

Acknowledgements

While in Grenoble, the second author received a grant by Gdr PMHC and GEO. Their support is gratefully acknowledged. The authors also wish to express their gratitude to A.L. Combe, who provided the data on minimum and maximum densities of the sands used in this study.

References

1. H. G. B. Allersma: Optical analysis of stress and strain in photoelastic particle assemblies. PhD Thesis, Delft University of Technology, the Netherlands (1987).

2. M. Ammi: Analyse de quelques propriétés des milieux granulaires modèles. Thèse de Doctorat es Sciences, Université de Rennes, France (1987).
3. J. F. R. Arthur, T. Dunstan, Q.A.J. Al–Ani, A. Assadi: Géotechnique, **27**, (1977).
4. J. F. R. Arthur, T. Dunstan: 'Rupture layers in granular materials'. In: *IUTAM Conference on Deformation and Failure of Granular Materials, Delft, the Netherlands, 1982* (Balkema, Rotterdam, 1982) pp. 453–459.
5. D. Bideau, J. P. Troadec, J. Lemaître, L. Oger: 'Dense packing of hard grains: effects of grain size distribution'. In *Physics of finely divided matters.* ed. by N. Boccara and M. Daoud (Springer 1985) pp. 76–81.
6. F. Calvetti, G. Combe, J. Lanier: Mechanics of Cohesive–Frictional Materials, **2**, 2 (1997).
7. A.L. Combe: 'Comportement du sable d'Hostun S28 au triaxial axisym'etrique. Comparaison avec le sable d'Hostun RF ' Rapport de stage, Laboratoire 3S, Grenoble (unpublished, 1998)
8. P. Dantu: Ann. Ponts et Chaussées, **IV** (1967).
9. J. Desrues: La localisation de la déformation dans les matériaux granulaires. Thèse de Doctorat es Sciences, USMG & INPG, Grenoble, France (1984).
10. J. Desrues, B. Duthilleul: Journal de Mécanique Théorique et Appliquée, **3**, 1 (1984).
11. J. Desrues, G. Viggiani: Mechanics of Cohesive–Frictional Materials (submitted for possible publication, 2000).
12. J. Desrues, R. Chambon, M. Mokni, F. Mazerolle: Géotechnique, **46**, 3 (1996).
13. R. J. Finno, W. W. Harris, M. A. Mooney, G. Viggiani: Géotechnique, **47**, 1 (1997).
14. E. Flavigny, J. Desrues, B. Palayer: Rev. Francaise Géotech., **53** (1990).
15. P. A. Gilbert, W. F. Marcusson: J. of Geotech. Engrg., ASCE, **114**, 1 (1998).
16. E. Guyon, L. Oger, T. J. Plona: Journal of Physics D: Appl. Phys., **20** (1987).
17. W. Hammad: Modélisation non linéaire et étude expérimentale des bandes de cisaillement dans les sables. Thèse de Doctorat de l'Université J. Fourier, Grenoble, France (1991).
18. W. W. Harris, G. Viggiani, M. A. Mooney, R. J. Finno: Geotechnical Testing Journal of Physics, ASTM, **18**, 4 (1987).
19. M. Mokni: Relations entre d'eformations en masse et déformations localisées dans les matériaux granulaires. Thèse de Doctorat de l'Université J. Fourier, Grenoble, France (1992).
20. H. B. Mühlhaus, I. Vardoulakis: Géotechnique, **37**, 3 (1987).
21. L. Oger: Etude des corrélations structure–propriétés dans les milieux granulaires modèles. Thèse de Doctorat es Sciences, Université de Rennes, France (1987).
22. K. H. Roscoe: Géotechnique, **20**, 2 (1970).
23. A. S. Saada, L. Liang, J. L. Figueroa, C. T. Cope: Géotechnique, **49**, 3 (1999).
24. I. Vardoulakis, J. Sulem: *Bifurcation Analysis in Geomechanics* (Blackie Academic & Professional, 1995)
25. T. Yoshida, F. Tatsuoka, Y. Kamegai, M. S. A. Siddiquee, C.–S. Park: 'Shear banding in sand observed in plane strain compression'. In: *Localization and Bifurcation Theory for Soils and Rocks, 3th International Workshop at Aussois, France, September, 1993*, ed. by R. Chambon, J. Desrues, I. Vardoulakis (Balkema, Rotterdam 1994) pp. 165–179.
26. Y. Yoshimi, F. Kuwabara, K. Tokimatsu: Soils and Foundations, **15**, 3 (1975).
27. A. Yudhbir, R. Abedinzadeh: Geotech. J. Testing Mat., GTJODJ, **14** (1991).

Micromechanics of the elastic behaviour of granular materials

N. P. Kruyt,[1] L. Rothenburg[2]

[1] Department of Mechanical Engineering, University of Twente, Enschede, The Netherlands

[2] Department of Civil Engineering, University of Waterloo, Waterloo, Ontario, Canada

Abstract. Micromechanics of granular materials deals with the relation between microscopic and macroscopic characteristics. This study focuses on elastic behaviour. An overview of micromechanics of granular materials is given. Two regimes are identified, uniform strain and uniform stress. It follows from extremum principles that these regimes correspond to upper and lower bounds for the moduli. A statistical theory is outlined that is based on maximum disorder. The results of Discrete Element simulations, based on isotropic assemblies of 50,000 particles, are used to gain insight in the relevant phenomena and to verify theoretical predictions. A linear relation between number of contacts and particle radius is observed. The statistical theory is valid for loose systems, while the uniform strain theory is valid for dense systems. In general, the relative displacements do not match the uniform strain assumption. Gaussian probability density functions are observed for the relative displacements. No equipartition of energy is found.

1 Introduction

Knowledge of the mechanical behaviour of granular materials is expressed by the constitutive relation that is usually based on continuum mechanics. Often phenomenological assumptions are made to obtain the constitutive relation.

Micromechanics forms a different approach to modelling the mechanical behaviour of granular materials. Contrary to the continuum viewpoint, it fully recognises the heterogeneous and discrete nature of granular materials. In micromechanics a granular material is modelled as an assembly of semi-rigid particles interacting by means of contact forces. The objective is to develop macroscopic (continuum) constitutive relations from the microscopic (contact) constitutive relations. Some recent micromechanical studies are [1–10, 14, 15, 19, 20, 25] [27–29, 33].

Here a detailed micromechanical study is made of the elastic behaviour of isotropic assemblies of two-dimensional, bonded and *non-rotating* particles. The results of this study suggest directions for further theoretical developments.

The outline of this paper is as follows. First an overview of micromechanics of granular materials is given. This is followed by a description of extremum principles and the Discrete Element simulation technique [11]. Finally, results are presented for the elastic moduli, magnitude and probability density function for relative displacements and energy.

2   Micromechanics

In this section a brief summary of micromechanics of quasi-static deformation of granular materials is given (see [21] for a more detailed account). Micromechanics of granular materials deals with the study of microscopic properties of particles, and their relation to overall characteristics. For granular materials, particles only exert forces on one another when they are in contact. Therefore the microscopic level is that of *contacts*. There the relevant physical quantities are contact force f_i^c and contact relative displacement Δ_i^c, which are related by some contact constitutive relation. The contact relative displacement Δ_i^{pq} between particles q and p is defined by

$$\Delta_i^{pq} = u_i^q - u_i^p \,, \tag{1}$$

where u_i^p is the displacement of the centre of particle p.

2.1   Branch and polygon vector

The contact geometry is depicted in Fig. 1: n_i^c and t_i^c are the normal and tangential vector of the contact and X_i^p is the centre of a particle p. The orientation of a contact θ^c is defined by the direction of its normal vector $n_i^c = \{\cos \theta^c, \sin \theta^c\}^T$. *Branch vectors* l_i^c are defined as the vectors that connect the centres of particles that are in contact. These branch vectors form closed loops, or networks, that define polygons (see Fig. 2). The vectors that connect the centres of adjacent polygons are the rotated polygon vectors g_i^c. The rotated polygon vectors also form closed loops that are shown in Fig. 2. The *polygon vectors* h_i^c are obtained by rotating the rotated polygon vectors counter-clockwise over 90 degrees. Note that the polygon vectors also form closed loops.

The branch and polygon vectors will be used in the micromechanical expressions for the stress and strain tensors.

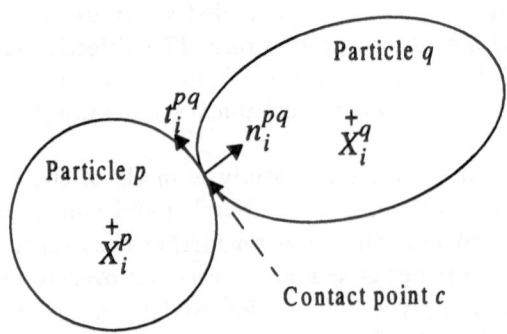

Fig. 1. Contact geometry

The graph formed by the rotated polygon vectors is the dual of the graph, see for example [22], of that formed by the branch vectors. Some properties of these graphs are discussed in [30].

Denote the number of particles by N_p, the number of contacts by N_c and the number of polygons (loops) by N_l. Euler's relation for large assemblies is, see for example [22]

$$N_p - N_c + N_l \cong 0 .\tag{2}$$

Denoting the average number of contacts per particle, or *average coordination number*, by Γ and the average number of sides per polygon by Ω, then it follows from (2) after noting that each contact is "shared" by two particles and two polygons

$$N_c = \frac{\Gamma}{2} N_p \;\; N_c = \frac{\Omega}{2} N_l \;\; \Omega \cong \frac{2\Gamma}{\Gamma - 2} .\tag{3}$$

A geometrical relation between branch and polygon vectors is [19]

$$\delta_{ij} = \frac{1}{S} \sum_{c \in S} l_i^c h_j^c ,\tag{4}$$

where δ_{ij} is the Kronecker symbol and the sum is over all contacts c in the region of interest with area S.

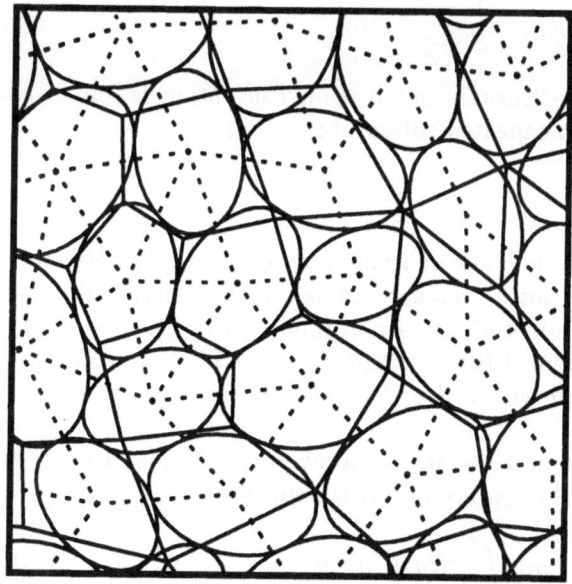

Network connecting polygon centres (rotated polygon vectors)

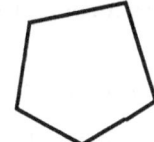

Network connecting particle centres (branch vectors)

Fig. 2. Networks of branch and rotated polygon vectors

2.2   Stress, strain and work

The quasi-static equilibrium conditions for particle p are

$$\sum_q f_i^{pq} + f_i^{p\beta} = 0 , \tag{5}$$

where the summation is over the particles q that are in contact with particle p and $f_i^{p\beta}$ is the force exerted by the boundary on particle p (if present).

The compatibility conditions for the relative displacements of particle centres are

$$\sum_s \Delta_i^{rs} + \Delta_i^{r\alpha} = 0 , \tag{6}$$

where the summation is over the sides s that form polygon r and $\Delta_i^{r\alpha}$ is the relative displacement between the boundary and contact r (if present).

The micromechanical expressions for the stress tensor σ_{ij} and the strain tensor ε_{ij} were derived from the equilibrium and compatibility conditions [19, 20] using a weighted averaging technique

$$\sigma_{ij} = \frac{1}{S} \sum_{c \in S} f_i^c l_j^c , \tag{7}$$

$$\varepsilon_{ij} = \frac{1}{S} \sum_{c \in S} \Delta_i^c h_j^c . \tag{8}$$

The work done at the boundaries must equal the internal work done at the contacts [20]. Hence

$$\sigma_{ij}\varepsilon_{ij} = \frac{1}{S} \sum_{c \in S} f_i^c \Delta_i^c . \tag{9}$$

Here and in the sequel, the Einstein summation convention is employed, which implies a summation over repeated subscripts.

2.3   Group averaging

For large assemblies that are random and spatially homogeneous, discrete quantities like those in (7) and (8) can be replaced by continuous distributions by grouping contacts with similar orientation θ. An example of this *group averaging* is the contact distribution function $E(\theta)$ [17]: $E(\theta)\Delta\theta$ is the fraction of contacts with orientation within the interval $(\theta, \theta + \Delta\theta)$. For the isotropic assemblies considered here $E(\theta) = 1/(2\pi)$.

For an arbitrary contact quantity α, its group average is denoted by $\overline{\alpha}(\theta)$, while an overall average over all orientations $\langle \overline{\alpha} \rangle$ is defined by

$$\langle \overline{\alpha} \rangle = m_S \int_{-\pi}^{\pi} \overline{\alpha}(\theta) E(\theta) \, d\theta , \tag{10}$$

where m_S is the contact density with respect to area $m_S = N_c/S$.

Since a contact can be "viewed" from two particles, the contact orientations θ and $-\theta$ are equivalent. Therefore the contact distribution function $E(\theta)$ and the normal and tangential components of the contact force f_n and f_t satisfy

$$E(-\theta) = E(\theta) \; ; \; \overline{f}_n(-\theta) = \overline{f}_n(\theta) \; ; \; \overline{f}_t(-\theta) = \overline{f}_t(\theta) \; . \tag{11}$$

Of course, similar relations hold for the branch, polygon and relative displacement vectors. It therefore suffices to consider only orientations θ in the interval $(0°, 180°)$.

In terms of overall averages the stress tensor (7) and strain tensor (8) are given by

$$\sigma_{ij} = \langle \overline{f_i l_j} \rangle \qquad \varepsilon_{ij} = \langle \overline{\Delta_i h_j} \rangle \; . \tag{12}$$

2.4   Contact constitutive relation

In the micromechanical approach the mechanical behaviour of the materials is specified at the microscopic (contact) level. Attention is focused here on the simple case of (linear) elastic behaviour of bonded, two-dimensional assemblies of *non-rotating* particles. The linear contact constitutive relation relates the normal and tangential forces f_n^c and f_t^c at the contact to the normal and tangential components Δ_n^c and Δ_t^c of the relative displacement of the two particles involved and to the stiffnesses k_n and k_t of springs in normal and tangential direction

$$f_n^c = k_n \Delta_n^c \qquad f_t^c = k_t \Delta_t^c \; . \tag{13}$$

The stiffness ratio λ is defined by $\lambda = k_t/k_n$.

The relation between force and relative displacement at the contact can also be expressed by a stiffness matrix S_{ij}^c

$$f_i^c = S_{ij}^c \Delta_j^c \qquad S_{ij}^c = k_n n_i^c n_j^c + k_t t_i^c t_j^c \; . \tag{14}$$

With this linear elastic contact constitutive relation, the macroscopic (continuum) constitutive relation is also elastic. The strain tensor ε_{ij} is related to the stress tensor σ_{ij} by the effective elastic stiffness tensor L_{ijkl} and compliance tensor M_{ijkl}

$$\sigma_{ij} = L_{ijkl} \varepsilon_{kl} \qquad \varepsilon_{ij} = M_{ijkl} \sigma_{kl} \; . \tag{15}$$

3   Extremum principles

Minimum potential energy and minimum complementary energy principles were derived in [20]. It was shown that a uniform stress field

$$f_i^c = \sigma_{ij} h_j^c \; , \tag{16}$$

leads to a lower bound to the moduli, while an upper bound follows from a uniform strain field

$$\Delta_i^c = \varepsilon_{ij} l_j^c \; . \tag{17}$$

The corresponding stiffness tensor L^ε_{ijkl} and compliance tensor M^σ_{ijkl} are

$$L^\varepsilon_{ijkl} = \langle S_{ik}\overline{l_j l_l}\rangle \; ; \qquad M^\sigma_{ijkl} = \langle S^{-1}_{ik}\overline{h_j h_l}\rangle \; . \qquad (18)$$

These expressions involve group averages of branch and polygon vectors. For isotropic assemblies we have

$$\overline{l_i}(\theta) = \overline{l_n}n_i(\theta) \qquad \overline{l_i l_j}(\theta) = \overline{l_n^2}n_i(\theta)n_j(\theta) + \overline{l_t^2}t_i(\theta)t_j(\theta)$$

$$\overline{h_i}(\theta) = \overline{h_n}n_i(\theta) \quad \overline{h_i h_j}(\theta) = \overline{h_n^2}n_i(\theta)n_j(\theta) + \overline{h_t^2}t_i(\theta)t_j(\theta) \; , \qquad (19)$$

where $\overline{l_n}, \overline{h_n}, \overline{l_n^2}, \overline{l_t^2}, \overline{h_n^2}$ and $\overline{h_t^2}$ are independent of orientation θ.

The corresponding uniform field moduli are

$$\frac{K^\varepsilon}{k_n} = \frac{m_S}{4}(\overline{l_n^2} + \lambda\overline{l_t^2}); \qquad \frac{G^\varepsilon}{k_n} = \frac{m_S}{8}(1+\lambda)(\overline{l_n^2} + \overline{l_t^2})$$

$$\frac{K^\sigma}{k_n} = \frac{\lambda}{m_S(\lambda\overline{h_n^2} + \overline{h_t^2})}; \quad \frac{G^\sigma}{k_n} = \frac{2\lambda}{m_S(1+\lambda)(\overline{h_n^2} + \overline{h_t^2})} \; , \qquad (20)$$

where K^ε and K^σ are bulk moduli and G^ε and G^σ are shear moduli. Expressions for K^ε and G^ε were already given in [4] for disks. Note that for disks as will be considered in the sequel $l_t \equiv 0$.

3.1   Statistical minimum potential energy theory

Here the statistical minimum potential energy theory of [20] is briefly recapitulated. In this theory potential energy is minimised subject to constraints. The constraints are prescribed strain (8) and the work relation (9). With the additional assumptions of Gaussian probability density functions for normal and tangential components of the relative displacements and maximum disorder, the effective stiffness tensor was obtained as

$$L_{ijkl} = \frac{L^\varepsilon_{ijkl} + L^\sigma_{ijkl}}{2} \; . \qquad (21)$$

4   Discrete Element simulations

Discrete Element simulations [11] are computer simulations in which the motion of a large assembly of particles (the discrete elements) is computed. The movement of the particles and the forces at the contacts, induced by stress or strain at the boundary, are computed from the equations of motion. This information would be very hard (if possible at all) to obtain from (photoelastic) experiments (for example [16, 18, 23]).

There are three main aspects to Discrete Element simulations (i) particle shape and particle size distribution (ii) contact constitutive relation (iii) numerical method for solving equations of motion. The numerical method employed here for quasi-static motion is the time-stepping method of [11] with underrelaxation instead of damping.

4.1 Particle size distribution

Various particle shapes have been employed in Discrete Element simulations: (i) disks, as originally used by [11] and also used here; (ii) ellipses [28, 31], or (iii) polygons [13].

The geometry of the particles is completed by specifying the particle size distribution. The selected particle size distribution for the particle radius R is a lognormal distribution that is frequently used for granular materials (see for example [26]). Its probability density function p is given by

$$p(R; R_{\text{med}}, \tau) = \frac{1}{\sqrt{2\pi \ln(1+\tau^2)} R} \exp\left(-\frac{1}{2} \frac{[\ln(R) - \ln(R_{\text{med}})]^2}{\ln(1+\tau^2)}\right) , \qquad (22)$$

where R_{med} is the median (i.e. the value where the cumulative probability function equals $1/2$) and τ is ratio between standard deviation $\text{StdDev}(R)$ and mean R_{avg}

$$R_{\text{avg}} = R_{\text{med}} \sqrt{1+\tau^2} ; \ \text{StdDev}(R) = \tau R_{\text{avg}} . \qquad (23)$$

The employed particle size distribution is fairly wide with $\tau = 0.25$. Note that the mean radius R_{avg} only determines the size of the area under consideration. Particles were divided into 100 groups. The radii of the groups were selected in such a manner that the resulting cumulative probability function closely matched the lognormal cumulative probability function. The ratio between the radius of the largest disk in the simulations to that of the smallest is 3.7.

4.2 Assemblies

The assemblies of 50,000 disks that were used for the Discrete Element simulations were created by slow uniform compression until the required average coordination numbers Γ were obtained. The assemblies generated in this manner were isotropic. Note that the average coordination number is a measure of the density of the packing. An approximate relation between packing density η and average coordination number Γ was derived in [19], $\eta \cong (\pi/\Gamma)/(\tan(\pi/\Gamma))$.

At higher average coordination numbers Γ there may be a significant deformation, or "overlap", of the particles. Then $R_1^c + R_2^c > l^c$, where R_1^c and R_2^c are the radii of the particles in contact c and l^c is the length of the branch vector. Since at the start of the elastic deformation all displacements and forces were reset to zero, the only effect of this "overlap" on the simulations is that it may obscure the relation between particle size distribution and branch vector at higher average coordination numbers Γ.

4.3 Discrete Element simulations

Discrete Element simulations were performed for nine different isotropic assemblies with average coordination numbers Γ in the range $4 - 6$, and for each average coordination number for nine different stiffness ratios λ in the range

0.05 − 1.0. Periodic boundaries [12] were employed to minimize the disturbing influence of boundaries.

The assemblies were subjected to two strain paths in order to compute bulk modulus K and shear moduli G from

$$\sigma_{11} + \sigma_{22} = 2K(\varepsilon_{11} + \varepsilon_{22}) \;;\; \sigma_{11} - \sigma_{22} = 2G(\varepsilon_{11} - \varepsilon_{22}) \,. \tag{24}$$

The strain paths are

$$\varepsilon = \varepsilon \begin{bmatrix} -1 & 0 \\ 0 & -1 \end{bmatrix} \text{Compression;} \quad \varepsilon = \varepsilon \begin{bmatrix} -1 & 0 \\ 0 & 1 \end{bmatrix} \text{Shear} \,. \tag{25}$$

The loading rate in the simulations was low enough that dynamic effects were negligible. This was verified by checking the linearity of the response and by performing simulations with a lower loading rate.

4.4 Averaging

Orientation-dependent quantities like the contact distribution function $E(theta)$, were computed by dividing the range of orientations $(0°, 180°)$ into 180 bins and computing the group averages for these bins. Probability density functions were computed by binning the relevant data into 100 bins.

5   Results from Discrete Element simulations

Here the results from the Discrete Element simulations will be discussed. The results involve characteristics of geometry, overall moduli, magnitude and probability density functions of relative displacements and energy.

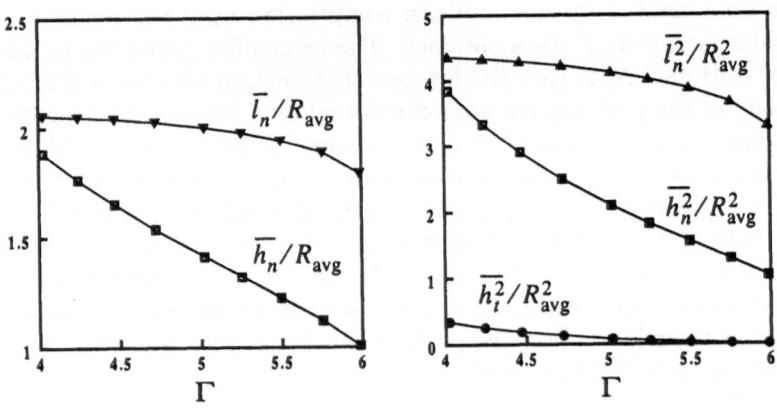

Fig. 3. Variation of mean and second moment of normal component of the branch and polygon vectors and tangential component of the polygon vector with average coordination number Γ

5.1   Geometry

The variation of normal components of branch and polygon vectors with average coordination number Γ, as well as the averages of squares (second moment) are shown in Fig. 3. The averages of squares of geometrical vectors are shown instead of standard deviations, since the squares are related to the bulk and shear moduli according to uniform strain and stress, see (20). The standard deviation for l_n is fairly constant with Γ, while the standard deviation of h_n decreases rapidly with increasing Γ, and becomes very small at $\Gamma = 6.0$.

A qualitative explanation for the decrease in standard deviation of the polygon vector is that as $\Gamma \to 6$, the polygons form triangles ($\Omega \to 3$, see (3)). Since the polygons then consist of three branch vectors with fairly small standard deviation of their lengths, the standard deviation in the position of the centre of polygons (and hence in the polygon vectors) becomes even smaller.

A remarkable observation from Fig. 3 is that at average coordination number $\Gamma = 4.0$, the average branch vector $\overline{l_n}$ is *larger* (by 2.8 %) than the average particle diameter $2R_{\mathrm{avg}}$, while one would expect it to be *smaller* due to the "overlap" between particles. That for higher Γ $\overline{l_n}$ no longer is larger than $2R_{\mathrm{avg}}$ is not surprising, since at higher Γ the "overlap" between particles leads to a reduction in $\overline{l_n}$. In micromechanical studies the assumption $\overline{l_n} = 2R_{\mathrm{avg}}$ is regularly made, see for example [6, 33].

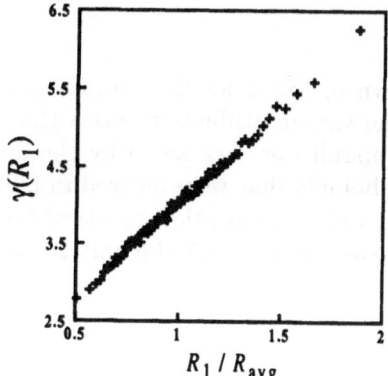 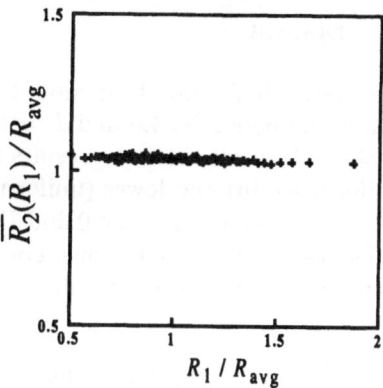

Fig. 4. Relation between particle radius and radius-dependent coordination number, and relation between radii of particles in contact. Data for the case of average coordination number $\Gamma = 4.0$

By investigating the relation between particle radius R and coordination number $\gamma(R)$, see Fig. 4, an explanation for $\overline{l_n} > 2R_{\mathrm{avg}}$ was found. Apparently there is a linear relation between the radius of a particle and its average number of contacts. A geometrical interpretation is that the larger the circumference of a particle, the more contacts are formed. The relation between the radii of the two particles involved in a contact was computed in order to investigate whether

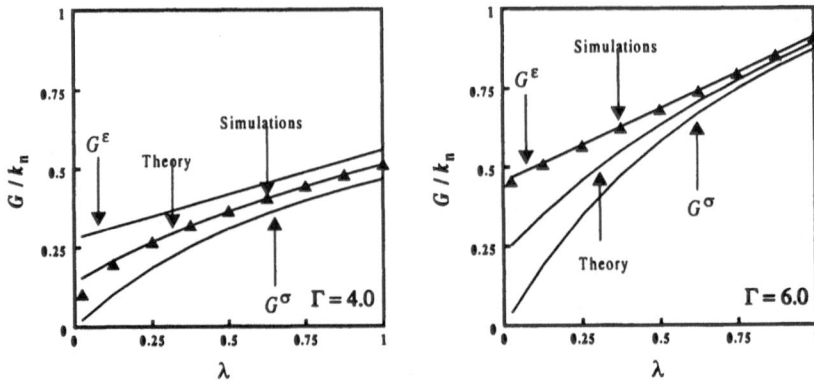

Fig. 5. Comparison between theoretical, uniform field moduli and moduli according to simulations in shear

large particles tend to form contacts with other large particles. For all particles with radius R_1, the average over all contacts of the radius of the second particle involved, $\overline{R_2}(R_1)$, is computed. The result is given in Fig. 4, which shows that the two radii are uncorrelated (correlation coefficient of -0.009). For consistency $\overline{R_2}(R_1)$ equals the average branch vector $\overline{l_n}/2$.

5.2   Moduli

The overall bulk and shear moduli are shown in Fig. 5 for the average coordination numbers $\Gamma = 4.0$ and $\Gamma = 6.0$ for the various stiffness ratios λ that are considered. In all cases the results for the moduli were bracketed by the upper (uniform strain) and lower (uniform stress) bounds that were derived in [20].

For loose systems $\Gamma = 4.0$ and $\Gamma = 5.0$ (not shown here) the statistical theory (21) gives accurate predictions. For the dense system $\Gamma = 6.0$, the uniform strain assumption gives correct results.

5.3   Relative displacements

In this section the results on the relative displacements are given. First the variation with average coordination number Γ and stiffness ratio λ of the magnitude of the relative displacements is presented. Then the probability density functions for the normal and tangential components of the relative displacements are given.

Generalised uniform strain   The direction-dependent averages of relative displacements can be expressed by a generalised uniform strain field

$$\overline{\Delta}_i = \zeta \varepsilon_{ij} \cdot \overline{l}_j \; . \tag{26}$$

This assumption is often made in the literature, for example [4–8, 10, 15, 27, 28], [33]. Note that for uniform strain $\zeta = 1$.

With the strain paths according to (25), this gives for isotropic assemblies

$$\overline{\Delta_n} = \zeta_n^K \varepsilon \overline{l_n}; \qquad \overline{\Delta_t} = 0 \qquad\qquad \text{Compression}$$

$$\overline{\Delta_n} = \zeta_n^G \varepsilon \overline{l_n} \cos 2\theta; \quad \overline{\Delta_t} = -\zeta_t^G \varepsilon \overline{l_n} \sin 2\theta \quad \text{Shear} . \tag{27}$$

These expressions match the orientation-dependent relative displacements that were observed in the simulations.

The direction-dependent averages of relative displacements are condensed into three numbers, two for normal (ζ_n^K for compression and ζ_n^G for shear) and one for tangential displacements (ζ_t^G for shear), that give the magnitude of the relative displacements.

The factors ζ were computed from the simulations for all average coordination numbers Γ and all stiffness ratios λ. Different ζ are obtained for normal and tangential displacements and for different strain paths. The results are shown Fig. 6.

From Fig. 6 it is clear that the (generalised) uniform strain assumption is incorrect. Only at high average coordination number Γ it is roughly correct. Remarkable is that $\zeta_t^G > 1$, for which no explanation has been found yet.

Assuming that relative displacements and polygon vectors are *independent*, $\overline{\Delta_i h_j} \cong \bar{\Delta}_i \bar{h}_j$, as well as branch vectors and polygon vectors, $\overline{l_i h_j} \cong \bar{l}_i \bar{h}_j$, it follows from (27) that for consistency with the expression (12) for the strain tensor

$$\zeta_n^K \cong 1 ; \quad \frac{1}{2} \left[\zeta_n^G + \zeta_t^G \right] \cong 1 . \tag{28}$$

Deviations of an order of magnitude of 15% from these expressions, see Fig. 6, are primarily caused by the limited validity of the first assumption of independence (in particular $\overline{\Delta_t h_t} \neq \bar{\Delta}_t \bar{h}_t$ as $\bar{h}_t = 0$).

Probability density functions The probability density functions for normal and tangential components of relative displacements were computed from the simulations for compressive and shearing loading. Since the distributions are direction-independent in compression, all contacts were used to compute the probability density function. Typical examples of the probability density functions are shown in Fig. 7, together with the corresponding Gaussian distribution. The probability density function for both normal and tangential distributions are nearly Gaussian, as was also found in [20]. In shear the probability density functions are also approximately Gaussian, although they more "noisy" due to the smaller number of contacts used in their computation.

The small difference between the computed and the corresponding Gaussian distribution can not be attributed to lack of accuracy due to insufficient data.

Some probability density functions for displacements for *non*-elastic behaviour are given in [2, 24].

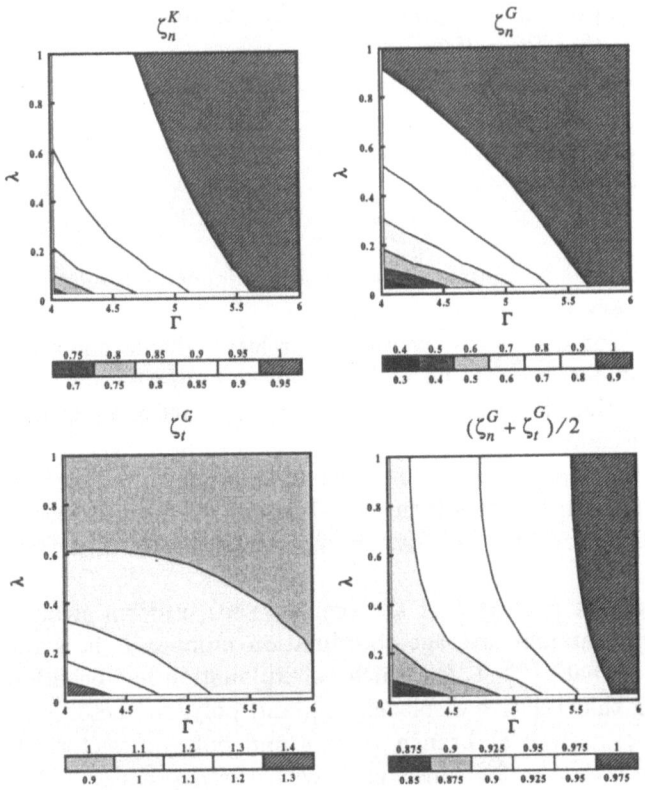

Fig. 6. Values of ζ from simulations

5.4 Energy distribution

The energies E_n and E_t in normal and tangential modes are defined by

$$E_n = \frac{1}{2}k_n\langle\overline{\Delta_n^2(\theta)}\rangle \; ; \qquad E_t = \frac{1}{2}k_t\langle\overline{\Delta_t^2(\theta)}\rangle \; . \tag{29}$$

Statistical-mechanical studies of ideal gases, see for example [32], lead to an equipartition of energy between the various modes. In analogy, the distribution of energy between normal and tangential modes has been computed here for all considered average coordination numbers Γ and stiffness ratios λ for compressive and shearing loading.

The ratio between the energies in the tangential and normal modes is shown in Fig. 8 for compression and shear. It is clear that generally there is no equipartition between normal and tangential modes. Only in shear for high stiffness ratio λ are the energies approximately equal. Surprising is the large contribution of the tangential mode at low average coordination number Γ and low stiffness ratio λ in shear. As expected, essentially no energy is present in the tangential mode in compression.

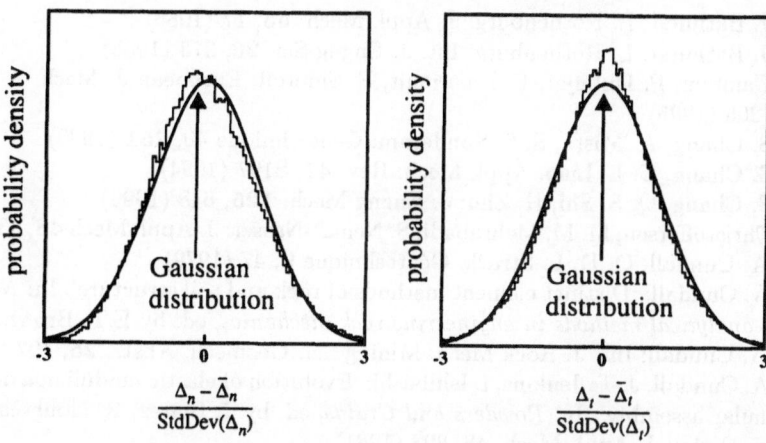

Fig. 7. Probability density functions for normal and tangential components of relative displacements together with the normalised Gaussian distribution in compression for average coordination number $\Gamma = 5.0$ and stiffness ratio $\lambda = 0.5$

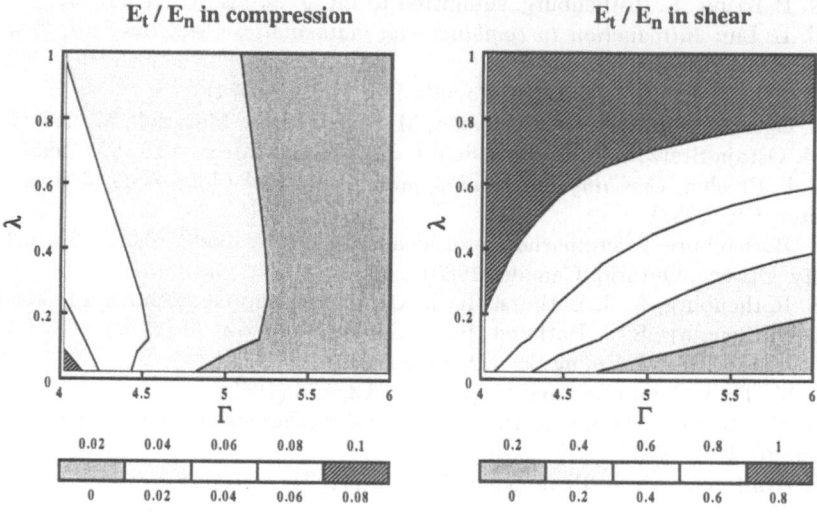

Fig. 8. Energy ratio E_t/E_n in compression and in shear

References

1. K. Alzebdeh, M. Ostoja-Starzewski: J. Appl. Mech. **66**, 172 (1999)
2. K. Bagi: 'On the definition of stress and strain in granular assemblies through the relation between micro- and macro-level characteristics'. In: *Powders and Grains '93*, 1993
3. K. Bagi: Mech. Materials 22, **165** (1995)

142   N. P. Kruyt, L. Rothenburg

4. R. J. Bathurst, L. Rothenburg: J. Appl. Mech. **55**, 17 (1988)
5. R. J. Bathurst, L. Rothenburg: Int. J. Engng Sci. **26**, 373 (1988)
6. B. Cambou, P. Dubujet, F. Emeriault, F. Sidoroff: European J. Mech. A / Solids **14**, 255 (1995)
7. C. S. Chang, A. Misra, S. S. Sundaram: Géotechnique 40, **251** (1990)
8. C. S. Chang, C. L. Liao: Appl. Mech. Rev. **47**, S197 (1994)
9. C. S. Chang, Q. S. Shi, H. Zhu: J. Engng Mech. **125**, 648 (1999)
10. J. Christoffersen, M. M. Mehrabadi, S. Nemat-Nasser: J. Appl. Mech **48**, 339 (1981)
11. P. A. Cundall, O. D. L. Strack: Géotechnique **9**, 47 (1979)
12. P. A. Cundall: 'Distinct element methods of rock and soil structure'. In: *Numerical and analytical methods in engineering rock mechanics*, ed. by E.T. Brown, 1986
13. P. A. Cundall: Int. J. Rock Mech. Mining Sci. Geomech. Abstr. **25**, 107 (1988)
14. P. A. Cundall, J. T. Jenkins, I. Ishibashi: 'Evolution of elastic moduli in a deforming granular assembly'. In: *Powders and Grains*, ed. by J. Biarez, R. Gourvès, 1989
15. P. J. Digby: J. Appl. Mech. 48, 803 (1981)
16. A. Drescher, G. de Josselin de Jong: J. Mech. Phys. Solids **20**, 337 (1972)
17. M. R. Horne: Proc. Royal Soc. London A **286**, 62 (1965)
18. G. de Josselin de Jong, A. Verruijt: Cahier Groupe Français Rheologie 2, 73 (1969)
19. N. P. Kruyt, L. Rothenburg: J. Appl. Mech. **63**, 706 (1996)
20. N. P. Kruyt, L. Rothenburg: Int. J. Engng Sci. **36**, 1127 (1998)
21. N. P. Kruyt, L. Rothenburg: submitted to Int. J. Solids Struct. (1999)
22. C. L. Liu: *Introduction to combinatorial mathematics* (McGraw-Hill, New York, NY, USA, 1968)
23. M. Oda, J. Konishi: Soils and Foundations **14**, 25 (1974)
24. L. Oger, S. B. Savage, D. Corriveau, M. Sayed: Mech. Materials **27**, 189 (1998)
25. M. Ostoja-Starzewski, K. Alzebdeh, I. Jasiuk: Acta Mech. **110**, 57 (1995)
26. C. L. Prasher: *Crushing and grinding process handbook* (John Wiley & Sons, Chichester, UK, 1987)
27. L. Rothenburg: Micromechanics of idealised granular materials. Carleton University, Ottawa, Ontario, Canada (1980)
28. L. Rothenburg, R. J. Bathurst, R. J.: Computers and Geotechnics **11**, 315 (1991)
29. L. Rothenburg, R. J. Bathurst, R. J.: Géotechnique **42**, 79 (1992)
30. M. Satake: Int. J. Engng Sci. **30**, 1525 (1992)
31. J. M. Ting: Computers and Geotechnics **13**, 175 (1992)
32. R. C. Tolman: *The principles of statistical mechanics* (Oxford University Press, Oxford, UK, 1938)
33. K. Walton: J. Mech. Phys. Solids **35**, 213 (1987)

On sticky-sphere assemblies

J. D. Goddard

Department of Mechanical and Aerospace Engineering, University of California, San Diego, 9500 Gilman Drive, La Jolla, CA 92093-0411, USA

Abstract. As the simplest model of a frictional-cohesive granular material, a variant on the (Baxter 1968) sticky hard-sphere of statistical mechanics is proposed. The present model is characterized by a hard-sphere diameter σ, a coefficient of sliding friction μ and a adhesive contact force T. It is argued that, depending on the initial density and disorder of the packing and on the magnitude of $T/p_o\sigma^3$, where p_0 is a confinement pressure, assemblies of such particles should exhibit purely inelastic behavior, ranging from that of a coherent solid to that of a noncohesive granular media. Recent computer simulations of a similar model for rock strength (Huang 1999) suggest that a three-dimensional variant of the classical Coulomb model with cohesion may apply.

The most important contribution of this brief article is perhaps the proposal for an organized multi-investigator collaboration on the computer simulation of the mechanics of various idealized granular media.

Modeling considerations

A long-standing problem in the physics of granular materials is the development of continuum constitutive models, adequate for macro-level mechanics. This problem is common to the statistical mechanics of molecular solids and fluids, and the power of modern computers allows for the possibility of testing continuum models directly against computer simulation based on more or less idealized molecular or micromechanical models.

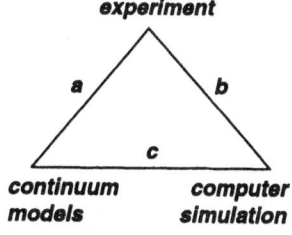

Fig. 1. Modeling paradigm (after Gray & Gubbins 1984)

Within the paradigm of Fig. 1 (Gray and Gubbins 1984, Goddard 1998), this provides an attractive theoretical alternative (route "c") to the classical approach ("a"). While some may see this as obviating the practical need for continuum models, (i.e. via route "b" in Fig. 1), the author subscribes to the view of Truesdell (1966) that a continuum description is essential to the mathematical physics, if not the actual physical understanding of various phenomena. At any rate, route "c" allows one to explore the validity of particular continuum models for simple micromechanical models before addressing more complex and realistic models or before proceeding to more costly experimentation on real systems.

In the field of molecular physics considerable progress has been made on the development of useful thermostatic equations of state by means of computer simulations based on certain standard idealized molecular models, e.g. the well-known Lennard-Jones fluid. In order to gain theoretical insight and to investigate parametric sensitivity, workers in the field often resort to simpler limiting forms of more complex and realistic models, examples being the hard-sphere fluid (or solid) and the sticky hard-sphere variant, apparently first proposed by Baxter (Baxter 1968, Berne 1977). We recall the latter is represented by the limit

$$\epsilon \to 0 \quad \text{and} \quad \phi \to \infty, \quad \text{with} \quad \frac{1}{\epsilon} e^{-\beta\phi} = \text{const.} \tag{1}$$

of the square-well interparticle potential represented by Fig. 2, where σ is the hard-sphere diameter, ϵ the width of the square well, representing a "breaking strain", and β the Boltzmann factor (inverse temperature). It is evident that this model involves an infinite force of separation and, hence, is unsuitable for models of quasi-static granular mechanics.

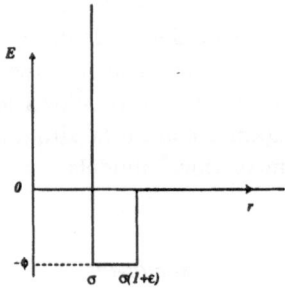

Fig. 2. Interparticle potential for Baxter's (1968) sticky hard-sphere

1   Cohesive materials

Previous work by the present author and coworkers have explored the quasi-static mechanics of rigid, non-cohesive frictional sphere assemblies (Didwania & Goddard, 1998, Goddard 1998, Didwania et al. 2000), using a combination of numerical simulation and continuum models based on mean-field micromechanics.

The simulations are based on frictional-elastic spheres with nearly rigid elastic contacts, such that

$$P = p_o\sigma/K << 1 \tag{2}$$

where p_o is confining pressure and K is the (Hookean) elastic constant for a contact.

While it has not been possible to determine all functions involved from limited numerical simulations, a general constitutive model is suggested, involving a symmetric second-rank "fabric" tensor $\mathbf{A}$ and evolution equations of the form

$$\overset{\circ}{\mathbf{A}} = \mathbf{g}(\mathbf{A}, \mathbf{S}) \tag{3}$$

where $\mathbf{g}$ is homogeneous degree-one in the local deviatoric rate of deformation $\mathbf{S}$, and where $\overset{\circ}{\mathbf{A}}$ denotes Jaumann rate. Given a rate-independent yield condition of the form

$$\mathcal{F}(\mathbf{A}, \mathbf{T}, \mathbf{S}) = 0, \tag{4}$$

homogeneous degree-zero in $\mathbf{S}$, one has a prescription for stress tensor $\mathbf{T}$ in terms of strain history. In the case of rigid non-cohesive spheres, it can be deduced that (4) is also homogeneous degree-zero in $\mathbf{T}$.

The addition of cohesion to the above model can be expected to have far-reaching consequences, for, depending on the magnitude of cohesive force, one can anticipate behavior ranging from coherent elastic solid through friable solid to non-cohesive granular mass. Indeed, the classical work of Budiansky & O'Connell (1976) on cracked-filled elastic solids suggests that the transition from coherent solid to granular material corresponds to the percolation threshold of a crack network, beyond which the shear (plastic-yield) strength is governed by dilatancy and friction and, hence, is directly proportional to the confining pressure p_o (Goddard & Didwania 1998). Subsequent works on the statistical physics of failure, such as that of Sahimi & Goddard (1986), indicate that disordered networks of breakable elastic springs can exhibit failure ranging from ductile to brittle, depending on the distribution of breaking strengths.

For purposes of conceptual exploration, let us consider a sticky elastic-sphere model for cohesive granular materials. A plausible alternative to the Baxter model is provided by the linear "sticky-spring" potential illustrated schematically in Fig. 3, with

$$\phi = \frac{1}{2}K\sigma^2\epsilon^2 \equiv \frac{1}{2}T\sigma\epsilon \text{ and } T = K\sigma\epsilon \tag{5}$$

where K represents contact stiffness (spring constant) and T contact tensile strength. If one adds tangential stiffness and (shear) strength, with (post-failure) Coulomb friction μ, one obtains a model similar to that employed by Huang (1999) for numerical simulations aimed at modeling rock strength. Some aspects of the model are also represented by the friable spring network of [9], which shows that strength depends on network disorder, hence, on density and disorder of a granular packing. However, in contrast to both those models, where broken contacts or springs undergo *irreversible* loss of adhesion, the model proposed here involves *reversible* adhesion.

146   J. D. Goddard

With either reversible or irreversible adhesion, it is clear that the quantities ϵ and $p_o\sigma^3/\phi$, connected to (2) by

$$P \equiv \epsilon(p_o\sigma^2/T) \equiv \frac{\epsilon^2}{2}(p_o\sigma^3/\phi), \qquad (6)$$

must now enter into hypothetical relations of the form (3-4), where the stress tensor $\mathbf{T}$ can, without loss of generality, be replaced by $\mathbf{T}/p_o$, with $p_o \equiv -\mathrm{tr}\{\mathbf{T}\}$ (Goddard & Didwania 1998).

To obtain a simpler model appropriate to rigid grain, consider the limit:

$$\epsilon \to 0 \quad \text{and} \quad K \to \infty, \quad \text{with} \quad T = K\sigma\epsilon = \text{const.} \qquad (7)$$

in (5), from which it is seen that $\phi \to 0$. Hence, one arrives at the important conclusion that an assembly of sticky rigid spheres with finite adhesive force must be purely *inelastic* in quasi-static deformations. In the case of irreversible adhesion, this inelastic deformation corresponds to true "failure", whereas with reversible adhesion it represents ductile plastic yielding. In either case, it is evident from (6) that the limiting forms of relations like (3-4) for $\epsilon \to 0$ will generally involve the parameter $T/p_o\sigma^2$ and must reduce to those of [4-6] for non-cohesive spheres when this parameter vanishes.

Pending a more detailed investigation, the (two-dimensional) computer simulations of Huang (1999) lead to the following conjectures for the above sticky hard-sphere assemblies:

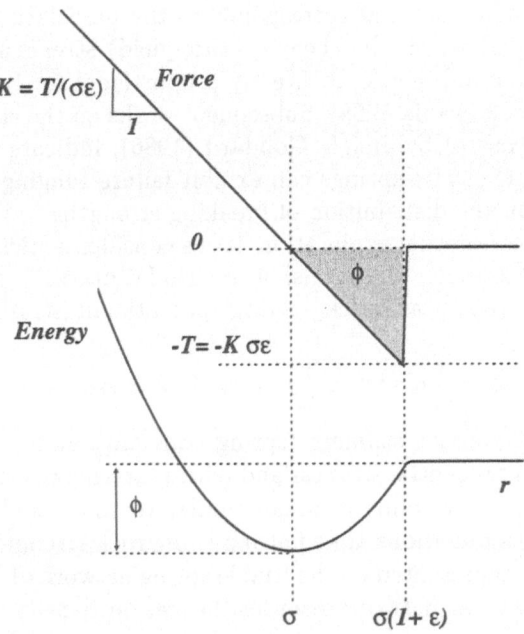

Fig. 3. The sticky Hookean sphere

1. The magnitude of uniaxial tensile strength is given approximately by T/σ^2, and

2. the yield surface is given roughly by the conical surface appropriate to the non-cohesive limit with "cohesive" shift along the negative pressure axis given by $T/p_o\sigma^3$, i.e. by an appropriate three-dimensional generalization of classical Coulomb surface with cohesion.

The first conjecture is suggested by the standard formula for granular stress (cf., Goddard 1998) and by the results presented in Huang (1999, Figs. 2-8 & 2-9, where the abscissa KR/T corresponds to $1/\epsilon$); whereas the second represents an inference from simulations of non-cohesive media (Goddard & Didwania 1998) together with a consideration of the Coulomb-like yield locus in Huang's Fig. 2-17.

2   Conclusions and recommendations

The sticky hard-sphere model proposed above is thought to provide an attractive model for theoretical investigation of cohesive effects in granular media. It is evident that it it represents a special case of a purely inelastic continuum, for which plausible conjectures for the form of the yield surface have been made above.

Following a successful and well-established tradition in the field of molecular fluids, it is recommended that the granular materials community adopt a similar strategy for development of continuum constitutive equations for various idealized granular media, such as that just proposed. This would involve a collaborative multi-investigator effort, with complementary simulations employed to develop and test continuum equations for idealized granular assemblies subject to various homogeneous strain histories, e.g. monotonic and cyclic "cubical triaxial" strains as well as more complex deformations involving rotation of principal-strain axes. The author regards this as an indispensable step in the development of continuum models for real materials.

Acknowledgments

Partial support from the U.S. National Aeronautics and Space Administration (Grant NAG3-1888) and the National Science Foundation (Grant CTS-9510121) is gratefully acknowledged.

References

1. R. J. Baxter: 'Percus-Yevick Equation for Hard Spheres with Surface Adhesion'. J. Chem. Phys. **49**, 2770-4 (1968)
2. B. J. Berne: 'Molecular dynamics of the rough sphere fluid. II. Kinetic models of partially sticky spheres, structural spheres, and rough screwballs'. J. Chem. Phys. **66**, 2821-30 (1977)

3. B. Budiansky, R. J. O'Connell: 'Elastic moduli of a cracked solid'. Int. J. Solids Structures **12**, 81-97(1976)
4. A. K. Didwania, K. Ledniczky,J. D. Goddard: 'Kinematic Diffusion in Quasi-static Granular Deformation'. submitted (2000).
5. J. D. Goddard:'Continuum Modeling of Granular Assemblies, - Quasi-Static Dilatancy and Yield'. In *Physics of Dry Granular Media, Proc. NATO ASI, Cargèse, 1997*, ed. by H. Herrmann et al., (Kluwer Academic, Amsterdam 1998) pp.1-24
6. J. D. Goddard,A. K. Didwania: 'Computations of Dilatancy and Yield Surfaces for Assemblies of Rigid Frictional Spheres'. Quart. J. Mech. Appl. Math. **51**,15-43 (1998)
7. C. G. Gray, K. Gubbins: *Theory of Molecular Fluids.* (Oxford University Press, Oxford 1984), Appendix E.
8. Haiying Huang: Discrete Element Modeling of Tool-Rock Interaction. PhD Thesis. University of Minnesota, Department of Civil Engineering, Minneapolis, Minnesota (1999)
9. M. Sahimi, J. D. Goddard: 'Elastic Percolation Models for Cohesive Mechanical Failure in Heterogeneous Systems'. Phys. Rev. B **33**, 7848-51 (1986)
10. C. Truesdell: *Six Lectures on Modern Natural Philosophy.* (Springer, New York, Berlin 1966)

Cohesive granular texture

F. Radjaï, I. Preechawuttipong, R. Peyroux

LMGC, CNRS-Université Montpellier II,
CC 048, Place Eugène Bataillon,
F-34095 Montpellier cedex 5, France

Abstract. We investigate the textural properties of cohesive 2D granular packings simulated by means of the molecular dynamics and contact dynamics methods involving simple contact laws with adhesion. We find that, while tensile forces appear naturally in response to external tensile loading, they appear only for a strong adhesion when the applied load is compressive in all directions. We introduce an adhesion index which represents the extent of activation of attractive forces compared to an external loading. The evolution of the tensile pressure and the average coordination number with the adhesion index suggests a transition between two regimes. In a first regime, the adhesion gives mainly rise to a geometrical rearrangement of the texture. In particular, the coordination number increases with the adhesion index. In a second regime, the adhesion entails a force reorganization involving tensile force chains and a partial crystallization of the contact network. Finally, we analyze the contribution of the tensile and compressive forces to shear stress and fabric anisotropy. An interesting result is that tensile forces play the same role in the stability of a cohesive packing as weak compressive forces.

1 Introduction

This paper is concerned with the textural properties of a cohesive 2D granular packing. By *texture* here we mean both force and contact networks when multiple contacts are present, e.g. in static equilibrium. Cohesion is the ability of a granular material to support tensile loads with an upper bound which defines the cohesion (threshold) of the material[1]. Cohesion stems from attractive surface forces acting at interparticle contacts. These adhesive forces, which may have various physico-chemical origins, together with hard core repulsive forces allow a pair of touching particles to support compressive forces (up to particle crash) and tensile forces with a threshold depending upon the nature and the geometry of the contact.

While cohesive granular materials, such as fine powders and clays, show a rich variety of behaviors, their textural properties have remained nearly unexplored. An important issue is how tensile forces appear and how they share the overall stresses carried by the medium. In a noncohesive granular medium, the transmission of shear stresses is insured by a particular organization of the texture involving two distinct yet complementary contact subnetworks[2]. Does this picture still hold in a cohesive medium? Do novel mechanisms appear as the cohesion is increased?

In this paper, we address some of these issues within two different numerical schemes, namely molecular dynamics and contact dynamics. We first discuss the contact laws with adhesion used in the present study. We introduce an "adhesion index" that represents the extent of activation of attractive forces as compared to the applied load. Then, we briefly present some of the investigated behaviors regarding force maps, average coordination number and tensile pressure as a function of the adhesion index. We also study the contribution of the tensile and compressive contacts to shear stress and fabric anisotropy.

2   Simple contact laws with adhesion

In molecular dynamics or most popular *discrete element* simulation methods, the particles are handled as rigid bodies moving according to Newton's equations of motion[3, 4]. The kinematic variable describing the contact between two touching particles is the depth δ of the overlap zone between two particles. This corresponds to the relative displacement of the particle centers from the "just at contact" position; see Fig. 1. Let r be the effective radius of curvature of the interface for $\delta = 0$. At first order in δ/r, the radius of the contact zone is given by $a = \sqrt{r\delta}$. It happens that δ, via the contact radius a, and its time derivative $\dot{\delta}$ are the relevant variables for the description of several behaviors including elastic repulsion, viscous dissipation and adhesion[5–8].

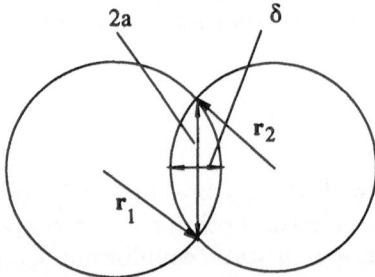

Fig. 1. Geometry of the contact between two disks

Although for the simulation of a given granular material it is desirable to implement the specific force laws, if available from experiments, governing particle interactions, a multitude of numerical simulations suggest that only very short-time dynamics and small-strain behaviors are influenced by the details of these interactions. Many observed granular phenomena have been reproduced successfully through simple force laws such as linear elastic repulsion with viscous damping or, more interestingly, with the contact dynamics method which makes no use of δ[9–11]. For this reason, we believe that numerical simulations with simple contact laws, incorporating basic aspects of particle interactions such as hard core repulsion and dissipation, should provide a useful guide for the investigation of many flow and failure properties of granular media. Obviously, this approach does not elude the numerical study of subtle effects of more complex contact laws as soon as the results with a simpler law are at hand.

We are interested here in the simplest force laws with adhesion for molecular dynamics simulations. A force law with adhesion in a broad sense involves that a pair of contiguous particles resist normal separation, but also relative sliding and rolling. When these conditions are fullfilled, the interparticle contact is *cemented*, i.e. the relative degrees of freedom of the particles are frozen within the range of supportable forces and moments. However, we will not consider here rolling friction and stiffness so that adhering particles may freely roll on one another with no bending moments acting between them. Then, with the convention of positive sign for compressive forces, the interparticle cohesion is characterized by a negative threshold $-N_c$, i.e. the "pull-off" force necessary for separating two particles, that can not be exceeded by the normal force N acting at the contact point, whereas sliding is resisted by a friction force with an upper limit T_c depending both on N and N_c.

Assuming that adhesion and repulsion have different origins and that they are functions only of δ, the normal force is simply the superposition of a repulsive force N^+ and an attractive force $-N^-$:

$$N(\delta) = N^+(\delta) - N^-(\delta) \tag{1}$$

The threshold $-N_c$ is reached at $\delta = \delta_c$ such that $-N_c = N^+(\delta_c) - N^-(\delta_c)$ and $dN/d\delta(\delta_c) = 0$. It is worth noting that, since $N^+(\delta_c) > 0$, the threshold N_c is not the largest value of N^-. We also note that for $\delta < \delta_c$, the contact is unstable, that is $dN/d\delta < 0$. The simplest form of (1) is when $N^-(\delta) = -N_c$, i.e. a constant attractive force $-N_c$ is exerted between all touching particles:

$$N(\delta) = -N_c + N^+(\delta) \tag{2}$$

This form implies instantaneous sticking upon collision between two particles. Let us remark that the attractive force $-N^-$, though acting along the contact normal, is merely a *contact* force and not a center-to-center force. This is a basic source of weakness of a cohesive granular medium compared to molecular aggregates, and implies that in the absence of external loading, and independently of the force law, the equilibrium contact force is zero just as is the case for nonadhesive particles. The equilibrium position δ_{eq} corresponds to $N^+(\delta_{eq}) = N^-(\delta_{eq})$.

For the sake of consistency we may require that the friction force threshold (or sliding threshold) T_c is a function of δ, too. According to Coulomb's law of friction, T_c is given by a coefficient of friction μ times the normal force. The latter, however, cannot be N because of the effect of adhesion on the contact area. The relevant force N' is a force that following the repulsion law $N^+(\delta)$ creates as much contact area, or equivalently δ, without adhesion as does N with adhesion[12]. In other words, assuming that $N(\delta)$ can be inverted (no hysteresis) to yield $\delta(N)$, we have

$$T_c(N) = \mu N'(N) = \mu N^+[\delta(N)] \tag{3}$$

Given the force law (2), i.e. with a constant adhesion force, (3) implies

$$T_c = \mu(N + N_c) \tag{4}$$

where the friction force varies linearly with N. This is nothing but the basic Coulomb yield function for a cohesive-frictional granular material[1]. With the more general equation (1), T_c can vary only nonlinearly with N.

There are two more general points worth discussing in connection with (1). First, whatever the value of N_c, the actual values of the normal contact forces, including tensile forces, in a granular medium are dictated by the external forces exerted on the medium. In fact, we will show below that a cohesive granular packing might contain no tensile force at all.

Another point which we would like to emphasize is that, even in the absence of tensile forces in a packing, the adhesion may be strongly sollicitated. In other words, the average $\langle N^- \rangle$ may be of the same order of magnitude or larger than N_c even thougth $N^+ > N^-$ at all contacts. In order to quantify this *activation* of adhesion in comparison to the average compressive force $\langle N^+ \rangle$, it is convenient to use the dimensionless quantity η defined by

$$\eta = \frac{\langle N^- \rangle}{\langle N^+ \rangle} \tag{5}$$

that we will refer to as the "adhesion index". It can be very small even with strong adhesion if the compressive loading is large in comparison to the adhesion threshold. In other words, η is an indicator of the *effective* cohesion of a granular medium at a given stage of its evolution. From (1) and (5) we get

$$\eta = 1 - \frac{\langle N \rangle}{\langle N^+ \rangle} \tag{6}$$

According to the expression of stress tensor in terms of contact forces, the average normal force $\langle N \rangle$ has the same sign as the average pressure p[13, 14]. Equation (6) shows that if the average pressure is positive, i.e. for a system globally in compression, then $0 < \eta < 1$, whereas for a system globally in tension we have $1 < \eta$. In the absence of external forces, where $N = 0$ for all contacts, and more generally if $p = 0$, we have $\eta = 1$. The maximum value of η depends on N_c and should tend to ∞ if N_c tends to ∞. It goes without saying that η may vary in space due to force gradients. For instance, in a wet sandpile the largest value of η is reached at the free surface where particle agglomerates are formed and play an important role in the avalanche process.

Now, let us consider the simple linear repulsion force $N^+ = k\,\delta$, where k is a spring constant, and an attractive force obeying a power law $N^- = \gamma\,r^{1-\alpha}\,\delta^\alpha$, where γ represents the interface energy. For an equilibrium state to exist, this form implies $\alpha < 1$. The limit case $\alpha = 0$ corresponds to (2) with $N_c = \gamma\,r$. The qualitative behavior is the same for all other values of α. We implemented this contact law with $\alpha = 0.5$, i.e.

$$N = k\,\delta - \gamma\sqrt{r\delta} \tag{7}$$

Figure 2 shows the evolution of N as a function of δ. The maximum value of η within this model is 2. It can easily be checked that the adhesion and friction

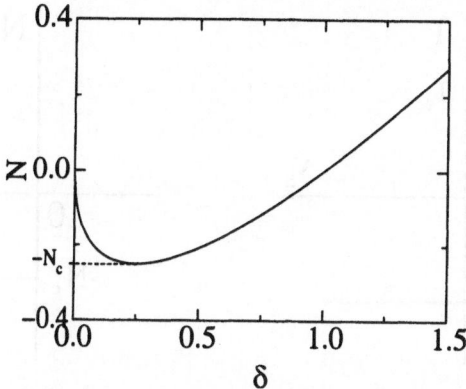

Fig. 2. Normal force N as a function of overlap depth δ between two particles according to (7)

thresholds are

$$N_c = \frac{r\gamma^2}{4k} \tag{8}$$

and

$$T_c = \mu N' = \mu N \left\{ 1 + 2\frac{N_c}{N} + 2\frac{N_c}{N}\sqrt{1 + \frac{N}{N_c}} \right\} \tag{9}$$

We see that the expression of T_c involves no other material parameter than N_c. For large positive values of N it tends to μN while in the limit $N = -N_c$ we have $T_c = \mu N_c$.

Although the derivation of the expression of T_c in (3) and (9) was based on a consistency argument, one may still rely on the simpler expression 4 of T_c without losing the basic features of the Coulomb friction law. Our requirement of a simple and robust contact law in a numerical approach justifies this choice.

The implementation of (7) within the molecular dynamics method is straightforward. It should, however, be completed by a new term accounting for dissipation. A common approach is to add a viscous damping term $\gamma_n \sqrt{m_{eff}}\dot{\delta}$, where γ_n is a normal damping constant and m_{eff} is the reduced mass, to the right hand side of (7). This term results in incomplete restitution upon a binary collision. It can easily be shown that with $\gamma \neq 0$, the coefficient of restitution is strictly zero for a finite range of relative head-on collision velocities.

The implementation of Coulomb's law in the molecular dynamics method, whatever the expression of T_c, poses a basic technical problem[5, 15]. In fact, the integration of the equations of motion requires a "smooth" force law such that the friction force T is a function of the sliding velocity v_s. However, according to Coulomb's law of friction, T and v_s belong to a set of permissible values which cannot be represented as a (single-valued) function. This "nonsmooth" relation is shown as a graph in Fig. 3(a)[10]. The only way to circumvent this difficulty in the framework of the molecular dynamics method is to adopt a "regularized"

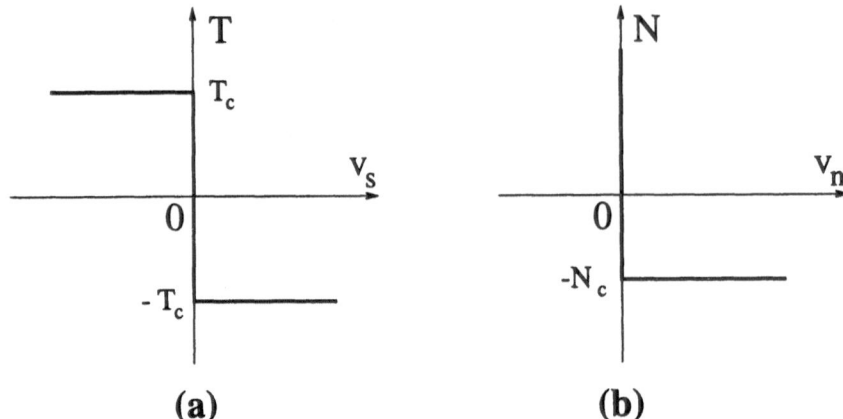

Fig. 3. Set of permissible pairs (N, v_n) and (T, v_s) defining a nonsmooth contact law with adhesion; see text

form of the exact Coulomb law. The simplest regularized form widely used for the simulation of granular materials is

$$T = \min\{|\gamma_s v_s|, T_c\}.sign(v_s) \tag{10}$$

where γ_s is a tangential damping constant. The latter should be given a value large enough to avoid numerical artefacts[5, 7].

We focused so far on the molecular dynamics method and examples of contact laws that account in a simple way for adhesion. The contact dynamics method, which can be qualified as a *nonsmooth* distinct element approach, is a conceptually different method based on a model of the bounded motions of a set of perfectly rigid bodies subject to constraints of impenetrability and friction[9]. This method allows, by construction, to treat Coulomb's law of friction without regularization and to handle contact unilaterality with no use of a repulsive potential. In order to calculate the contact percussions (integrated force over one time step) and velocity changes, which can be discontinuous due to collisions and the nonsmooth feature of the friction law, the equations of motion are transformed into a set of equations relating contact percussions ($N\Delta t$ and $T\Delta t$, where Δt is the time step) and relative velocities at the end of the time step (thus an implicit-type differencing). An iteration process over these equations allows to converge to a solution satisfying the contact laws defined by a set S of permissible pairs (N, v_n) and (T, v_s), where v_n is the normal separation velocity between two touching particles.

Within this framework, the implementation of adhesion needs only a consistent definition of the set S. Figure 3 shows the simplest example of a nonsmooth contact law with adhesion. This set involves an adhesion threshold N_c and an expression for T_c such as (4). The relation between N and v_n shown in Fig. 3(b) has the following physical properties:

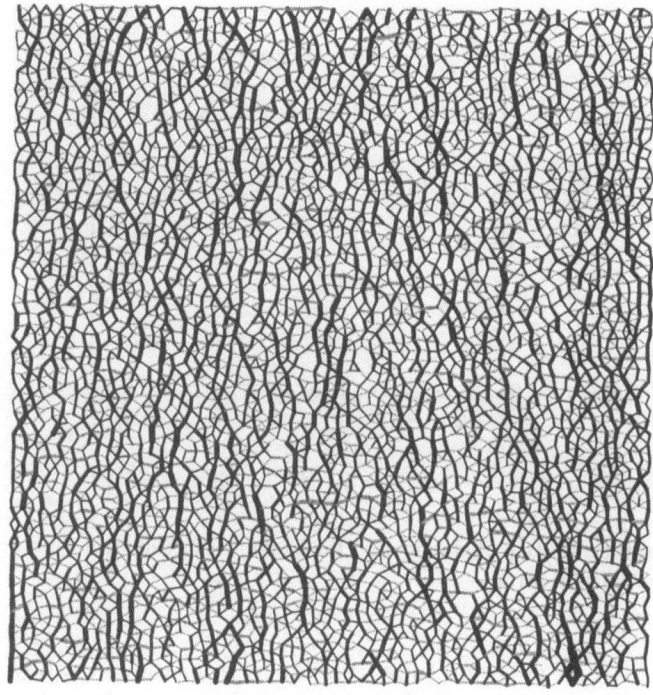

Fig. 4. Normal contact forces in a cohesive packing in compression vertically and in tension horizontally. The line thickness is proportional to normal force. The black and gray lines show compressive and tensile forces, respectively

1. The separation velocity of two touching particles can not be negative (no interpenetration).
2. When a contact yields ($v_n > 0$), it is the locus of the limit tensile percussion $N_c\Delta t$.
3. At a persisting contact ($v_n = 0$) the normal force N belongs to the range $[-N_c, \infty]$.

Both the force components (N, T) and the relative velocities (v_n, v_s) at the end of a time step are determined by the dynamics. The solution is unique for a binary collision. In a *multicontact* system, the solution is degenerate due to the nonsmooth feature of the contact laws. However, our simulations show that, as a result of geometric disorder, the extent of this degeneracy is so narrow that the solution is practically unique. In other words, the problem is more to find *one* solution than the risk of missing the right solution. In our view, this is an important property of a granular medium that the physical behavior is determined essentially by mutual exclusion of particles, the Coulomb-like friction law and dissipation rather than the details of the contact laws.

We see that the contact dynamics approach involves no *a priori* distinction between the attractive and repulsive contributions (N^- and N^+) of normal

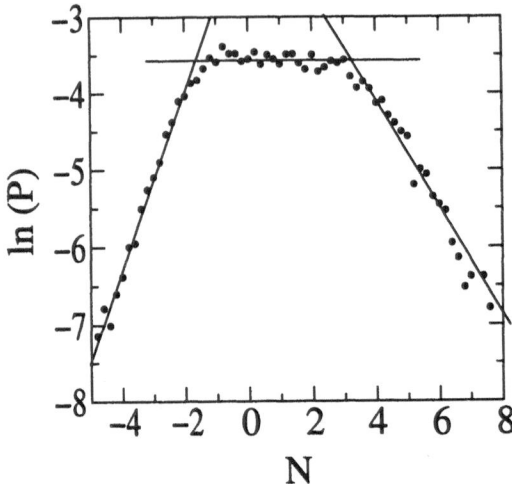

Fig. 5. Probability density of the normal forces shown in Fig. 4. Forces are normalized by the average normal force

forces. How then an adhesion index similar to η, introduced above for smooth contact laws, can be defined within this model? Given a force law such as (7), the repulsion term $N^+(\delta)$ can be expressed as a function of N (instead of δ) by simply replacing δ with its expression as a function of N. This is nothing but $N'(N)$ defined by (3) and whose explicit expression as a function of N and N_c is given by (9) for the contact law (7). The force $N''(N) = N'(N) - N$ represents the corresponding attractive part, so that the adhesion index is given by $\eta = \langle N'' \rangle / \langle N' \rangle = 1 - \langle N \rangle / \langle N' \rangle$.

3   Examples of observed behaviors

In this section we present examples of cohesive granular textures in static equilibrium and their statistical properties. The investigated systems consist of several thousand disks confined inside a rectangular box with a fixed vertical wall, a fixed horizontal wall, a vertical wall free to move horizontally and a horizontal wall free to move vertically. Particle radii are uniformly distributed in a range such that the ratio of the largest to the smallest radius is two. An equilibrium state is reached following a biaxial compression with zero adhesion. Then, the adhesion is switched on and the sample is allowed to evolve towards a new equilibrium state under the action of the forces applied on the two free walls.

Figure 4 shows the map of normal forces in a granular system simulated by means of the contact dynamics method. The confining box is subjected to a compressive vertical force and a tensile horizontal force. The major principal axis of the average stress tensor σ is vertical.

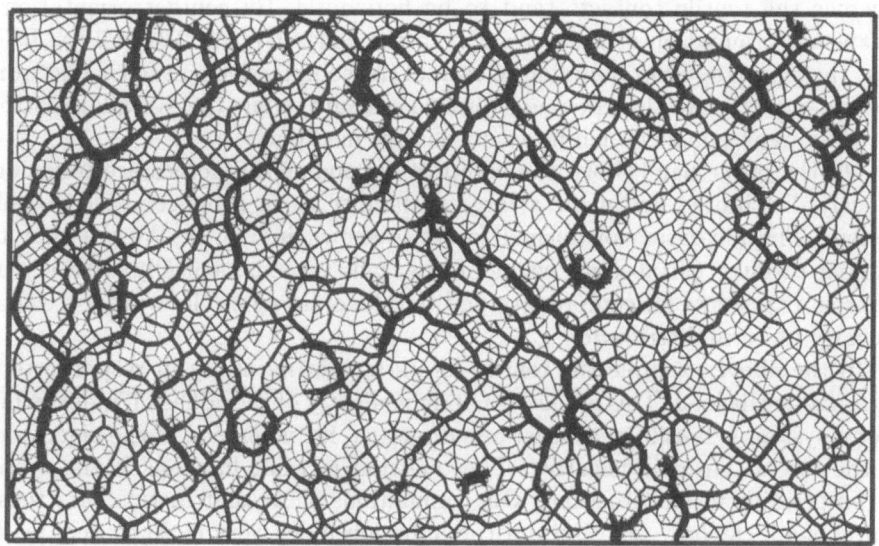

Fig. 6. Normal contact forces in a strongly cohesive packing in compression in all directions. The line thickness is proportional to normal force. The black and gray lines show compressive and tensile forces, respectively

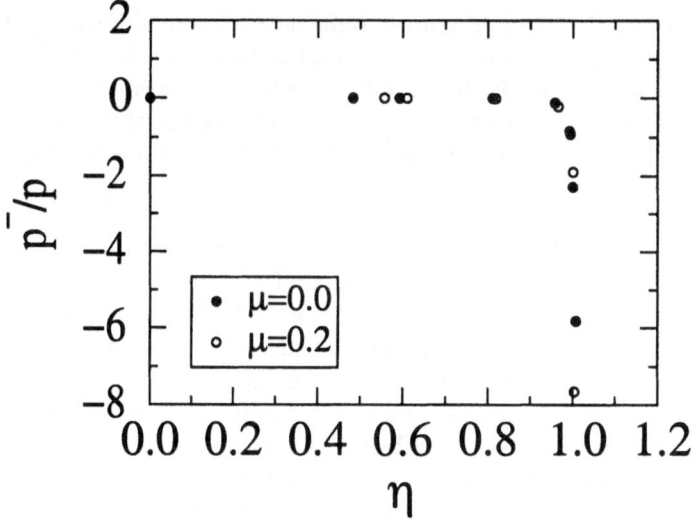

Fig. 7. The tensile pressure p^- as a function of the adhesion index η

With the sign convention adopted here, the eigenvalues σ_1 and σ_2 are such that $\sigma_1 > 0$, $\sigma_2 < 0$ and $p = (\sigma_1 + \sigma_2)/2 > 0$. The adhesion index in this example is 0.6. As expected, the compressive contacts are on average vertically oriented, whereas the tensile contacts tend to be horizontal. By contact orientation is meant the orientation of the normal n to the contact plane. We will see below that the anisotropy correponding to each of the compressive and tensile networks is quite large (larger than in a noncohesive medium) while the overall anisotropy is rather weak (smaller than in a noncohesive medium).

On the other hand, both the compressive and tensile forces are inhomogeneously distributed and show a large variability as in a noncohesive granular system. The histogram of normal forces is shown in Fig. 5. We see that the number of large (in absolute value) forces falls off exponentially with slightly different exponents for compressive and tensile forces while the weak forces have a nearly uniform distribution.

In the above example the tensile forces appear naturally in response to an external tensile load. This is the case, independently of the adhesion index, whenever a system is tensile in at least one direction, i.e. when one of the eigenvalues of the stress tensor is negative. The question is what happens if the system is compressive in all directions. Our simulations show that in this case tensile contacts appear only for values of the adhesion index very close to one, i.e. for very strong adhesion or for very weak compressive loading. Figure 6 shows an example of the force map in such a system simulated by means of the molecular dynamics method. We see that the networks of tensile and compressive forces are strongly correlated and the orders of magnitude of the largest tensile and compressive forces are quite comparable. The corresponding force histogram is similar to the one shown in Fig. 5.

Since the tensile contacts appear quite late as a function of η, the average negative pressure p^- (due to tensile forces) remains nearly equal to zero up to the vicinity of $\eta = 1$. This is shown in Fig. 7 where p^-, normalized by the average pressure p, is plotted versus η. Each point on this plot is the result of a separate simulation. The average pressure p has the following expression[14]:

$$p = n_c \langle \ell\, f \rangle_V \tag{11}$$

where n_c is the number density of contacts, ℓ is the length of the branch vector ℓ joining the centers of touching particles and f is the magnitude of the contact force f. The average $\langle \cdots \rangle_c$ is taken over the contacts inside a control volume V that may cover the whole packing. The tensile pressure p^- is given by the same expression when restricted to the tensile contacts.

This quasi-absence of tensile contacts in a system subjected only to compressive stresses means that the equilibrium of each particle is more easily (in the probabilistic sense) and more stably achieved by means of compressive forces. Indeed, a tensile force exerted on one side of a given particle can not be balanced by a compressive force unless the latter is exterted on the same side; see Fig. 8. This mechanism, however, is strongly obstructed by steric exclusion between two adjacent particles in contact with a third particle. This picture implies that a particle subjected to a tensile force is more efficiently equilibrated if another tensile force is exerted on the opposite side of the particle. In other words, if tensile forces are activated, they should appear mostly in *chain*. This is what one observes on the force map shown in Fig. 6.

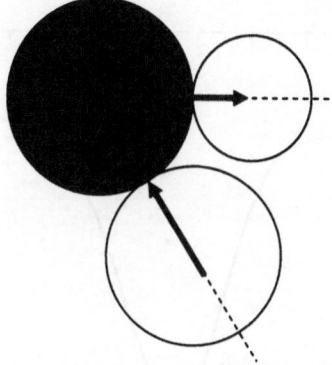

Fig. 8. A tensile and a compressive force exerted on the same side of a particle

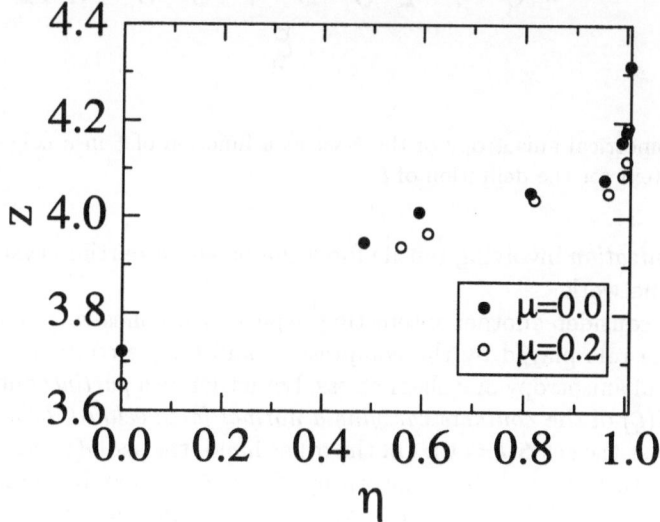

Fig. 9. Average coordination number z versus the adhesion index η for two different coefficients of friction μ

Another consequence of the foregoing picture is that tensile forces are favored by larger coordination numbers z. Figure 9 shows the evolution of z as a function of η. We see that z begins to increase with η to stagnate around 4 before increasing further beyond $z = 4$. This suggests a transition to a dense and partially crystallized texture where tensile forces begin to take part in the force landscape.

On these grounds, we may distinguish two regimes as a function of the adhesion index. In a first regime, the adhesion gives mainly rise to a *geometrical rearrangement* of the texture. In particular, the coordination number increases in this regime with adhesion index. In a second regime, the adhesion entails a

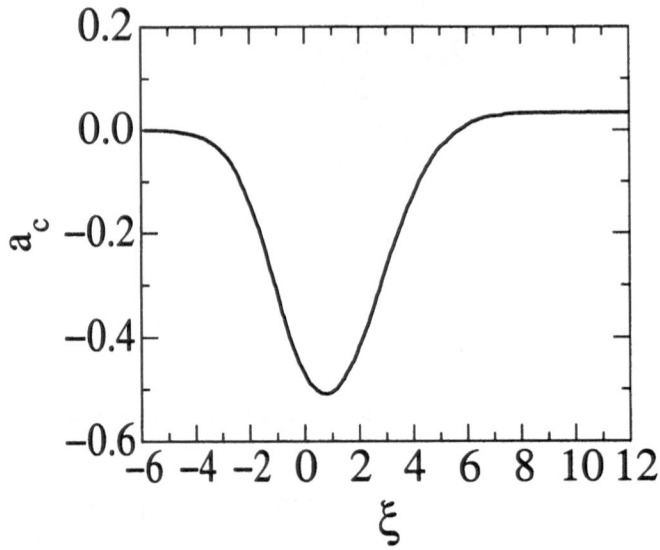

Fig. 10. Geometrical anisotropy of the ξ-set as a function of ξ in a cohesive granular packing; see text for the definition of ξ

force reorganization involving tensile force chains and a partial crystallization of the contact network.

We now consider another interesting aspect of a cohesive granular texture regarding the role played by the compressive and tensile contacts with respect to geometrical anisotropy and shear stress. Let us define a *partial* contact network by the set $S(\xi)$ of the *contacts carrying a normal force below* $\xi \langle N \rangle$. The tensile network is just the set $S(\xi = 0)$. On the other hand, the set $S(\xi = \infty)$ represents the whole contact network. The anisotropy of the "ξ-set" can be determined from the corresponding fabric tensor $\phi(\xi)$ whose components $\phi_{ij}(\xi)$ are given by

$$\phi_{ij}(\xi) = \langle n_i n_j \rangle_{S(\xi)} \tag{12}$$

where n_i and n_j are the components of the contact normal $\boldsymbol{n}$. The average is taken over the contacts belonging to $S(\xi)$. The anisotropy $a_c(\xi)$ of this set is just two times the deviator of $\phi(\xi)$[2].

Figure 10 displays the evolution of a_c with ξ in the packing shown in Fig. 4. By definition, the anisotropy a_c is a positive quantity. A negative value of a_c in Fig. 10 means simply that the major principal axis of $\phi(\xi)$ is horizontal while that of $\phi(\xi = \infty)$, i.e. the fabric tensor of the whole packing, is vertical.

Figure 10 shows that the total anisotropy $a_c(\xi = \infty)$ is very weak (about 0.03), whereas $a_c(\xi = 1)$ is quite large in absolue value (about -0.5). In other words, the set of tensile contacts and those compressive contacts which carry a force below the average normal force provides an important *negative contribution* to $a_c(\xi = \infty)$. The complementary set gives a large positive contribution $a_c(\xi = \infty) - a_c(\xi = 1)$

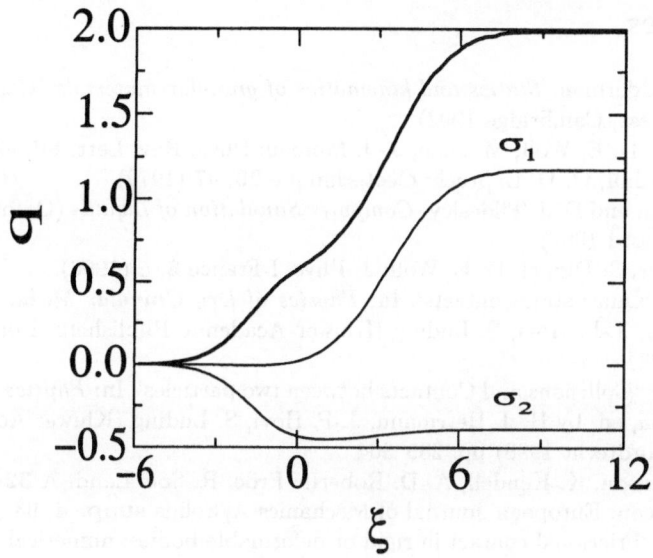

Fig. 11. The shear stress q and principal stresses σ_1 and σ_2 of the ξ-set as a function of ξ in a cohesive granular packing

Let us point out that the same property holds in a noncohesive granular medium where the set $\mathcal{S}(\xi = 1)$, called *weak network* in the case of a noncohesive packing, contains only the (compressive) contacts carrying a force below the average normal force [2]. The observation of the same phenomenon in a cohesive packing sheds a new light on the matter. The stability of a granular packing is related in a lesser extent to its property of building up a large anisotropy than to its capacity of organizing itself into two complementary networks with large anisotropies. This capacity increases with adhesion due to the activation of tensile forces which seem to play the same role as the weak network of compressive forces. Hence, the average geometrical quantity which best represents the strength of the granular texture is the anisotropy of the *strong network*, complementary set of the weak network.

This picture is reinforced by the analysis of partial stresses as a function of ξ. A partial stress tensor $\sigma(\xi)$ can be defined by simply restricting the usual expression of stress tensor to the ξ-set. Then, its components $\sigma_{ij}(\xi)$ are given by

$$\sigma_{ij}(\xi) = \langle f_i \ell_j \rangle_{\mathcal{S}(\xi)} \tag{13}$$

Figure 11 shows the evolution of the principal stresses σ_1 and σ_2, as well as the normalized shear stress, $q = (\sigma_1 - \sigma_2)/(\sigma_1 + \sigma_2)$ with ξ. We see that σ_2 reaches its maximum in absolute value at $\xi = 1$. In other words, the strong network gives nearly no contribution to the negative stress, the latter coming both from tensiles forces and weak compressive forces.

References

1. R. M. Nedderman: *Statics and kinematics of granular materials* (Cambridge University Press, Cambridge 1992)
2. F. Radjai, D. E. Wolf, M. Jean, J.-J. Moreau: Phys. Rev. Lett. **80**, 61 (1998)
3. P. A. Cundall, O. D. L. Stack: Geotechnique **29**, 47 (1979)
4. M. P. Allen and D. J. Tildesley: *Computer Simulation of Liquids* (Oxford University Press, Oxford 1987)
5. J. Schaefer, S. Dippel, D. E. Wolf: J. Phys. I France **6**, 5 (1996)
6. S. Roux: 'Quasi-staic contacts'. In: *Physics of Dry Granular Media*, ed. by H. J. Herrmann, J.-P. Hovi, S. Luding (Kluwer Academic Publishers, Dordrecht 1998) pp. 267–284
7. S. Luding: 'Collisions and Contacts between two particles'. In: *Physics of Dry Granular Media*, ed. by H. J. Herrmann, J.-P. Hovi, S. Luding (Kluwer Academic Publishers, Dordrecht 1998) pp. 285–304
8. K. L. Johnson, K. Kendell, A. D. Roberts: Proc. R. Soc. Lond. A **324**, 301 (1971)
9. J. J. Moreau: European Journal of Mechanics A/Solids **supp. 4**, 93 (1994)
10. M. Jean: 'Frictional contact in rigid or deformable bodies: numerical simulation of geomaterials'. In: *Mechanics of Geomaterial Interfaces*, ed. by A. P. S. Salvadurai, J. M. Boulon (Elsevier, Amsterdam 1995) pp. 463–486
11. F. Radjai, J. Schäfer, S. Dippel, D. E. Wolf: J. Phys. I France **7**, 1053 (1997)
12. H. M. Pollock: 'Surface forces and adhesion'. In: *Fundamentals of friction: Macroscopic and Microscopic Processes*, ed. by I. L. Singer, H. M. Pollock (Kluwer Academic Publishers, Dordrecht 1992) pp. 77–94
13. F. Radjai, S. Roux, J.-J. Moreau: Chaos **9**, 544 (1999)
14. F. Radjai, D. E. Wolf: Granular Matter **1**, 3 (1998)
15. F. Radjai: 'Multicontact dynamics'. In: *Physics of Dry Granular Media*, ed. by H. J. Herrmann, J.-P. Hovi, S. Luding (Kluwer Academic Publishers, Dordrecht 1998) pp. 305–311

Micro-mechanisms of deformation in granular materials: experiments and numerical results

J. Lanier

Laboratoire Sols, Solides, Structures, BP 53, 38041 Grenoble, Cedex 9, France

Abstract. To study experimentally the micro-mechanisms of deformation in granular materials a special shear apparatus was designed. The tested material is an assembly of wooden rods. Experimental results are concerned with macroscopic behavior (boundary conditions) and local kinematics (namely, rods displacements and rotations). In a first part, these local variables are compared with the continuum mechanics predictions. In a second part *Contact Dynamics method* is used for numerical simulations of *homogeneous* or *non-homogeneous* tests. Comparison with experimental results shows a good accuracy of the numerical code to describe local kinematics if the initial configuration of grains is chosen exactly as the experimental one.

Introduction

This paper is concerned by a study of micro-mechanisms of deformation in granular materials. From a mechanical point of view, these materials are generally considered as a continuum whose behavior is governed by a phenomenological constitutive equation to relate stress and strain. More recently increasing interest is devoted to micro-mechanical approach: grains are assumed as quasi-rigid bodies and the local kinematics is described by grains displacements and rotations. The correlation between these two approaches are studied theoretically (homogenization theories) or numerically with discret elements methods, but very few experimental results are published on this subject. Our paper will be divided in two parts: the first one describes experimental results obtained with a special shear apparatus. The material is a 2D-material (circular wooden rods). The second one is devoted to numerical simulations with a *Contacts Dynamics Method* code [4, 5] and it is shown that numerical simulations are able to reproduce very well local mechanism of deformation like shear band.

1 Experimental results

1.1 Experimental procedure

To obtain experimental measurements concerned with local mechanisms of deformation in a granular assembly, a special shear apparatus was build [1, 2]. In fact this apparatus is a vertical rectangular frame filled with 2D material (wooden rods with three sizes: 28, 18, 13 mm diameter). This frame can be deformed as parallelogram (see figure 1). As a consequence, general plane strain conditions

Fig. 1. Special shear apparatus used to study micromechanisms of deformation in granular assembly

are applied at the boundary $(\epsilon_x, \epsilon_y, \gamma)$. The sample is prepared inside the frame by pilling the rods one by one. During a test, successive pictures are taken and by digitalizing each of them, successive positions and orientations of each rods can be measured. Displacements field of rods centers and rotations are deduced by comparison of two pictures.

In the following, some experimental results are presented concerned with center displacements and rotations of rods. As far as possible an attempt of correlation with continuum mechanics analysis is presented.

1.2   Displacements field of rods centers

In our experiments the deformation of the sample is obtained by the boundary conditions imposed by the loading frame. These conditions allow the definition of an homogeneous deformation in continuum mechanics whose displacement vector field is defined by three constants α, β, ϱ:

$$u_x \;=\; \alpha.x + \beta.y \tag{1}$$
$$u_y \;=\; \varrho.y \tag{2}$$

The streamlines associated with this vector field are defined by the differential equation:

$$\frac{\mathrm{d}x}{u_x} = \frac{\mathrm{d}y}{u_y} \tag{3}$$

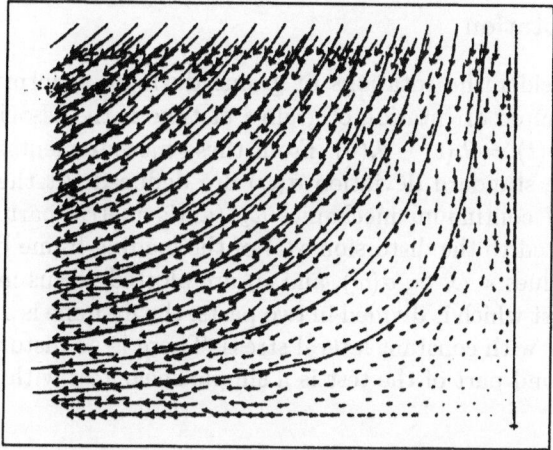

Fig. 2. Measured displacements field in a vertical compression test

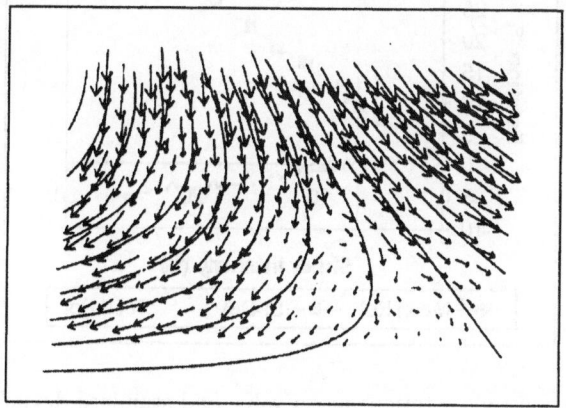

Fig. 3. Measured displacements field in a complex strain path test

On the other hand, the measured displacements of rods centers are obtained experimentally. On figures 2 and 3 these two approaches are plotted. The real displacement fields are plotted as arrows (with a scale factor) and the continuum mechanics streamlines appear as superposed lines. Two experiments are presented:

- a vertical compression ($\epsilon = 12\%$) with constant lateral stress (figure 2).
- a complex strain path with: $\epsilon_x = -2,24\%$ (horizontal), $\epsilon_y = +1,53\%$ (vertical), $\gamma = 3,5\%$ (distorsion) (figure 3).

These two examples show that continuum mechanics predictions give the general tendency for the displacements of rods centers. This is not a suprise: in dense granular assembly mobility of grains is strongly constraint by the no-overlapping conditions and the scheme of deformation is essentially defined by the boundary conditions.

1.3   Grains rotation

Let us now consider the rotations of grains during a deformation process. If $\theta_i(t)$ is the measured orientation of grain i at time t, the associated rotation ω_i is defined by: $\omega_i(t) = \theta_i(t) - \theta_i(0)$. From these measurements, the mean value $< \omega_r >$ and the standard deviation $\sigma(\omega)$ are deduced. At the same time, the rigid rotation of continuum mechanics ω_s (antisymetric part of displacement gradient) is related to the distorsion $\gamma(t)$ of the loading frame by: $\omega_s = \gamma/2$. On figure 4 these values $< \omega_r >$, $\sigma(\omega)$, and ω_s are plotted versus macroscopic shear intensity for a test which is divided in two parts: the first one is a classical vertical compression test with constant lateral stress. There is no distorsion ($\gamma = 0$) and $\omega_s = 0$. The second part of the test is a simple shear test with distorsion γ.

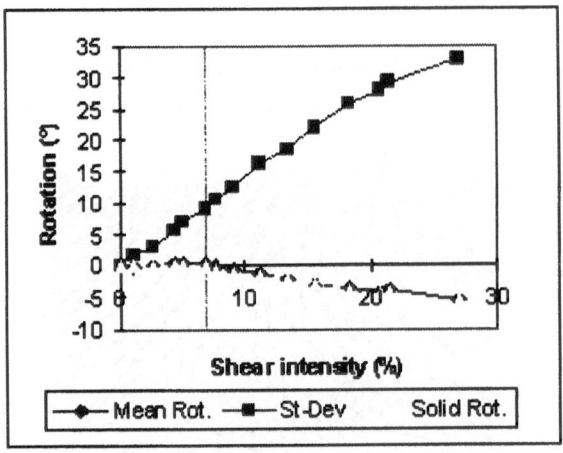

Fig. 4. Evolutions of rods rotations mean value and standard deviation during a test. Comparison with solid rotation defined by continuum mechanics

The main fact to observe is that the mean value of rods rotations remains very close to the continuum mechanics prediction all along the test, whereas the standard deviation is always increasing up to 35° without noticeable changes in its rate during the two parts of the test. Once more we can conclude that the discrete nature of the material is expressed by the dispersion around a mean value which can be predicted by continuum approach.

1.4   Rolling without sliding

Let us consider two contacting grains in an initial configuration and suppose that this contact is persitent during deformation. We can define the two oriented arcs lengths $a = \overrightarrow{C'C_1}$ and $b = \overrightarrow{C'C_2}$ where C' is the contact point in the deformed configuration and C_1, C_2 are the two materials points (belonging to each grains) located at the contact point in the initial configuration. The rolling without

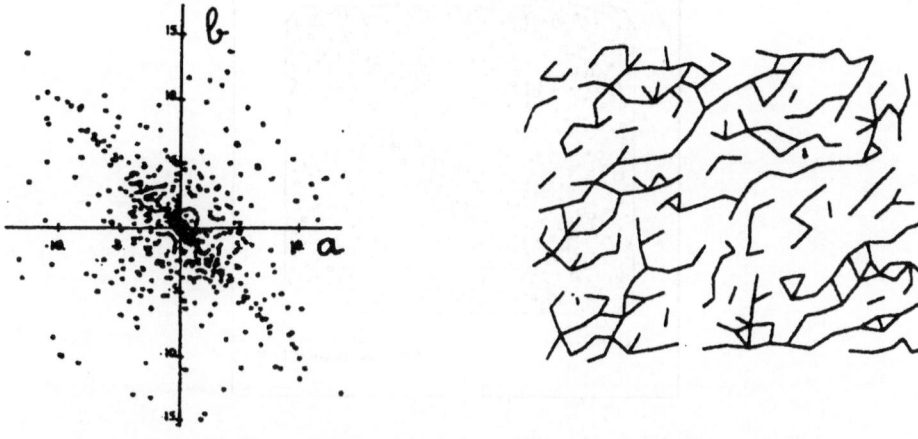

Fig. 5. Rolling without sliding for a simple shear test. (a) plane (a,b), the rolling without sliding condition is expressed by $a + b = 0$; (b) Contacts chains which verify the rolling without sliding condition

sliding condition is expressed by:

$$a + b = 0 \qquad (4)$$

This condition means that the grains behave like gears and there is no dissipation of energy in this mechanism of deformation. From our measurements of center displacements and grains rotation the two quantities a and b can be evaluated. They are plotted on figure 5a for a simple shear test. It is clear that there is a lot of contacts which verify the rolling without sliding condition (4). The figure 5b precises where these contacts are located in the sample: they are organized in chains whose orientations are close to 45 degrees which can be related to the principal axe of compression in the simple shear test.

1.5   Local deformation and shear band

To study the local mechanisms of deformation in the sample we defined for each grain a local strain measure which is deduced from the measured displacements of grains in the immediate vicinity of the considered grain [1]. This definition implies elementary cells whose sizes are at the scale of 3 grains; it cannot be used as a macroscopic deformation. Figure 6 presents a map of such a local deformation for a vertical biaxial compression test: for each grain of the sample (722 grains) the local strain is evaluated and the shear intensity $E = \epsilon_1 - \epsilon_2$ is deduced ($\epsilon_1 > \epsilon_2$ are the principal value of local strain). The size of each square on the map is proportional to local E-value. This example shows that the global mechanism of deformation in this test occurs with shear banding inside the sample.

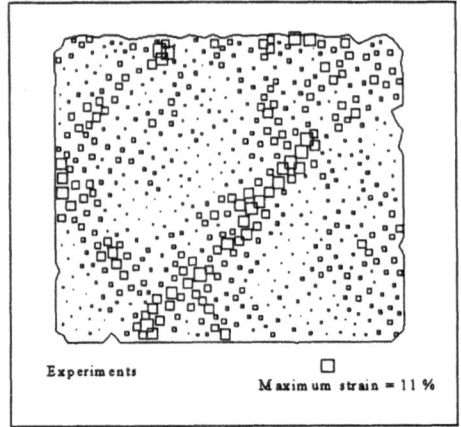

Fig. 6. Map of local shear intensity for a vertical compression test The macroscopic vertical strain is 3%

2   Numerical simulations

Numerical simulations are performed with *Contact dynamics method* (CD) elaborated by JJ.Moreau and M.Jean since 10 years ago [3–5]. It differs essentially from the classical *Distinct Element method* by the following hypothesis:

- grains are assumed to be rigid bodies without any stiffness at the contact: the contact law is the no-overlapping condition for the normal part and the Coulomb friction for the tangential one. These conditions are presented on figure 7.
- the integrating scheme is implicit.
- a shock rule can be implemented. In the following, inelastic shock is assumed.

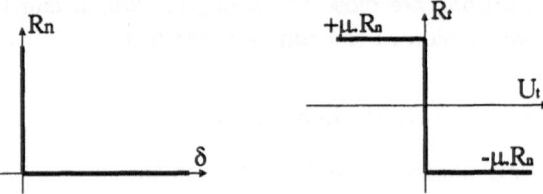

Fig. 7. Contacts rules in *CD method*. δ is the gap between two particules in potential contact, R_n, R_t are the normal and tangential contact forces, μ is the friction coefficient

The solving algorithm is an iterative process whose unknows are the contact forces, constrained by the contact rules. Then the Newton law applied for each grain gives the movement (velocity and position of grains). A detailed presentation of the method can be found in [2]. Our code is written for 2D problems and is able to take into account assemblies of circular or rectangular elements.

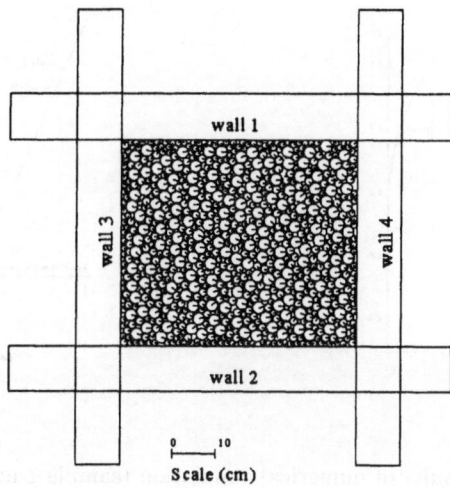

Fig. 8. Numerical sample for simulation of biaxial tests

2.1   Numerical simulations of biaxial tests

Two biaxial tests (vertical compression with constant horizontal stress) were numerically simulated. The sample is an assembly of 726 disks inside a rectangular frame consisting in 4 walls (figure 8). The friction coefficients are chosen such that: disk-disk contacts $\mu = 0.53$ (measured value), disk-wall contact $\mu = 0.0$ (no-friction at the boundary). The two tests differ only by the sample preparation. For the sample 1 the initial configuration (position and radii of disks) is exactly the measured experimental configuration in the test analysed in figure 6. Sample 2 is prepared directly in the frame from sample 1 according to the following numerical procedure: first, all the radii are decreased by 0.5 mm, then gravity is applied on disks and wall 1. As a result a completely new sample is obtained with less contacts than the sample 1.

The results obtained with these two samples may be classically analysed in terms of macroscopic behavior, that to say stress-strain curve and volume change, and compared with measured experimental results (figure 9). Macroscopic deformation is deduced from the lengths changes between wall 1-2 and 3-4. Vertical and horizontal stress are obtained by the resultant contact forces between disks and walls.

From these simulations the following conclusions can be drawned:

- initial behavior is very sensitive with the sample preparation. Sample 1 behaves like a rigid plastic body whereas mechanisms of deformation in sample 2 is different: it looks like successive instabilities. The real experiments lies between these two behaviors.

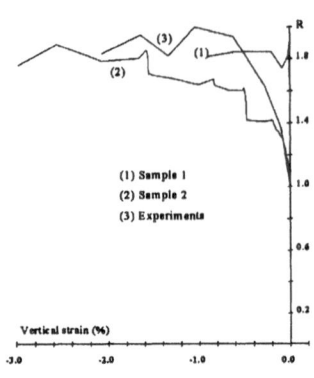
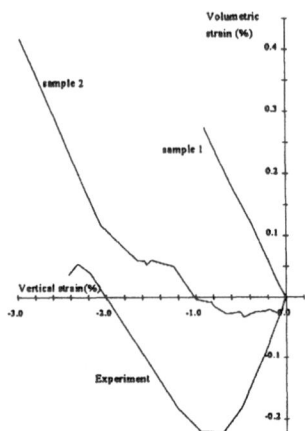

Fig. 9. Macroscopic results of numerical simulation (sample 1 and 2) and comparison with real experiments

- the maximum strength (R is the stress ratio) is quite similar for the three curves.
- the volume change curves show similar dilatancy when plastic flow occurs but initial compaction is not found in numerical simulations.

2.2   Local mechanisms of deformation

As the initial configuration of sample 1 is exactly the experimental one we may compare local mechanism of deformation in this numerical sample with experimental results presented on figure 6. The map of local shear intensity is presented on figure 10.

On this figure it is clear that shear banding is present with the same position and orientation as in real experiment (figure 6). Results for sample 2 (not displayed here), whose preparation is obtained by disturbing sample 1, are completly different. The conclusion is that the local mechanisms of deformation, especially initiation of shear bands, are strongly dependent of the geometry. If and only if the exact configuration of the sample is considered, numerical simulations are able to reproduce the macroscopic scheme of deformation.

2.3   Numerical simulation of pull-out test

The previous experimental or numerical tests with granular materials were conducted with boundary condition which may be considered as homogeneous in the framework of continuum mechanics. This last example is a non-homogeneous one: a rectangular plate is buried horizontally in a disks assembly with three diameters 28, 18, 13 mm (figure 11) and is pulled out in vertical translation. During the real experiment, pictures are taken which allow as previously the measure

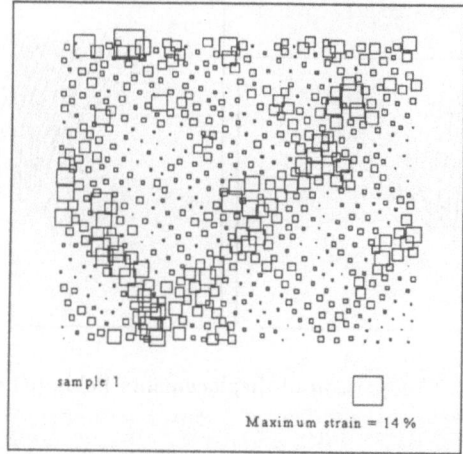

Fig. 10. Map of local shear intensity in sample 1 (numerical simulation to compare with figure 6)

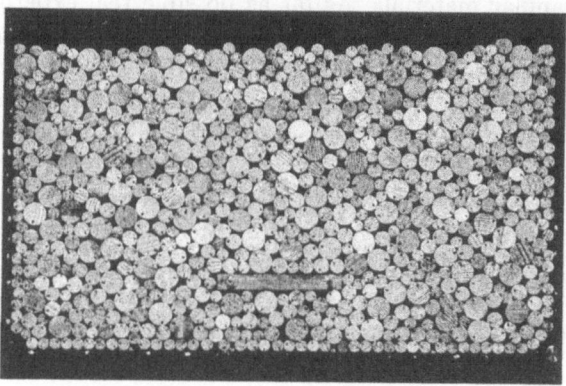

Fig. 11. Pull out test in a granular assembly. Initial configuration

of rods displacements. To perform numerical simulation, the first picture is used to define the initial numerical configuration (disks and plate positions, radii).

The comparison of measured and numerical displacements fields is presented on figure 12 for a 3 cm-vertical translation of the plate. Once more numerical simulation seems to reproduce the observed mechanisms of deformation, namely:

- the movement of the plate induced local deformation of the upper part of the sample while the lower part remains at rest.
- a rigid corner on the top of the plate moves without deformation (rigid zone)
- the mechanism of deformation is non-symmetric. This point can only be explained in numerical simulation by the fact that real configuration is considered.
- a void is created below the plate and some disks fall in.

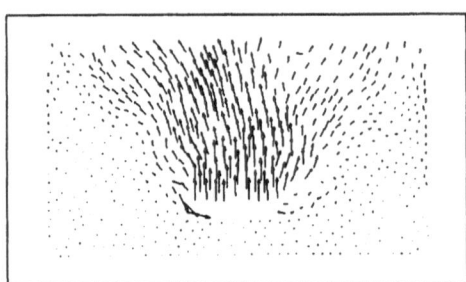 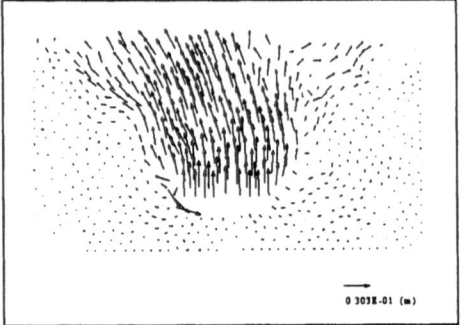

Fig. 12. Pull out test. Comparison of displacements fields (**a**) experimental, (**b**) numerical

3   Conclusion

The two parts of this paper are concerned with analysis of mechanisms of deformation in granular materials. As far as possible real experiments are taken as a basis for analysis. In the first part, the kinematics of granular material is described by displacement and rotations fields. It is shown that these fields are strongly connected with continuum mechanics predictions. The discret nature of the material is only expressed by dispersion around a predicted mean value. In the second part numerical simulations with *CD method* are presented. Comparisons with experimental results show that the local mechanisms of deformation (kinematics variables) are pretty well reproduced if the real experimental configuration is taken as initial numerical data. As far as stress-strain behavior is concerned, numerical simulations appear very sensitive to the sample preparation, but plastic flow (maximum strength and dilatancy) is correctly obtained.

References

1. Calvetti F. and Combe G. and Lanier J., "Experimental micromechanical analysis of a 2D granular material: relation between structure evolution and loading path" Mech. Cohes. Frict. Mat. **2**, 121-163 (1997).
2. Lanier J. and Jean M., "Experiments and numerical simulations with 2D disks assembly", Powder Technology **109**, 206-221 (2000).
3. Moreau JJ., "Some numerical methods in multibody dynamics: application to granular materials", Eur. J. Mech. A/Solids 13/4, 93-114 (1994).
4. Moreau JJ., "New computation methods in granular dynamics", Proc. Powders and Grains **93**, Balkema, C. Thornton (ed.), 227-232 (1993).
5. Jean M., "Frictional contact in collections of rigig or deformable bodies: numerical simulation of geomaterial motion", Mechanics of geomaterial interfaces, Boulon and Salvadurai (eds.), Elsevier, 463-486 (1995).

Scaling properties of granular materials

T. Pöschel,[1] C. Salueña,[2] T. Schwager[1]

[1] Humboldt-University Berlin - Charité, Institute for Biochemistry, Hessische Str. 3-4, D-10115 Berlin, Germany. http://summa.physik.hu-berlin.de/~kies/
[2] Humboldt-University Berlin, Institute of Physics, Invalidenstr. 110, D-10115 Berlin, Germany.

Abstract. Given an assembly of viscoelastic spheres with certain material properties, we raise the question how the macroscopic properties of the assembly will change if all lengths of the system, i.e. radii, container size etc., are scaled by a constant. The result leads to a method to scale down experiments to lab-size.

1 Introduction

It is frequently desired to investigate large scale phenomena in granular systems experimentally, such as geophysical effects or industrial applications. To this end one has to rescale all lengths of the system to a size which is in accordance with the laboratory size, i.e. big boulders in the original system become centimeter sized particles in the experiment. Of course, one wishes that the effects which occur in the large system occur equivalently in the scaled system too. We will show that naive scaling will modify the properties of a granular system such that the original system and the scaled system might reveal quite different dynamic properties. To guarantee equivalent dynamical properties of the original and the scaled systems we have to modify the material properties in accordance with the scaling factor and we have to redefine the unit of time.

In a simple approximation, a granular system may be described as an assembly of spheres of radii R_i, $i = 1, \ldots, N$. The interaction between spheres i and j at positions r_i and r_j of radii R_i and R_j is given by a pairwise force law

$$F_{ij} = \begin{cases} F_{ij}^n n_{ij} + F_{ij}^t t_{ij} & \text{if} \quad \xi_{ij} \equiv R_i + R_j - |r_i - r_j| > 0 \\ 0 & \text{else}, \end{cases} \qquad (1)$$

with the unit vector in normal direction $n \equiv \frac{r_j - r_i}{|r_j - r_i|}$ and the respective unit vector in tangential direction t.

The dynamics of the system can be found, in principle, by integrating Eq. (1) for all particles $i = 1, \ldots, N$ simultaneously with appropriate initial conditions. Eventually, external forces as, e.g. gravity, may also act on the particles. In practice the dynamics is found by integrating Eq. (1) numerically. Molecular Dynamics techniques (e.g. [1]) exploit this idea, which has provided much insight into the properties of granular systems over the last decade (see e.g. [2] and many references therein).

The detailed formulation of the normal and tangential forces F^n and F^t depends on the grain model; several models have been studied in [3]. It is known that the tangential force F^t is essential to explain many macroscopic effects, in particular if static properties of the granular material become important [4]. With this formulation we mean that at least for a certain time interval the static hindrance with respect to rotation of contacting spheres is important for the dynamics of the system. Nevertheless there exist phenomena for which the rotational degree of freedom of the particles is less important as, e.g., in the case of highly agitated granular system as granular gases [5]. In other systems neglecting the tangential force F^t still might be a good approximation. In the present study we first assume that the tangential force F^t may be neglected to describe the dynamics of the system, i.e., $\boldsymbol{F}_{ij} = F^n \boldsymbol{n}_{ij}$. The incorporation of tangential forces is discussed briefly in section 6.

Assume we know the dynamics of a certain granular system S. Will the dynamics change if we rescale all sizes by a constant factor α, i.e., $R_i' \equiv \alpha R_i$, but leaving the material properties constant? Here and in the following we mark all variables which describe the scaled system S' with a prime. If the scaling affects the system properties, how do we have to modify the material properties to assure that the systems S and S' behave identically?

2   The normal force F^n

The normal force F^n can be subdivided into two parts, the elastic part F_{el}^n and the dissipative part F_{dis}^n:

$$F^n = F_{el}^n + F_{dis}^n \,. \tag{2}$$

(For simplicity of notation we drop the indices ij of the variables which refer to pairs of particles.)

The elastic force for colliding spheres is given by Hertz's law [6]

$$F_{el}^n = \frac{2Y}{3(1-\nu^2)} \sqrt{R^{eff}} \xi^{3/2} \equiv \rho \xi^{3/2} \,, \tag{3}$$

with $R^{eff} = R_i R_j / (R_i + R_j)$ and Y, ν being the Young modulus and the Poisson ratio. Equation (3) also defines the prefactor ρ which we will need below.

The formulation of the dissipative part of the force F_{dis}^n depends on the mechanism of damping. The most simple mechanism is elastic interaction ($F_{dis}^n = 0$). The second simplest type is viscoelastic damping, which we will focus on. More complicated mechanisms are plastic deformation or brittle deformation. The last two damping types are more complicated since the shape of the particles changes due to collisions. Therefore, the simultaneous assumption of plastic deformation and spherical shape of the particles is inconsistent (although frequently applied in simulations of granular material). It is not even clear that the spherical shape of particles which undergo plastic deformation is conserved *on average* [7].

We assume the most simple nontrivial interaction between colliding grains, which is viscoelastic interaction. It implies that the elastic part of the stress tensor is a linear function of the deformation tensor, whereas the dissipative part of the stress tensor is a linear function of the deformation rate tensor. It is valid if the characteristic velocity (the impact rate g) is much smaller than the speed of sound c in the material and the viscous relaxation time τ_{vis} is much smaller than the duration of the collisions τ_c (quasistatic motion).

The range of our assumption is, hence, limited from both sides: the collisions should not be to fast to assure $g \ll c$, $\tau_{\mathrm{vis}} \ll \tau_c$, and not too slow to avoid influences of surface effects as adhesion. For spheres the dissipative part of the normal force reads [8]

$$F_{\mathrm{dis}}^n = A \frac{d\xi}{dt} \frac{d}{d\xi} F_{\mathrm{el}}^n \qquad (4)$$

$$= \frac{3}{2} A \rho \sqrt{\xi} \frac{d\xi}{dt} \qquad (5)$$

with

$$A = \frac{1}{3} \frac{(3\eta_2 - \eta_1)^2}{3\eta_2 + 2\eta_1} \frac{(1 - \nu)(1 - 2\nu)}{Y \nu^2}. \qquad (6)$$

The dissipative material constant A is a function of the viscous constants $\eta_{1/2}$, the Young modulus Y and the Poisson ratio ν. The functional form of Eq. (5) was guessed (but not derived) by Kuwabara and Kono before [9] and has been derived independently by Brilliantov et al. [8] and Morgado and Oppenheim [10] using very different approaches. However, only the strict analysis of the viscoelastic deformation in Ref. [8] yields the prefactors A and ρ as functions of the material properties. The knowledge about these prefactors is crucial for the derivation of the scaling properties. We want to remark that the general equation (4) is not limited to spheres but holds true for any smooth (in mathematical sense) interaction of viscoelastic bodies [11].

Combining the forces (3) and (5) one obtains the equation of motion for colliding viscoelastic spheres

$$\frac{d^2\xi}{dt^2} + \frac{\rho}{m^{\mathrm{eff}}} \left(\xi^{3/2} + \frac{3}{2} A \sqrt{\xi} \frac{d\xi}{dt} \right) = 0 \qquad (7)$$

$$\xi|_{t=0} = 0$$

$$\frac{d\xi}{dt}\bigg|_{t=0} = g,$$

with $m^{\mathrm{eff}} = m_i m_j / (m_i + m_j)$.

3   Scaling properties

First we want to write down the equation of motion (7) in a dimensionless form [12]. To this end we need a characteristic length and a characteristic time of the system. A reasonable inherent length of the problem of colliding viscoelastic spheres is the maximal compression ξ_0 for the equivalent undamped (elastic) problem. It can be found by equating the kinetic energy of the impact $m^{\mathrm{eff}} g^2 / 2$ with the elastic energy at the instant of maximal compression:

$$m^{\mathrm{eff}} \frac{g^2}{2} = m^{\mathrm{eff}} \frac{\rho}{m^{\mathrm{eff}}} \frac{2}{5} \xi_0^{5/2}, \qquad (8)$$

yielding

$$\xi_0 \equiv \left(\frac{5}{4}\frac{m^{\text{eff}}}{\rho}\right)^{2/5} g^{4/5}. \tag{9}$$

As characteristic time we define the time in which the distance between the particles changes by the characteristic length just before the collision starts:

$$\tau_0 \equiv \xi_0/g. \tag{10}$$

Note that up to a numerical prefactor the timescale τ_0 is equal to the duration of the undamped collision [6], which would be an alternative (and equivalent) choice of the timescale. Using the definitions (9) and (10) we find the rescaled length, velocity, and acceleration

$$\hat{\xi} = \xi/\xi_0 \tag{11}$$

$$\frac{d\hat{\xi}}{d\tau} = \frac{1}{g}\frac{d\xi}{dt} \tag{12}$$

$$\frac{d^2\hat{\xi}}{d\tau^2} = \frac{\xi_0}{g^2}\frac{d^2\xi}{dt^2} \tag{13}$$

and rewrite the equation of motion (7) in dimensionless form

$$\frac{d^2\hat{\xi}}{d\tau^2} + \frac{5}{4}\hat{\xi}^{3/2} + \frac{3}{2}\left(\frac{5}{4}\right)^{3/5} A\left(\frac{\rho}{m^{\text{eff}}}\right)^{2/5} g^{1/5}\sqrt{\hat{\xi}}\,\frac{d\hat{\xi}}{d\tau} = 0 \tag{14}$$

$$\hat{\xi}\Big|_{\tau=0} = 0$$

$$\frac{d\hat{\xi}}{d\tau}\Big|_{\tau=0} = 1.$$

We see that the only term which depends explicitly on the system size and on material properties is the prefactor in front of the third term of Eq. (14). Scaling the system, therefore, can only affect the dynamics of a granular system if it affects the value of this term. In other words, two systems will behave identically (in the scaled variables) if its value is conserved.

Expanding our abbreviations we obtain

$$A\left(\frac{\rho}{m^{\text{eff}}}\right)^{2/5} g^{1/5} = A\left(\frac{2Y\sqrt{R^{\text{eff}}}}{3(1-\nu^2)m^{\text{eff}}}\right)^{2/5} g^{1/5} \tag{15}$$

$$= A\left(\frac{Y\sqrt{\frac{R_iR_j}{R_i+R_j}}}{2\pi(1-\nu^2)\phi\frac{R_i^3R_j^3}{R_i^3+R_j^3}}\right)^{2/5} g^{1/5} \tag{16}$$

$$\sim AY^{2/5}\phi^{-2/5}\left(1-\nu^2\right)^{-2/5} F(R_i, R_j)\, g^{1/5}, \tag{17}$$

with ϕ being the material density. The function $F(R_i, R_j)$ in Eq. (17) collects all terms containing R_i and R_j. The third line Eq. (17) equals the second line, Eq. (16), up to the constant $(2\pi)^{-2/5}$ which is not relevant for the scaling properties.

The function $F(R_i, R_j)$ is directly affected by scaling the radii $R'_i = \alpha R_i$, $R'_j = \alpha R_j$. Let us see how this function scales:

$$F(R'_i, R'_j) = F(\alpha R_i, \alpha R_j) = \frac{\left(\frac{\alpha R_i \alpha R_j}{\alpha R_i + \alpha R_j}\right)^{1/5}}{\left(\frac{\alpha^3 R_i^3 \alpha^3 R_j^3}{\alpha^3 R_i^3 + \alpha^3 R_j^3}\right)^{2/5}} = \alpha^{-1} \frac{\left(\frac{R_i R_j}{R_i + R_j}\right)^{1/5}}{\left(\frac{R_i^3 R_j^3}{R_i^3 + R_j^3}\right)^{2/5}}$$

$$= \alpha^{-1} F(R_i, R_j). \tag{18}$$

4   Scaling large phenomena down to "lab-size" experiments

We have already seen that simple scaling of the system in general affects the prefactor (15) already via the scaling properties of $F(R_i, R_j)$ (see Eq. (18)), i.e., in general the original system and the scaled system might reveal quite different dynamic properties. More explicitly, one can show that naively scaling the system by a factor $\alpha < 1$ will lead to a comparatively more damped dynamics.

Therefore, to guarantee equivalent dynamical properties of the systems we have to modify the material properties in a way to assure that the prefactors (15) of the original system and the scaled system are identical.

One of the few things which cannot be modified in the experiment with reasonable effort is the constant of gravity G. That implies that going from S to S' not only G but all other accelerations must remain unaffected too:

$$\left(\frac{d^2 x}{dt^2}\right)' = \frac{d^2 x}{dt^2}$$

$$\frac{d^2(\alpha x)}{d(t')^2} = \frac{d^2 x}{dt^2}$$

$$t' = \sqrt{\alpha}\, t. \tag{19}$$

Hence, scaling all lengths $x' = \alpha x$ implies that times scale as $t' = \sqrt{\alpha} t$ if we request that the gravity constant stays unaffected. Thus, our clock in the laboratory should run by a factor $\sqrt{\alpha}$ faster or slower than the clock in the original system. In other words, if in the original system we observe a phenomenon at time $t = 100$ sec, we will find the same effect in the scaled system at time $t' = \sqrt{\alpha}\, 100$ sec. Scaling of time is a direct consequence of the spatial scaling if the constant of gravity has the same value in both systems.

In order to obtain the necessary conditions for the material properties of the scaled system, we require the scaled equation of motion to be exactly equivalent to its counterpart in the unscaled system. This ensures that the trajectories in the scaled system are exactly the same (after reversing the scaling procedure) as in the unscaled system. In the scaled system the equation of motion during a binary collision or a permanent contact reads

$$\frac{d^2\xi'}{dt'^2} + \frac{\rho'}{(m^{\text{eff}})'}(\xi')^{3/2} + A'\frac{\rho'}{(m^{\text{eff}})'}\sqrt{\xi'}\,\frac{d\xi'}{dt'} = 0. \tag{20}$$

If we apply our scaling relations which were introduced above, i.e.

$$\xi' = \alpha\xi \tag{21}$$

$$\frac{d\xi'}{dt'} = \sqrt{\alpha}\frac{d\xi}{dt} \tag{22}$$

$$\frac{d^2\xi'}{dt'^2} = \frac{d^2\xi}{dt^2} \tag{23}$$

we obtain

$$\frac{d^2\xi}{dt^2} + \alpha^{3/2}\frac{\rho'}{(m^{\text{eff}})'}\xi^{3/2} + \alpha A'\frac{\rho'}{(m^{\text{eff}})'}\sqrt{\xi}\,\frac{d\xi}{dt} = 0. \tag{24}$$

Since the scaling does not affect the physical meaning of a given equation of motion, all systems whose equation of motion can be transformed into each other by simply scaling all lengths and the time can be considered to be equivalent. Therefore, our scaled system is equivalent to the unscaled system if and only if

$$\alpha^{3/2}\frac{\rho'}{(m^{\text{eff}})'} = \frac{\rho}{m^{\text{eff}}} \tag{25}$$

$$\alpha A'\frac{\rho'}{(m^{\text{eff}})'} = A\frac{\rho}{m^{\text{eff}}}. \tag{26}$$

If we choose material constants which obey Eqs. (25) and (26) we will obtain the original equation of motion after scaling the system back to original size, i.e. the two systems are equivalent.

Equations (25) and (26) can be further simplified yielding

$$\frac{\rho'}{(m^{\text{eff}})'} = \alpha^{-3/2}\frac{\rho}{m^{\text{eff}}} \tag{27}$$

$$A' = \sqrt{\alpha}\,A. \tag{28}$$

The last equation shows that A, which is essentially the viscous relaxation time of the spheres involved in the contact, has to behave exactly as any other time.

To check the validity of our considerations we will now study two fundamental characteristics of a binary collision whose scaling behaviour is known. These are the coefficient of restitution, which describes the ratio of the normal relative velocity of the two particles after and before the collision

$$\epsilon = \frac{g_{\text{after}}}{g_{\text{before}}}, \tag{29}$$

and the duration of the collision. Both values depend on the impact velocity –
for experimental evidence see e.g. [13–17]. For viscoelastic spheres the restitution
coefficient has been derived rigorously from Eq. (5) [12, 18]:

$$\epsilon = 1 - C_1 \frac{3}{2} A \left(\frac{\rho}{m^{\text{eff}}} \right)^{2/5} g^{1/5} + C_2 \left(\frac{3}{2} A \right)^2 \left(\frac{\rho}{m^{\text{eff}}} \right)^{4/5} g^{2/5} \mp \dots \tag{30}$$

For the duration of a collision we find [18]

$$T_c = \left(\frac{\rho}{m^{\text{eff}}} \right)^{-2/5} g^{-1/5} \left(D_0 + D_1 \frac{3}{2} A \left(\frac{\rho}{m^{\text{eff}}} \right)^{2/5} g^{1/5} \right.$$
$$\left. + D_2 \left(\frac{3}{2} A \right)^2 \left(\frac{\rho}{m^{\text{eff}}} \right)^{4/5} g^{2/5} + \dots \right). \tag{31}$$

To be physically consistent we have to require that the restitution coefficient
is invariant and that the duration of the collision scales as any other time, i.e.

$$\epsilon'(g') = \epsilon(g) \tag{32}$$
$$T_c'(g') = \sqrt{\alpha}\, T_c(g). \tag{33}$$

The symbols ϵ' and T_c' denote the coefficient of restitution and collision time
for spheres in the scaled system, i.e., where the relevant material properties are
ρ', $(m^{\text{eff}})'$ and A'.

Eq. (32) is easily verified by noting that

$$A' \left(\frac{\rho'}{(m^{\text{eff}})'} \right)^{2/5} (g')^{1/5} = \sqrt{\alpha}\alpha^{-3/5}\alpha^{1/10} A \left(\frac{\rho}{m^{\text{eff}}} \right)^{2/5} g^{1/5} \tag{34}$$

$$= A \left(\frac{\rho}{m^{\text{eff}}} \right)^{2/5} g^{1/5}. \tag{35}$$

To check Eq. (33) due to property (35) we only have to verify

$$\left(\frac{\rho'}{(m^{\text{eff}})'} \right)^{-2/5} (g')^{-1/5} = \sqrt{\alpha} \left(\frac{\rho}{m^{\text{eff}}} \right)^{-2/5} g^{-1/5}, \tag{36}$$

which is done by inserting Eqs. (27) and (28).

The discussion of the coefficient of restitution and the duration of collision
shows that our scaling procedure is physically consistent.

The replacement of our abbreviations with material parameters yields

$$\frac{\rho}{m^{\text{eff}}} = \frac{2Y\sqrt{R^{\text{eff}}}}{3(1-\nu^2)m^{\text{eff}}} = \frac{Y}{2\pi(1-\nu^2)\phi} [F(R_1, R_2)]^{5/2} \tag{37}$$

$$\frac{\rho'}{(m^{\text{eff}})'} = \frac{Y'}{2\pi \left(1-(\nu')^2\right)\phi'} [F(R_1', R_2')]^{5/2}. \tag{38}$$

With Eqs. (18), (37) and (38) we obtain

$$\frac{Y'}{\phi'\left(1-(\nu')^2\right)} = \alpha \frac{Y}{\phi\left(1-\nu^2\right)}. \tag{39}$$

Note that in a simple approximation one can identify $\sqrt{Y/\phi}$ with the speed of sound in the material which is strictly valid only for gases. If one neglects the scaling of the Poisson ratio, one discovers that the speed of sound scales as any other velocity, namely with the factor $\sqrt{\alpha}$ (see Eq. (22)).

Hence, simple scaling of lengths by a constant factor α and the request that the gravity constant is conserved leads to a scaling of the elastic and dissipative material properties and of the time if we wish that original and rescaled systems behave identically.

	original system	scaled system
all lengths	x	αx
time	t	$\sqrt{\alpha}\,t$
elastic constant	$\dfrac{Y}{\phi\left(1-\nu^2\right)}$	$\alpha\,\dfrac{Y}{\phi\left(1-\nu^2\right)}$
dissipative constant	A	$\sqrt{\alpha}A$

If one scales down an experiment by a factor α, therefore, one has to change the material as well according to the scaling relations given in the table, in order to find the same effects as in the original system. Moreover one has to scale time, i.e. an effect which is observed at time t in the original system will occur at time $\sqrt{\alpha}\,t$ in the rescaled system.

We want to give an example: Assume in the original system one deals with steel spheres ($Y = 20.6 \cdot 10^{10}$ Nm^{-2}, $\nu = 0.29$ and $\phi = 7,700$ kg m^{-3}) of average radius $\bar{R} = 10$ cm and system size of $L = 10$ m. The property whose scaling behaviour is known is $Y/(\phi(1-\nu^2)) = 2.92 \cdot 10^7$ m^{2}sec^{-2}. One wishes to know (to measure) a certain value at time $t = 100$ sec. In the lab we perform the experiment with an equivalent system of size $L' = 1$ m, i.e. we scale the system by the factor $\alpha = 0.1$, including all radii. From the scaling relations we see that we have to find a material whose scaled property is $Y'/(\phi'(1-\nu'^2)) \approx 0.3 \cdot 10^7$ m^{2}sec^{-2}. From tables [19] we see that we can use plexiglass ($Y = 0.32 \cdot 10^{10}$ Nm^{-2}, $\nu = 0.35$ and $\phi = 1,200$ kg m^{-3}) in order to obtain this value. Therefore, we have to perform the experiment with plexiglass spheres and have to measure the value of interest at time $t' = 31.6$ sec.

One can imagine that not for all scaling factors α one will find a proper material, however, nowadays it is possible to manufacture materials which can meet demanding requirements, such as high softness along with a custom-designed damping constant.

5   Bouncing ball

One of the most simple experiments one can think of is a ball which falls from height h due to gravity and collides with another ball of the same material which is attached at the ground. For a highly sophisticated experiment on this system see [20]. Assume that in our original system which is to be simulated a steel ball of radius $R = 10$ cm falls from height $h = 20$ cm. The material parameters are $Y = 20.6 \cdot 10^{10}$ Pa, $\phi = 7700$ kg/m^3, $\nu = 0.29$, $A = 10^{-4}$ sec, and gravity is $G = 9.81$ m/sec^2. Of course, the value of A is fictious since there are no tabulated values available.

In Fig. 1 we have drawn the distance between the spheres over time for the original system (left) and a system which is scaled down by a factor $\alpha = 0.25$ (right), i.e., the sphere has a radius of $R' = 2.5$ cm and is dropped from $h' = 5$ cm. As can be seen in Fig. 1 both trajectories are virtually indistinguishable if one

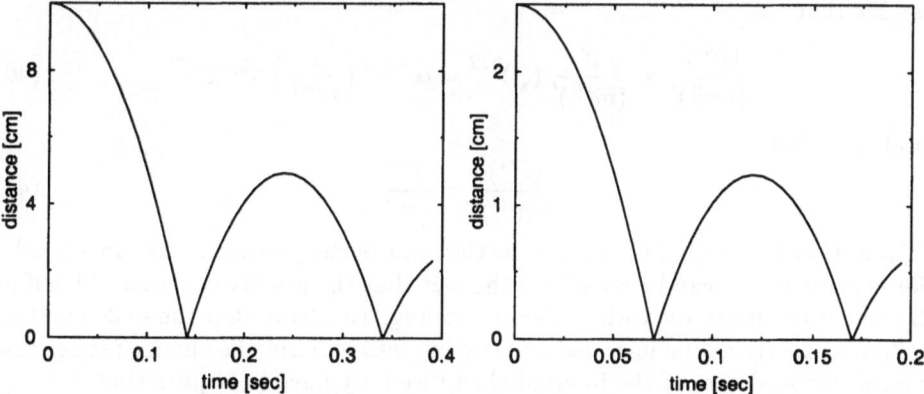

Fig. 1. Numerical results for the distance between the colliding balls over time. One of them is fixed to the ground – unscaled system (left) and the scaled system (right). For explanation of the simulation parameters see text.

takes into account that the distance between the balls is scaled by 0.25 and the time by 0.5 in accordance with our scaling scheme.

In Fig. 2 we show the distance between the spheres during the first collision in the unscaled and scaled systems. Negative values mean that the particles are compressed, i.e. the distance between their centres is smaller than the sum of their radii. Again it can be seen that both systems behave identically.

6   Consideration of the tangential force

In order to incorporate the tangential force into the analysis we have to require that it scales exactly as the normal force, or gravity respectively. From Eq. (25)

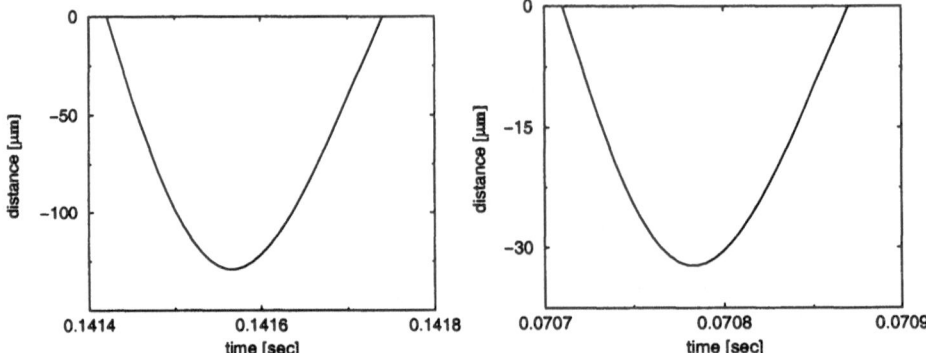

Fig. 2. The trajectory of the same system as in Fig. 1 during a collision shown with higher resolution in space and time. Again the unscaled system is shown on the left, the scaled one on the right.

we see that

$$\frac{(F^n)'}{(m^{\text{eff}})'} = \frac{\rho'}{(m^{\text{eff}})'} (\xi')^{3/2} = \alpha^{-3/2} \left(\frac{\rho}{m^{\text{eff}}}\right) \alpha^{3/2} \xi^{3/2} \tag{40}$$

and thus find

$$\frac{(F^n)'}{(m^{\text{eff}})'} = \frac{F^n}{m^{\text{eff}}} . \tag{41}$$

We see that the normal force scales as the mass of the particles. The same is valid for gravity, which can be seen from the fact that the gravity constant G itself is, by definition, invariant with respect to scaling. To ensure that the scaled system behaves exactly as the unscaled one (taking into account the changed timescale, i.e the changed rate of the internal clock) we only have to require that

$$\frac{(F^t)'}{(m^{\text{eff}})'} = \frac{F^t}{m^{\text{eff}}} . \tag{42}$$

This requirement has to be met by appropriately scaling the material constants, particularly the friction constant. This will give us an additional scaling equation. Its form depends on the underlying friction model, i.e. on the functional dependence of the tangential force on the geometry and the material properties as well as on the compression and the relative velocity. For instance if we assume the most simple tangential force law

$$F^t = \mu F^n , \tag{43}$$

with μ being the friction coefficient we can conclude that the friction coefficient has to be invariant with respect to scaling

$$\mu \equiv \text{const.} \tag{44}$$

Other friction models will lead to different scaling properties. However, a thorough discussion of the possible friction models is beyond the scope of this study.

7   Conclusion

In the present article we have shown that by means of a straightforward scaling procedure one can construct for any given system of viscoelastic spheres an equivalent system of scaled size. Along with a change of size of the system and all of its constituents, one has to modify the timescale (rate of the internal clock) and the material properties in a predefined way (see Eqs. (28) and (39)).

This scaling scheme has a number of useful applications. The most notable one is the possibility to scale down real world systems, e.g. geophysical or industrial granular systems, to sizes where laboratory experiments can be performed. If one scales down such a granular system one has to replace the original material by a material which meets the scaling requirements discussed in the text.

One can further apply the scaling scheme in Molecular Dynamics simulations. Here it can be desirable to change the values of the material parameters in order to achieve more accurate results at a given value of the integration time step. A study of the impact of the scaling on the accuracy of numerical simulations is subject of further studies [21].

References

1. M. P. Allen and D. J. Tildesley, *Computer Simulations of Liquids*, Clarendon Press (Oxford, 1987).
2. H. J. Herrmann, J.-P. Hovi, and S. Luding (eds.), *Physics of Dry Granular Media*, Kluwer (Dordrecht, 1998).
3. J. Schäfer, S. Dippel, and D. E. Wolf, *Force schemes in simulations of granular materials*, J. Physique **6**, 5-20 (1996).
4. T. Pöschel and V. Buchholtz, *Static friction phenomena in granular materials: Coulomb law versus particle geometry*, Phys. Rev. Lett. **71**, 3963-3966 (1993).
5. T. Pöschel and S. Luding (Eds.), *Granular Gases*, Springer (Berlin, 2000).
6. H. Hertz, *Über die Berührung fester elastischer Körper*, J. f. reine u. angewandte Math. **92**, 156-171 (1882).
7. W. F. Busse and F. C. Starr, *Change of viscoelastic sphere to a torus by random impacts*, Am. J. Phys. **28**, 19-23 (1960).
8. N. V. Brilliantov, F. Spahn, J.-M. Hertzsch, and T. Pöschel, *Model for collisions in granular gases*, Phys. Rev. E. **53**, 5382-5392 (1996).
9. G. Kuwabara and K. Kono, *Restitution coefficient in a collision between two spheres*, Jpn. J. Appl. Phys. **26**, 1230-1233 (1987).
10. W. A. M. Morgado and I. Oppenheim, *Energy dissipation for quasielastic granular particle collisions*, Phys. Rev. E **55**, 1940-1945 (1997).
11. N. V. Brilliantov and T. Pöschel, *Contact of viscoelastic bodies*, in preparation (2000).
12. R. Ramírez, T. Pöschel, N. V. Brilliantov, and T. Schwager, *Coefficient of restitution of colliding viscoelastic spheres*, Phys. Rev. E **60**, 4465-4472 (1999).
13. F. G. Bridges, A. Hatzes, and D. N. C. Lin, *Structure, stability and evolution of Saturn's rings*, Nature **309**, 333-335 (1984).
14. R. M. Brach, *Rigid body collisions*, J. Appl. Mech. **56**, 133-138 (1989).
15. S. Wall, W. John, H. C. Wang, and S. L. Goren, *Measurement of kinetic energy loss for particles impacting surfaces*, Aerosol Sci. Tech. **12**, 926-946 (1990).

184   T. Pöschel *et al.*

16. W. Goldsmith, *Impact: The Theory and Physical Behaviour of Colliding Solids*, Edward Arnold (London, 1960).
17. P. F. Luckham, *The measurement of interparticle forces*, Powder Techn. **58**, 75-91 (1989).
18. T. Schwager and T. Pöschel, *Coefficient of restitution of viscous particles and cooling rate of granular gases*, Phys. Rev. E, **57**, 650-654 (1998).
19. H. Kuchling, *Physik*, VEB Fachbuchverlag Leipzig (Leipzig, 1989).
20. E. Falcon, C. Laroche, S. Fauve, and C. Coste, Eur. Phys. J. **B 3**, 45-57 (1998).
21. T. Pöschel, C. Salueña, and T. Schwager, *in preparation*.

Discrete and continuum modelling
of granular materials

H.-B. Mühlhaus,[1] H. Sakaguchi,[1] L. Moresi,[1] M. Fahey[2]

[1] CSIRO Exploration and Mining, PO Box 437, Nedlands, WA 6009, Australia
[2] Department of Civil and Resource Engineering, The University of Western
   Australia, Nedlands, WA 6009, Australia

Abstract. We give an outline of discrete element and continuum models for granular
flows involving large deformations, and arbitrary particle shapes. A symbolic solution
in the form of a finite element based, particle in cell formulation is also presented. The
theories and methods are illustrated by examples such as silo flow, simulated triaxial
tests, and trapdoor problems.

1   Introduction

Of the currently unresolved problems in the area of continuum theories for granular materials, three in particular stand out: the crossovers between fluid, solid
and gas like behaviour, the consideration of particle shapes, and a computational tool capable of modelling all of the complexities of granular flows. In our
paper we address the latter two problems. We apply two different numerical approaches — a Lagrangian Particle Finite Element Method [14] and the Discrete
Element Method — to problems of flow alignment in connection with oriented
(pencil shaped) particles and free surface flow and discharge of a granular material from a hopper. The granular material is assumed to be rigid-plastic, obeying
the Drucker-Prager yield criterion.

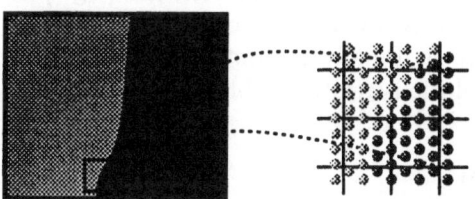

Fig. 1. Schematic of the LPM

In the Lagrangian Particle Method (LPM), one traces the position of a finite
number of material points, for instance, within a finite element mesh. Within
each time or loadingstep, the position of the particles is updated based on current position, nodal point velocities, shape functions and a suitable integration
procedure (Figure 1). Here we concentrate on slow or quasi-static flows. The

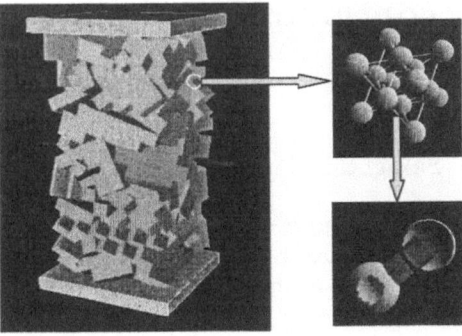

Fig. 2. Schematic of the DEM

particles can cross element boundaries and carry information pertinent to the problem, chemical and thermal history depending on the nature of the system.

In the Discrete Particle Model, solids are described as assemblies of bonded or unbonded spherical particles (Figure 2). The procedure for the establishment of the domains of each boundary value problem, and the procedure for the solution of the equations of motion are similar to the ones employed in the discrete element method originally developed by Cundall and Strack [4]. However, in the contact search algorithm, we take advantage of the fact that most of the contacts within the macro particles are permanent (see [12] for details).

The paper is structured as follows: We first give brief outlines of the governing equations for the continuous and the discrete models. In the continuum description we assume the framework of a Cosserat Continuum and point out the relationship to Director Continua (e.g. Chandrasekar [2]). The use of a director as the basic variable rather than the rotation tensor as in the Cosserat theory may be advantageous in the case of oriented particles such as those in Figure 3. We next discuss aspects of the constitutive behaviour of granular media with an emphasis on the characterisation of oriented (e.g. pencil or penny shaped particles). This is followed by an outline of the Lagrangian Particle Method as we have implemented it. Finally the potential of the methods is illustrated by means of example solutions related to block-caving and hopper flows.

2   Formulation

2.1   Continuum model

In the following we give a brief outline of the salient balance equations of the Cosserat continuum. We use rectangular, Cartesian coodinates x_i, $(i = 1, 2, 3)$, and v_i and ω_i^c are the translational and rotational velocities of a material point at position x_i and time t. In general ω_i^c is not equal to the spin

$$\omega_i = -\frac{1}{2}\epsilon_{ikl}W_{kl} \tag{1}$$

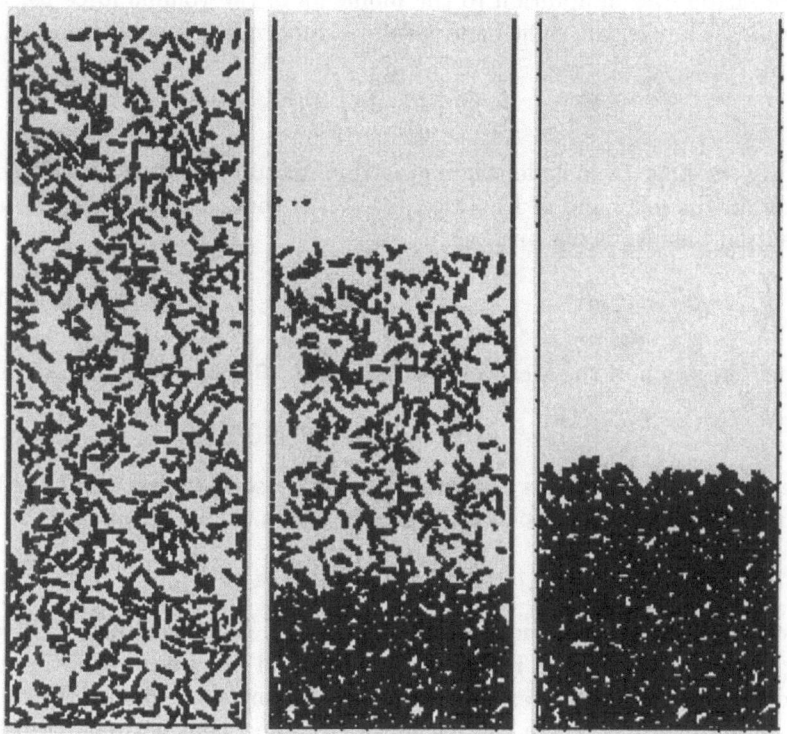

Fig. 3. Assembly of non-spherical particles. Each macro-particle is represented by an assembly of bonded, disc-shaped particles.

where

$$W_{kl} = \tfrac{1}{2}(v_{k,l} - v_{l,k})$$

of an infinitesimal element of the continuum. In Equation (1), ϵ_{ikl} designates the permutation symbol and $(\cdot)_{,k} = \partial(\cdot)/\partial x_k$ are partial derivatives. The mass balance equation

$$\frac{\partial \rho}{\partial t} + (\rho v_i)_{,i} = 0 \tag{2}$$

and the balance of linear momentum

$$\sigma_{ij,j} + \rho(b_i - \dot{v}_i) = 0 \tag{3}$$

of the standard continuum and the Cosserat continuum are identical. (ρ is the density.) In the Cosseart theory the angular momentum balance of the standard continuum

$$\frac{d}{dt} \int_V \rho \epsilon_{ikl} x_k v_l dV = \int_S \epsilon_{ikl} x_k t_l dS + \int_V \rho \epsilon_{ikl} x_k b_l dV \tag{4}$$

is extended in two repects. Firstly, a volume moment ρm_i^V and a surface moment m_i^S are introduced in addition to the moments of the volume force ρb_i and the traction t_i. That is, the right hand side of Equation (4) is supplemented by

$$\int_V \rho m_i^V dV + \int_S m_i^S dS \tag{5}$$

Secondly, we have to include the angular momentum ρd_i of the particles in the integral on the left hand of Equation (4). With these assumptions, the angular momentum balance is obtained as

$$\frac{d}{dt} \int_V \rho(\epsilon_{ikl} x_k v_l d_i) dV = \int_S (\epsilon_{ikl} x_k t_l + m_i^S) dS + \int_V \rho(\epsilon_{ikl} x_k b_l + m_i^V) dV \tag{6}$$

For the derivation of the local form of Equation (6), we use the relationships

$$t_i = \sigma_{ij} n_j \text{ and } m_i^S = m_{ij} n_j \tag{7}$$

where σ_{ij} and m_{ij} are stress and couple stress tensors, respectively. Application of Gauss' theorem and consideration of Equations (2) and (3) yields

$$m_{ij,j} - \epsilon_{ikl} \sigma_{kl} + \rho(m_i^V - \dot{d}_i) = 0 \tag{8}$$

We conclude that the Cauchy stress tensor σ_{kl} is nonsymmetric in general, though symmetry could be retained if m_i^V, $\dot{d}_i$ and $m_{ij,j}$ were to form an equlibrated system by themselves. Thus the non-symmetry might not be of importance in applications where the exchange of momentum between translational and rotational freedoms is weak.

For a more complete outline of the Cosserat theory as applied to granular material we refer to Mühlhaus & Hornby [9] [10]. The papers address among other things, statistical aspects, granular heat and the energetics of granular flow.

For finite element computations we require the weak form of the momentum and angular momentum balance equations. The weak form is obtained in the ususal way multiplying (3) and (8) with the test functions w_i and φ_i respectively. Integration over the volume and application of Gauss' theorem yields:

$$\int_V \sigma_{ij}(w_{i,j} \epsilon_{ijk} \varphi_k) m_{ij} \varphi_{i,j} dV = \int_V \rho \left((b_i - \dot{v}_i) w_i + (m_i^V - \dot{d}_i) \varphi_i \right) dV$$
$$+ \int_A (t_i w_i + m_i^S \varphi_i) dA \tag{9}$$

Equation(9) must hold for all kinematically admissible w_i, φ_i vanishing on those parts of the boundary where displacements and/or rotations are prescribed.

If the granulate is composed of oriented particles (rod or disc shaped) it is sometimes advantage to use the director $\boldsymbol{a}$, with $|\boldsymbol{a}| = 1$ instead of the rotation tensor $\boldsymbol{R}^c$, where $\boldsymbol{W}^c = \dot{\boldsymbol{R}}^c \boldsymbol{R}^T$ to describe the micro-kinematics of the granulate. In case of rods or discs, $\boldsymbol{a}$ signifies the averaged local rod axis on the average unit normal of the disc surfaces.

A theory based on v, grad v, a, grad a was proposed by Chandrasekar [2] (see also [5]) in the context of a continuum theory for liquid crystals. It should be noted that in director theories the torsional moment about the director axis usually vanishes. For 2D problems, however, Cosserat and director theories are fully equivalent. In section 3 we propose a continuum model for 2D flows of rod-shaped particles within the framework of a director theory.

2.2   Discrete element model

We consider two granules i and j having translational velocities v^i and v^j and angular velocities s^i and s^j. Continuum versions of the following formulations have been derived by Mühlhaus & Oka [8] and Mühlhaus & Hornby [9]. The relative velocity at the contact is

$$g^{ji} = v^{ji}d^{ji} + n^{ji} \times s^{ji}, \tag{10}$$

where $v^{ji} = v^j - v^i$, $s^{ji} = s^j + s^i$, n^{ji} is the unit vector along the connecting line between the mass centres of the granules and d^{ji} is the distance between the mass centres.

In the traditional manner, we calculate the normal component of the contact force vector, F_n^{ji}, at each instant in time, from the overlap, $d^{ji} - D^{ji}$, where d^{ji} is the current distance between particles j and i and D^{ji} is the sum of their radii. Note that non-spherical particles are represented as assemblies of spherical "micro-particles" using the macro-particle generator of Sakaguchi and Mühlhaus [12]. Accordingly, F_n^{ji} is always parallel to n^{ji}.

The tangent force, F_t^{ji} and the contact moments are much less significant than F_n^{ji} for the macroscopic properties of the granulate, but unfortunately much more complicated to calculate. We write the corrotational rates of F_t^{ji} and M^{ji} as

$$\overset{\triangledown}{F_t^{ji}} = \dot{F_t^{ji}} - W^{ji}F_t^{ji}, \tag{11}$$

$$\overset{\triangledown}{M^{ji}} = \dot{M^{ji}} - W^{ji}M^{ji}, \tag{12}$$

$$W_{\alpha\beta}^{ji} = \frac{1}{d^{ji}} \left(v_\alpha^{ji} n_\beta^{ji} - v_\beta^{ji} n_\alpha^{ji} \right) \tag{13}$$

Where $\dot{n}^{ji} = W^{ji}n^{ji}$ and $\dot{t}^{ji} = W^{ji}t^{ji}$; t^{ji}, with $n^{ji} \cdot t^{ji} = 0$, is the unit vector in the direction of F_t^{ji}

Elastic contacts are described in the simplest possible way by

$$\overset{\triangledown}{F_{el}^{ji}} = K_s \left(g^{ji} - (g^{ji} \cdot n^{ji})n^{ji} \right) + K_n(g^{ji} \cdot n^{ji})n^{ji}, \tag{14}$$

$$\overset{\triangledown}{M_{el}^{ji}} = d^{ji^2} K_r(s^j - s^i), \tag{15}$$

where K_s, K_n and K_r are constant tangent, normal and rotational stiffnesses. Note that the above formulation is physically meaningful only if the magnitude of the relative displacements is infinitesimal.

The total contact forces and moments are in general made up of an elastic and a viscous part, so that

$$\boldsymbol{F}^{ji} = \boldsymbol{F}^{ji}_{\text{el}} + \boldsymbol{F}^{ji}_{\text{v}}, \tag{16}$$

$$\boldsymbol{M}^{ji} = \boldsymbol{M}^{ji}_{\text{el}} + \boldsymbol{M}^{ji}_{\text{v}}. \tag{17}$$

In analogy to (14) and (15) we assume

$$\boldsymbol{F}^{ji}_{\text{v}} = \eta_s(\boldsymbol{g}^{ji} - (\boldsymbol{g}^{ji} \cdot \boldsymbol{n}^{ji})\boldsymbol{n}^{ji}) + \eta_n(\boldsymbol{g}^{ji} \cdot \boldsymbol{n}^{ji})\boldsymbol{n}^{ji}, \tag{18}$$

$$\boldsymbol{M}^{ji}_{\text{v}} = d^{ji^2}\eta_r(\boldsymbol{s}^j - \boldsymbol{s}^i), \tag{19}$$

where η_s, η_n and η_r are constant tangent, normal and rotational viscosities.

The contact viscosities may have a real physical significance *e.g.* for viscoelastic contacts or in fast granular flows. In the latter case, the viscosities are defined by the normal and tangent coefficients of restitution [9]. In connection with the solution of quasi-static problems by dynamic relaxation, the significance of the viscosities is purely numerical and the values of the viscosities are chosen exclusively in view of numerical efficiency.

In the 2D analyses intended here, $\boldsymbol{F}^{ji}, \boldsymbol{g}^{ji}, \boldsymbol{s}^j$ and $\boldsymbol{M}^{ji}$ have components $(F^{ji}_s, F^{ji}_n, 0), (g^{ji}_s, g^{ji}_n, 0), (0, 0, s^j)$ and $(0, 0, M^{ji})$ respectively.

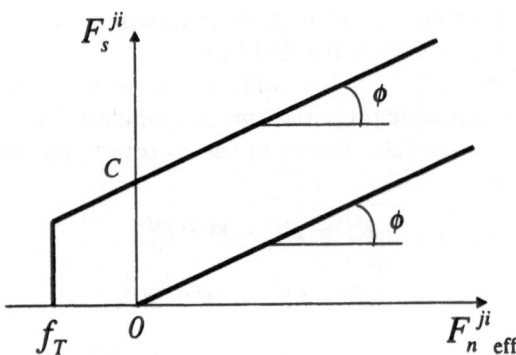

Fig. 4. Contact yield criterion

The contact yield criterion in the present model is illustrated in Figure 4, where C and f_T designate the cohesive force and tensile strength of the contact, ϕ is Coulomb's angle of friction and $F^{ji}_{n\text{eff}}$ is an effective normal force defined as

$$F^{ji}_{n\text{eff}} = F^{ji}_n + \frac{2\alpha^{ji}}{d^{ji}}|M^{ji}|, \tag{20}$$

where $\alpha^{ji} \leq 1$ is a dimensionless material parameter. Through the inclusion of the moment term in equation (20) we consider the influence of possible load eccentricities with respect to the mathematical centre of the contact (a similar criterion was used by van Mier *et al.*, [16] within the context of a lattice model for concrete fracture). Upon initial yield C and f_T are put equal to zero and

$$F_s^{ji} = \tan \phi F_{\text{neff}}^{ji}. \tag{21}$$

Separation of the particles i and j occurs if the effective normal force $F_{\text{neff}}^{ji} = 0$.

The configurations of the particle system are determined as usual by integration of the equations of motion

$$\sum_{\text{contacts}} F^{ji} + \gamma^i - m^i \dot{v}^i = 0 \tag{22}$$

and

$$\sum_{\text{contacts}} (\frac{D^i}{2} n^{ji} \times F^{ji} + M^{ji}) + \mu^i - \frac{1}{8} m^i D^{i^2} \dot{s}^i = 0, \tag{23}$$

where m^i and D^i are the mass and diameter of the i-th particle; γ^i and μ^i are external particle forces and particle moments respectively. Equations (22) and (23) are the discrete particle versions of (3) and (8) respectively. An algorithm for the growth arbitrarily shaped macro-particles is described in Sakaguchi and Mühlhaus(1997).

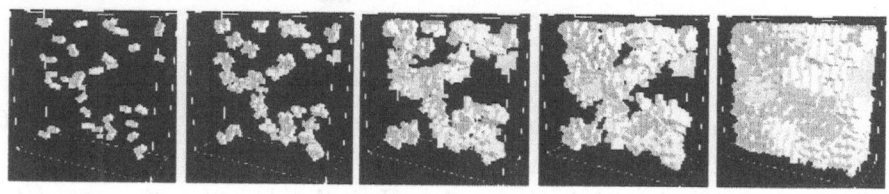

Fig. 5. Growth of three dimensional macro-particles. *see* Sakaguchi and Mühlhaus [12]

On the scale of grains and micro cracks the structure of rock appears highly disordered and complex in general. A major advantage of the present model is that such complexities can be easily dealt with. A simple algorithm for the generation of complex shaped macro-particles is described in Sakaguchi and Mühlhaus [12].

Initially, seed particles are generated in a given domain by a uniform random process in which coordinate pairs or triples are continuously determined until the total number of seed particles reaches a preset value. Once all the seed particles are positioned, macro-particles are grown according to a random process from seed particles to fill the space. The resulting particle shapes are usually irregular

(Figure 5) with a significant proportion of non-convex particle shapes. However, the proportion of convex particle shapes or in fact any particle shape can be controlled by using specific growth algorithms (see Sakaguchi and Mühlhaus [12]).

Assemblies of bonded macro-particles may represent, for instance, particular rock or ceramics fabrics, while assemblies of loose macro-particles are considered in fragmentation, comminution and flow problems involving angular granules.

3   Lagrangian particle method

Initially, particles are evenly distributed within each element so that the element mass can be portioned out equally to the individual particles: $m_p = \rho_0 V_e / N_{ep}$ where m_p is the mass of particle p, ρ_0 is the fluid density, V_e is the volume of element e, and N_{ep} is the number of particles in element e. The discrete density distribution is then represented by

$$\rho(x,t) = \sum_{p=1}^{N_e p} m_p \delta(x - x_p) \tag{24}$$

The shape functions for the underlying mesh are used to interpolate the nodal velocity and rotation fields to any location

$$v_i = \sum_{\alpha=1}^{N_{en}} N_\alpha v_i(\boldsymbol{x}_\alpha), \qquad \omega_i^c = \sum_{\alpha=1}^{N_{en}} N_\alpha \omega_i^c(\boldsymbol{x}_\alpha) \tag{25}$$

and also

$$w_i = \sum_{\alpha=1}^{N_{en}} N_\alpha w_i(\boldsymbol{x}_\alpha), \qquad \phi_i = \sum_{\alpha=1}^{N_{en}} N_\alpha \phi_i(\boldsymbol{x}_\alpha) \tag{26}$$

where N_{en} is the number of element nodes, N_α is the shape function associated with node α, and $\boldsymbol{x}_\alpha$ is the location of node α.

Specific stresses and moment stresses are defined as

$$s_{ij} = \frac{1}{\rho}\sigma_{ij}, \qquad \mu_{ij} = \frac{1}{\rho}m_{ij} \tag{27}$$

Subsituting (25),(27) into (9), and dropping acceleration terms as required by the infinite Prandtl number approximation,

$$\sum_{\text{elements}} \left(\int_{V_e} \rho \sum_{\alpha=1}^{N_{en}} (\mathbf{B}_\alpha \mathbf{W}_\alpha)^T s dV - \int_{V_e} \rho \sum_{\alpha=1}^{N_{en}} (N_\alpha \mathbf{W}_\alpha)^T g dV \right) -$$

$$\sum_{\text{elements}} \int_{\partial V_e \in \partial V} \sum_{\alpha=1}^{N_{en}} (N_\alpha \mathbf{W}_\alpha)^T s_s dA = 0 \tag{28}$$

The vectors and matrices appearing in (28) are defined as follows:

$$\mathbf{W}_\alpha^T = \{w_i, \phi_i\}_{x=x_\alpha} \qquad \mathbf{s}^T = \{s_{ij}, \mu_{ij}\} \tag{29}$$

$$\mathbf{g}^T = \{b_i, m_i^v\} \qquad \mathbf{s}_s^T = \{t_i, m_i^s\} \tag{30}$$

In 2D we have

$$\mathbf{B}_\alpha = \begin{bmatrix} N_{\alpha,1} & 0 & 0 \\ 0 & N_{\alpha,2} & 0 \\ N_{\alpha,2} & 0 & N_\alpha \\ 0 & N_{\alpha,1} & -N_\alpha \\ 0 & 0 & N_{\alpha,1} \\ 0 & 0 & N_{\alpha,2} \end{bmatrix} \tag{31}$$

and the $\mathbf{W}$ and $\mathbf{s}$ pseudo-vectors are defined as

$$\mathbf{W}_\alpha^T = \{w_1, w_2, \omega_3^c\}, \qquad \mathbf{s}^T = \{s_{11}, s_{22}, s_{12}, s_{21}, \mu_{31}, \mu_{32}\} \tag{32}$$

The definitions in (31) and (32) are restricted to 2D (x_1, x_2 plane); (28) – (30) are general.

This derivation is, so far, identical to that for a standard, grid-based finite element discretization. However, when (24) is inserted into (28), we obtain terms of the form

$$\mathbf{f}_\alpha^e = \int_{V_e} \left(\sum_{p=1}^{N_{pe}} m_p \delta(x - x_p) \mathbf{B}_\alpha^T \right) dV = \sum_{p=1}^{N_{pe}} m_p \mathbf{B}_\alpha^T \Big|_{x=x_p} \tag{33}$$

where $\mathbf{f}_\alpha^e$ is the element force vector referred to the grid but now computed using the particle locations and properties.

3.1 Lagrangian particles

In a particle-in-cell method, the grid remains fixed whereas particles move through the mesh during computations. The use of an Eulerian mesh allows the identification of the material time derivative with the time derivative itself:

$$\frac{D}{Dt} = \frac{\partial}{\partial t} \tag{34}$$

Particle velocities are interpolated from nodal velocities and then particle position is updated using a suitable integration scheme such as:

$$x_p^{t\Delta t} = x_p^t + \sum_{node} v_n \Delta t N_n(x_p) \tag{35}$$

where v is the nodal velocity and N are the shape functions associated with the nodes of the element in which the particle currently resides. In practice, a higher order scheme such as Runge-Kutta gives a more accurate result. Particle updates can be done in a predictor-corrector fashion, initially updating particle locations to obtain velocity solutions and then repeatedly correcting the final locations to obtain a converged velocity. We have found that the improvements in accuracy from iterating on the particle locations are generally too small to justify the increase in complexity, and, for non-linear rheological laws, the iterative scheme may become unstable. Our preference is therefore to reduce the timestep whenever higher accuracy in the time integration is required.

3.2   *Numerical integration*

At the heart of the finite element method is the concept that the weak form of the equations can be split up into a number of sub-integrals over individual elements, and that these integrals can be computed approximately using a suitable quadrature scheme.

$$\int_{\Omega^e} \psi d\Omega^e \approx \sum_{p=1}^{P} w_p \psi(x_p) \tag{36}$$

where Ω^e is an element volume, ψ is a field quantity which is evaluated at a number of sample points, x_p, and w_p are weights for each integration point.

In standard FEM, the locations of the quadrature points, x_p and their weights, are chosen to optimize the integration accuracy for a given set of interpolation functions. The criterion for choosing the quadrature scheme is usually computational efficiency: the minimum number of locations required to achieve exact integration of a specific degree polynomial.

We assume the sample points coincide with the particles notionally attached to the fluid which therefore move with respect to the mesh. The locations of the quadrature points are, as a consequence, given for each element, and it is necessary to vary the weights in order to obtain the correct integral for a given element. The procedure is similar to the determination of weights for any quadrature rule: deriving a set of constraints based on the requirement that polynomials of a certain order have to be integrated exactly, and then equating coefficients to obtain the set of w_p values.

If we consider the one dimensional case in which ψ in (36) is a polynomial, and the integral is over the range -1 to 1(a typical "master element").

$$\psi(x) = \alpha_0 \alpha_1 x \alpha_2 x^2 \cdots \alpha_n x^n \tag{37}$$

We integrate (37) algebraically over the domain, and equate coefficients with the quadrature expansion of the integral from (36) to obtain n constraints on the set of w_p values:

$$\sum_{p=1}^{n_{ep}} w_p = 2 \qquad \text{(constant terms)} \tag{38}$$

$$\sum_{p=1}^{n_{ep}} w_p x_p = 0 \qquad \text{(linear terms)} \tag{39}$$

$$\sum_{p=1}^{n_{ep}} w_p x_p^2 = \frac{2}{3} \qquad \text{(quadratic terms)} \qquad (40)$$

$$\sum_{p=1}^{n_{ep}} w_p x_p^3 = 0 \qquad \text{(cubic terms)} \qquad (41)$$

and so on.

We can envisage choosing a suitable number of constraints, and performing an inversion for the w_p values, but in practice this is very time-consuming, and may produce very poor results if the particles are unevenly distributed within an element, including the generation of negative w_p values. If we wish to associate the value of w_p attributed to a particle with the representative mass or volume of fluid which it occupies, then w_p should be positive.

Constant particle weights are not guaranteed to satisfy even the lowest order constraint once the particle's positions evolve to a general configuration (the numbers of particles per element may vary), however, the accuracy of the integration scheme is relatively good for simulations where material strain is not extreme (see Ref [7]). Using this observation, we store the values of w_p determined from the local volume occupied by the particle in the initial configuration. These reference values are adjusted to best fit the constraints up to a certain degree, and subject to the further constraint that no tracer should have a negative weight. For viscous flow using bilinear elements we expect optimal convergence rates, in the limit of an infinitely fine mesh, if the constant and the linear constraint terms are satisfied ([6]).

3.3 Element matrices and particle properties

In a standard finite element formulation, the stiffness matrix for an element, k^e is built up in a segregated form, often written as:

$$k_{ab}^e = \int_{\Omega^e} B_a^T D B_b d\Omega^e \qquad (42)$$

where B_a is a matrix comprising shape function derivatives (associated with node a) obtained from the constitutive relationship, and D is a matrix of material properties. For our formulation, the material property matrix is considered to be an attribute of an individual particle together with its current state and history and not a property of the mesh. Therefore, in the context of the numerical integration scheme discussed above,

$$k_{ab}^e = \sum_{p=1}^{n_{ep}} w_p B_a^T(x_p) D_p B_b(x_p) \qquad (43)$$

3.4 Particle splitting

Fluid flow close to a stagnation point produces an elongation in one direction and a corresponding shortening in the perpendicular direction. This has the effect of

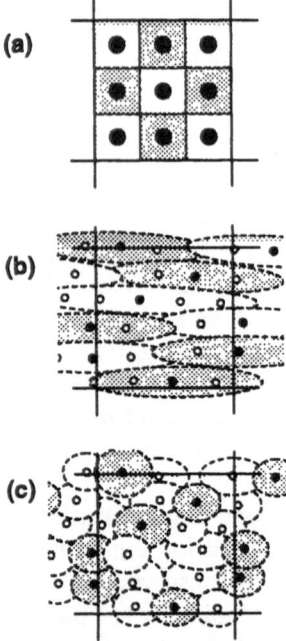

Fig. 6. Integration schemes become more complicated when large material strains produce elongated "local volumes" for particles. (a) Initial configuration, (b) flow near a stagnation point, (c) splitting particles as they distort and remapping the local volumes

distorting the original volume local to each tracer into a narrow filament (Figure 6 a and b). This may mean that the fluid initially associated with a particle lies in several different elements, while the integration scheme simply lumps the entire amount into the element containing the tracer itself. In the worst case this may leave some elements entirely empty of tracers and others considerably over represented.

The remedy is to ensure that the volume of fluid associated with a particle never becomes too distorted. We keep track of a local measure of strain associated with each particle and use this to generate new particles nearby which occupy the extremities of the distorted local volume. This is illustrated in Figure 6b where the heavily shaded particles sitting at the centroid of the salami-shaped local regions represent the original occupants of the volumes from Figure 6a. The material within the element is poorly represented by the particles which actually contribute to the element integrals. The lightly shaded particles are later additions which aim to correct this problem. Local volumes corresponding to the new particle distributions are indicated in Figure 6c.

When splitting particles, we attribute the same history variables to the copies as were held on the original particle. To ensure that this approximation is a reasonable one the particle splitting should occur when the distortion is relatively small.

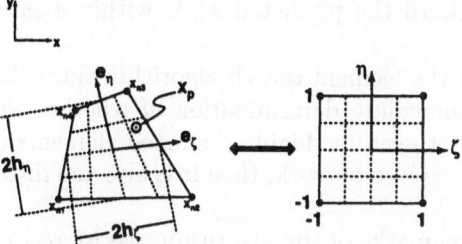

Fig. 7. Coordinate systems in global mesh, distorted element, and master element reference frames

3.5 *Element inverse mapping*

It is usual to change variables in the element integrals such as (42) to a regular master element. This greatly simplifies the computation, but in our case there is an additional step required before we can map to a master element. The particle positions are known in the global coordinate system and must first be mapped into the element local coordinate system. [17] found an algebraic mapping for bilinear quadrilateral elements, but a more general approach is required for higher order elements and for 3D.

The notation is given in Figure 7 for the 2D case — the extension to three dimensions is straightforward. e_ξ and e_η are unit vectors in the 'natural directions' of the distorted element which map to the ξ and η axes respectively in the master element. h_ξ and h_η are characteristic dimensions of the element in the appropriate directions. We wish to map the coordinates, x_p of the particle p to the coordinates in the master element, ξ_p. The procedure is iterative: we first guess the initial value of ξ_p and use this to predict the global coordinates x_p^0 as follows

$$\xi_p = (0,0) \tag{44}$$

$$x_p^0 = \left(\sum_{n=1}^{n_{en}} N_n(\xi_p)x_n, \sum_{n=1}^{n_{en}} N_n(\xi_p)y_n \right) \tag{45}$$

where n_{en} is the number of nodes in the element, and x_n are their coordinates. We compute ξ_p through a number of corrector steps:

$$\xi_p \leftarrow \xi_p \beta \left(e_{\xi x}x_p^i e_{\xi y}x_p^i \right) / h_\xi \tag{46}$$

$$\eta_p \leftarrow \eta_p \beta \left(e_{\eta x}x_p^i e_{\eta y}x_p^i \right) / h_\eta \tag{47}$$

$$x_p^i = \left(\sum_{n=1}^{n_{en}} N_n(\xi_p)x_n, \sum_{n=1}^{n_{en}} N_n(\xi_p)y_n \right) \tag{48}$$

i is the iteration index, β is a relaxation parameter which we chose to be 1.0 for the first iteration and 0.9 for subsequent iterations. When e_ξ and e_η are orthogonal, the iteration completes in one step, otherwise it is necessary to repeat the correction step until the predicted x_p^i is within a satisfactory tolerance of the known value.

We use ξ within the element search algorithm since the master element coordinates give an immediate determination of whether the particle lies inside or outside the element even for highly distorted elements (if the element is too distorted for this procedure to work, then it is also too distorted to use to create the element matrices)

For detailed benchmarks of the algorithm, see Moresi *et al* [7].

4   Examples

4.1   DEM model simulating a triaxial compression test

The discrete element method offers the opportunity to study the behaviour of virtual materials, virtual granulates in particular, under the perfectly controlled conditions of a numerical experiment (*e.g.* [15],[1]). Adjustment of the properties of the virtual materials to the properties of a real geo-material is performed conveniently by comparing real and simulated triaxial compression tests. For this purpose, we have established a virtual triaxial compression testing procedure.

The sample specimen can be generated in 3D either by a radius growth strategy for spherical particles or, as described earlier, by generating macro-particles for irregular shaped particles. Both methods can produce isotropic packing with controlled particle size distribution and initial void space.

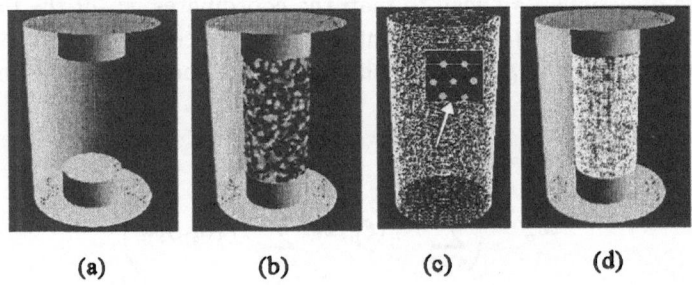

(a)   (b)   (c)   (d)

Fig. 8. Setup of virtual triaxial compression test. (a) Triaxial compression cell and platen, (b) Specimen under consolidation at isotropic confining pressure, (c) Membrane particles, (d) Specimen covered by membrane.

The virtual specimen is initially isotropically consolidated at a specified, constant confining pressure. During consolidation the radial and vertical displacements at the cylinder and top and bottom surfaces respectively of the specimen are kept uniform by servo control mechanism (Figure 8(a) and (b)). During the actual compression test the axial displacement (Figure 9) at the top surface is increased at constant confining pressure. During this phase the displacement control at the cylinder surfaces is replaced by membrane boundary elements which support the constant confining pressure during monotonically increasing axial compression (Figure 9).

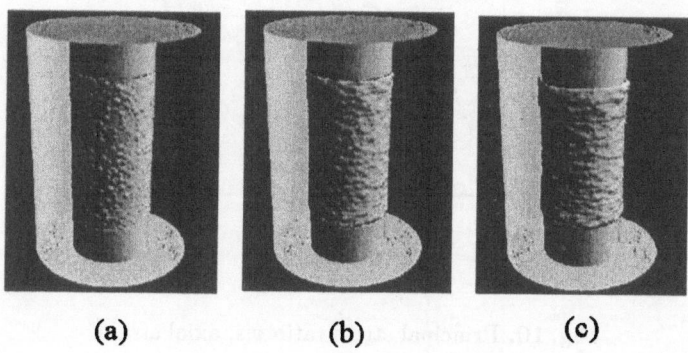

Fig. 9. Surface deformation represented by membrane elements. (a) Initial,(b) 1% axial strain, (c) 3% axial strain

The membrane elements (Figure 8(c)) are defined by particles which are initially positioned loosely around the cylinder surface and are arranged in a lattice of equal sided triangles. The triangles are connected by linear elastic springs. Contact during loading between the membrane particles and the specimen particles are detected and considered by standard discrete element contact detection techniques. The results presented here are snapshots of work in progress. The virtual servo control, the membrane elements, all work very well, however during the simulation the stiffness of the membrane was vastly overestimated (by about 10 times) so that the results correspond more what one would expect in an oedometer test (Figure 10) rather than a triaxial compression test. A re-run of these calculations will be presented shortly. Nevertheless the overall deformation pattern looks remarkably realistic. Note the wrinkles appearing with progressive deformation on the membrane surface.

4.2   DEM model of granular flow

Let us consider a rectangular domain, which has a trap door located at the centre of the bottom side, filled with randomly shaped, cohesionless particles (Figure 11(A)) generated by the macro-particles generator which briefly described above. This example could be a simplified model for mining of an ore body in block caving. Upon opening of a trap door centered at the lower side of the domain, granular flow set in and stops after a stable arch has formed above the trapdoor. Figure 11(B) shows the final equilibrium configuration. In the following example we demonstrate that the stability of the arch is determined by the ratio of the strength of the asperities of the macro-particles to the weight of the elemental

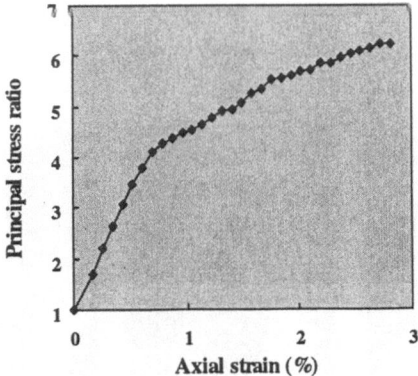

Fig. 10. Principal stress ratio *v.s.* axial strain.

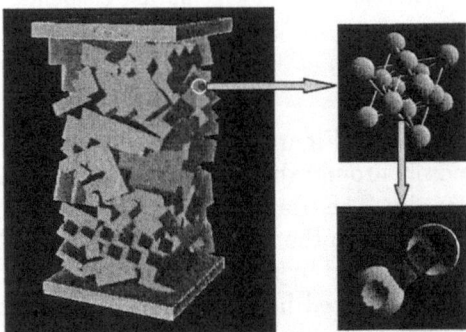

Fig. 11. Block caving simulation. (A) Initial state, (B) Stable equilibrium configuration, (C) Flow pattern for fracturing particles.

particles. The initial particle configuration and the strength and elasticity properties are the same as in the above example. However the weight of the elemental particles is ten times higher. In this case the intensity of the asperity and particle fracture process is high enough to accommodate sustained granular flow. A snap shot of the discharge process is shown in Figure 11(C).

4.3   LPM large deformation benchmark

We model driven convection in an unit square box (Fig. 12). The default boundary condition is free-slip everywhere except on the top between $x = 0.4$ and $x = 0.6$ where a horizontal velocity is applied towards the right. We track the motion of the material along the vertical mid-line through time (Fig. 12). The velocity solution is uniquely determined by the boundary conditions and is independent of time.

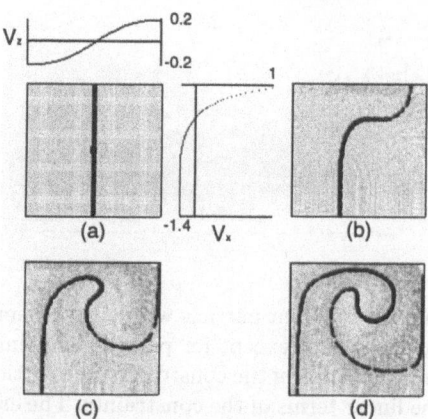

Fig. 12. Geometry of the convection model at initial timestep (a) with velocity profiles, timestep = 150 (b), 500 (c) and 800 (d).

The flow carries particles around the box through time, however, which will stir any material interfaces, alter the distribution of particles within elements, and consequently may disrupt the integration scheme. Intense deformation in the corners results in a need to introduce new particles resulting in an increasing number of particles as a function of time.

Errors were computed by comparison with a very fine mesh solution. This solution is obtained at the first timestep for a regular square mesh of 192×192 elements using a Gaussian integration scheme of 4 integration points per element. Our study model is a 48×48 elements with either Gaussian integration points or evenly distributed and weighted points. The error due to the use of a coarser mesh is about 2.8%.

Recomputing the particle weights is a computationaly costly step not required in the Material Point Method. Problems with extreme deformation do, however, require some recomputation. In Figure 13, we show the accuracy of integration for the driven convection problem when the weights are fixed, when they satisfy the constant terms of the constraints, and when the constant and linear terms are satisfied. We ran each model for several hundred timesteps with initially 16 particles per element and evaluated the accuracy of the results as well as the fluctuation of the solution. Fluctuation of the error was computed using a moving window of 15 point width to give the average error $(Ave(t))$ and its variance $(Var(t))$. Then we plotted for each case $Ave(t) + Var(t)$ and $Ave(t) - Var(t)$.

When we keep the particle weight constant so that the particle mass is conserved we observe relatively large errors with large fluctuations which becomes worse for very large deformation (Fig. 13.(a)) despite an ever increasing number of particles. Modifying the weights stabilizes the error through time with a mean value close to that of the Gaussian integration scheme. The variation in the error

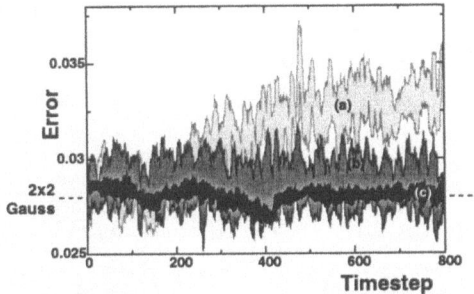

Fig. 13. Error versus time for different particle weighting schemes: (a) particle weights are conserved exactly through time (except for particle splitting), (b) particle weights are adjusted to fit the constant terms of the constraints, (c) particle weights are adjusted to fit the constant and the linear terms of the constraints. The error obtained in a single step from Gaussian quadrature is indicated by the dotted lines.

is considerably smaller when the constraint equation (38) for both constant and linear terms is applied (Fig. 13(b,c)).

We obtain reasonable results using four particles (initially) per element and adjusting their weights as they are moved with the flow. Sixteen particles are needed before we have the same number of degrees of freedom (and integration accuracy) as a four point Gauss scheme in 2D, allowing for the fact that some particles will be in the wrong position within an element to contribute to the integrals.

4.4   LPM model of discharging silo

We model a silo using a logically rectangular mesh of bilinear-velocity, constant-pressure quadrilateral elements. The silo shape is obtained by distorting the lower portion of the mesh. We assume symmetry about the mid-plane to save computational effort, and apply a reflecting boundary condition there. The sidewalls are given a no-slip boundary condition, and the bottom and top are unconstrained. Particles are initially distributed evenly within elements such that the silo is three-quarters full of material. A rectangular grid of dyed particles is used to measure the deformation during the run. The material is assumed to be rigid-plastic, obeying a Drucker-Prager yield criterion [11] Figure 14 shows a sample run in which the material had a friction coefficient of 0.7. The initial filling level is shown by a horizontal line, the light grey grid lines are the strain-markers, and the intensity of the shading indicates the strain rate. Interacting shear bands develop after very small deformation has occurred. As the deformation increases, the surface begins to slump, in particular above the opening. The longer-term behaviour shown in the inset is taken from a lower resolution simulation.

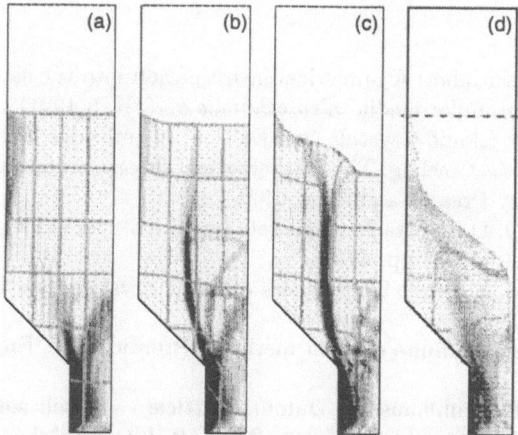

Fig. 14. Silo discharge model using the LPM (A) Initial state with grid of deformation markers, (B,C) discharge showing bands of high strain rate evolving with the free surface. (D) Result from lower resolution (fewer elements) simulation: Near complete discharge.

5 Concluding remarks

Classical field theories often describe the relations and interactions between macroscopic observables that arise from some sort of averaging over microscopic complexions of the system. It is also the case that many systems which are quite different at the microscopic scale possess essentially the same description at the macroscopic level leading to a relatively small number of field theories that model a wide variety of materials. By contrast, generalized continuum theories and discrete element methods of the type we have described provide a richer physical framework to describe and distinguish between different material microstructures. Starting from the virtual power expression for dense assemblies of spherical granules, expressions for the average stress tensor, momentum and angular momentum balances and corresponding variational boundary conditions can be derived. Turning to computational aspects, we derived a symbolic solution of a general, mixed boundary value problem for Cosserat materials within the framework of a finite element based particle-in-cell method. The potential of the theory and the particle-in-cell model is demonstrated by means of two dimensional silo flow and trapdoor problems.

Acknowledgement

The financial support of the second author by the Centre for Offshore Foundation System (COFS) is greatfully acknowledged.

204   H.-B. Mühlhaus *et al.*

References

1. J. P. Bardet, J. Proubet: A numerical investigation into the structure of persistent shear bands in granular media. *Géotechnique* 41 (4)L 5(1991) pp. 99–613
2. S. Chandrasekar: Liquid Crystals, 2nd Edition, (Cambridge University Press, 1992)
3. S. Chapman, T. G. Cowling: The mathematical theory of non-uniform gases, (Cambridge University Press, Cambridge, 1995)
4. P. A. Cundall, O. D. L. Strack: A discrete numerical model for granular assemblies. *Géotechnique.* 29 (1979) pp. 47–65.
5. P. G. de Gennes, J. Prost: The physics of liquid crystals, 2nd Edition, (Clarendon Press, Oxford, 1995)
6. T. J. R. Hughes: The finite element method, (Prentice Hall, Englewood-Cliffs, New Jersey, 1987)
7. L. Moresi, H.-B. Mühlhaus, F. Dufour: Particle - in -cell solutions for creeping viscous flows with internal interfaces, *Proc. 5th Int. Workshop on Bifurcation and Localisation in Geomechanics*, Perth, Australia (Balkema, 2000) in press.
8. H.-B. Mühlhaus, F. Oka: Dispersion and wave propogation in discrete and continuous models for granular materials. *Int. J. Solids Structures*, 33 (1996) pp. 2841–2858
9. H.-B. Mühlhaus, P. Hornby: On the reality of antisymmetric stresses in fast granular flows. *Proc. IUTAM Conf. on Granular and Porous Media*, eds. Fleck and Cocks, (Kluwer Academic Publishers, 1997) pp 299–311.
10. H.-B. Mühlhaus, P. Hornby: Polar continua and the micromechanics of granular materials. Chapter 2.2 *Mechanics of granular materials, an introduction*, eds. M. Oda and K. Iwashita, (A. A. Balkema: Rotterdam, 1999) pp. 86–106
11. H.-B. Mühlhaus, L. Moresi, R. Freij-Ayoub: Lagrangian particle modelling of granular flows, 2nd Australian Congress on Applied Mechanics, Canberra, 10-12, (February, 1999) pp. 482-487
12. H. Sakaguchi, H.-B. Mühlhaus: Mesh free modelling of failure and localization in brittle materials. In: *Deformation and progressive failure in geomechanics, Proc. IS-Nagoya '97* eds. Asaoka, A., Adachi, T. and Oka, F. (Pergamon, 1997) pp.15–21
13. S. B. Savage: Analysis of slow, high concentration flows of granular materials. *J. Fluid Mech.* 377 (1998) pp. 1–26
14. D. Sulsky, Z. Chen, H. L. Schreyer: A particle method for history dependent materials, *Comput. Methods Appl. Mech. Engrg.*, 118 (1994) pp. 179–196
15. C. Thornton: Microscopic approach contributions to constitutive modelling. In: *Constitutive modelling of granular materials* ed. D. Kolymbas (Springer-Verlag, 2000) pp. 193–208
16. J. G. M. van Mier, E. Schlangen, A. Vervuurt: Lattice type fracture models for concrete. In : *Continuum Models for Materials with Microstructure* ed. H.B.Mühlhaus (Wiley, 1995) pp.341–377
17. C. Zhao, B. E. Hobbs, H.-B. Mühlhaus, & A. Ord, A consistent point-searching algorithm for solution interpolation in unstructured meshes consisting of 4-node bilinear quadrilateral elements, Int J. Numer. Math. Engng, 45, 1509–1526. (1999)

Difficulties and limitation of statistical homogenization in granular materials

B. Cambou, Ph. Dubujet

Ecole Centrale de Lyon
Laboratoire de Tribologie et Dynamique des Systèmes
69131 Ecully Cedex, France

Abstract. A global constitutive model for granular material can be obtained from a change of scale using a statistical homogenization procedure. This kind of approach uses various operators for the change of scale. This paper presents the difficulties encountered to define all these operators and analyzes the validity of the relations proposed in the literature. These different analyses are based on the results obtained from a 2 D numerical simulation using the DEM.

1 Definition of statistical homogenization in granular materials

Homogenization is the term generally used when defining a constitutive law at a global level (for instance at the level of a mesh element in a finite element method analysis) from the local behaviour defined at a smaller scale (for example at the grain level in granular materials) and from knowledge of the fabric of the material in question [11].

This approach can be applied to two kinds of material: periodic materials in which a representative cell can be defined and described, and non-periodic materials in which the different phases are randomly distributed. Granular materials are generally considered to belong to the second kind of material. In this case the material cannot be described accurately at the micro level. Usually only a statistical description is available, in which case the procedure which can be used is called : "statistical homogenization". Usually the physical information available is rather poor, so the statistical description of the phases is not complete. This means that only bounds or estimates for the characteristics of such materials, can be derived [4] [10].

A statistical homogenization process applied to granular materials includes (Fig.1):

- a localisation operator,
- a local constitutive law,
- an averaging operator.

The global constitutive model can be obtained using two different homogenization routes using, for the first, a static localisation operator and a kinematic averaging operator, and for the second a kinematic localisation operator and a

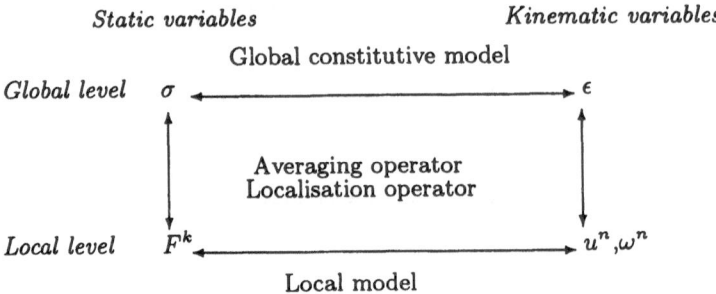

Fig. 1. Statistic homogenization procedure

static averaging process. The two routes usually produce different estimates of the real behaviour of the considered material. We will analyse hereunder the difficulties encountered to develop these kinds of approach.

2 Static averaging operator

This operator is well established [12] [6] and can be written as

$$\sigma_{ij} = \frac{1}{V} \sum_{k=1}^{N} F_i^k l_j^k \tag{1}$$

N : number of contacts in volume V.

The local variable proposed by Cambou et al [4] can also be considered

$$f_i(n_k) = P(n_k) F_i(n_k) N_o l_o \frac{\pi(d+1)}{3} \tag{2}$$

d : dimension of the space (2 or 3)
N_0 : number of contacts per unit volume
l_0 : mean value of the distance between particles in contact.

With this variable relation (1) can be written as

$$\sigma_{ij} = \frac{3}{4\pi} \int_{\Omega} f_i(n_k) n_j d\Omega \tag{3}$$

$d\Omega$: element of solid angle, oriented in direction n_i and Ω : all possible orientations.

3   Static localisation operator

3.1   General formulation

The definition of a realistic localisation operator is one of the main difficulties of a homogenization process. Two relationships proposed in the literature will be analysed. The static local variable which is considered in these approaches is the mean value of contact forces for a given contact orientation $\overline{F_i}(n_k)$. The grains in contact are considered to be cylindrical (2D) or spherical (3D) . The expression proposed by Chang [5] starts from

$$\overline{F_i}(n_k) = A\sigma_{ij}^* n_j \tag{4}$$

A : constant and σ_{ij}^* modified stress.

The expression of $\overline{F_i}(n_k)$ is introduced into the static averaging operator which leads to the definition of A and σ_{ij}^* and then to the following relationship

$$\overline{F_i}(n_k) = \sigma_{im}G_{jm}n_j l_o / N_0 \tag{5}$$

with $G_{ij}H_{jp} = \delta_{ip}$ and $H_{jp} = \langle n_j n_p \rangle = \int_\Omega P(n_k) n_j n_p d\Omega$
N_0 : number of contacts per unit volume

In this expression variable G_{ij} allows the fabric anisotropy to be considered, H_{ij} is the contact fabric tensor.

The expression proposed by Cambou et al [4] is based on theorem representation. $f_i(n_k)$ is defined in a more general expression depending isotropically on n_k, e_{ij}, and linearly on σ_{ij}. Moreover $f_i(n_k)$ has to satisfy the static averaging operator (3). Finally the expression of $f_i(n_k)$ depends on the 2 internal constants (μ and e_{ij})

$$f_i(n_k) = \mu\sigma_{ij}n_j + \frac{1-\mu}{2}\left[4n_u\sigma_{uv}n_v - \sigma_{kk}\right]n_i + \sigma_{kk}\left[n_u e_{uv}n_v n_i - \frac{1}{2}e_{ik}n_k\right] \tag{6}$$

e_{ij} is a deviatoric tensor which allows the fabric anisotropy of the material to be taken into account.

3.2   Analysis of the physical meanings
of internal parameters μ and e_{ij}

The internal parameter μ characterizes the orientation of the mean contact forces for a given orientation n_k: for $\mu = 0$, all the mean contact forces are normal, if $\mu = 1$ the mean contact forces take on the same orientation as the stress in a continuum ($\sigma_{ij}n_j$). As parameter μ is a scalar it can be taken to be independant of the fabric of the material. The influence of the geometrical fabric which evolves during the loading process is taken into account by tensor e_{ij}. It seems reasonable to suppose that e_{ij} is directly linked to H_{ij}, the contact fabric tensor. These parameters can be identified from a numerical simulation using the values of

σ_{ij}^N, s_{ij}^N which are the parts of σ_{ij}, and s_{ij}, linked to normal contact forces and given by

$$\sigma_{ij}^N = \frac{1}{V} \sum F_i^{kN} l_j^k \quad s_{ij}^N = \sigma_{ij}^N - \frac{1}{3}\delta_{ij}\sigma_{kk}^N \quad s_{ij}^T = \sigma_{ij} - \sigma_{ij}^N \tag{7}$$

At the beginning of a deviatoric loading ($\left|s_{ij}^T/\sigma_{kk}\right| \ll 1$), the internal state can be assumed to be isotropic ($e_{ij} = 0$), so it can be easily demonstrated that

$$\mu = 2\frac{s_{ij}^T}{s_{ij}} \tag{8}$$

For greater values of s_{ij}, e_{ij} ($\neq 0$) can be defined by

$$e_{ij} = 2\left(s_{ij}^N - s_{ij}^T - (1-\mu)\, s_{ij}\right)/\sigma_{kk} \tag{9}$$

Numerical simulations have been conducted to define these parameters μ and e_{ij}. The local contact behaviour is modelled in the analyzed simulations by a linear elastic model with a threshold defined by the local friction angle Ψ

$$F^N = k^N c^N \quad F^T = k^T c^T \quad \alpha = k^T/k^N \quad F^T \leq F^N \tan(\Psi) \tag{10}$$

The results of this numerical simulation show clearly (Fig.2) that parameter μ depends on the local contact law, in particular on Ψ. It is also clear from figure 3 that parameter e_{ij} depends only on H_{ij} and not on the local contact law. These results confirm the validity of the considered hypotheses on the two internal variables μ and e_{ij}. μ characterizes the local contact law and e_{ij} characterizes the fabric anisotropy.

3.3   Analysis of the capacity of different localisation operators from a numerical simulation

A two dimensional numerical simulation considering cylindrical particles and using a DEM code [7] is analysed.

In this simulation the distribution of local forces is analysed. The two projections in the normal direction of contact (f^N) and in the tangential direction of contact (f^T) of variables $f_i(n_k)$ defined by relation (2) are considered. These distributions have been drawn for the level of stress equal to

$$q/p = (\sigma_1 - \sigma_2)/(\sigma_1 + \sigma_2) = 0.69 \tag{11}$$

It is clear in figure 4 that formulation given by (6) is much more acceptable than the formulation given by (5). This result shows that a simple formulation for the static localisation considering only the anisotropy linked to the contact distribution cannot give a realistic local description. The description using two internal variables (μ and e_{ij}) allowing not only the internal fabric but also the contact conditions to be taken into account, is more complex but seems to improve greatly the local description of contact forces.

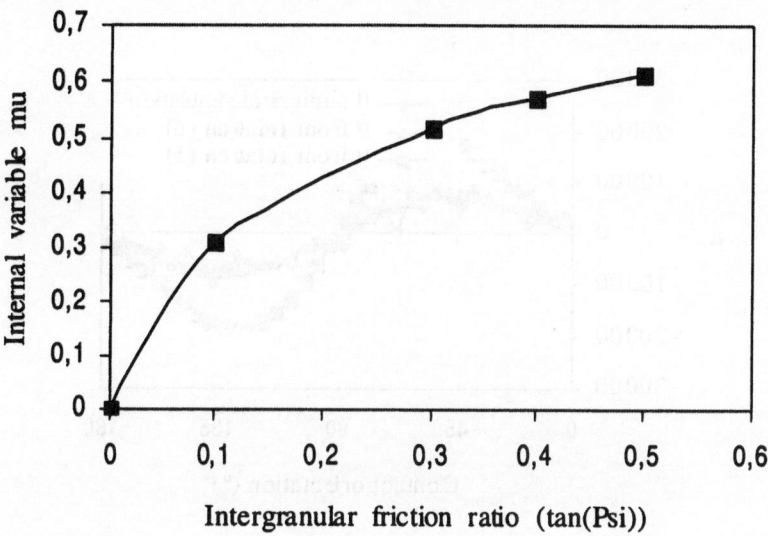

Fig. 2. Relation between internal parameter μ and the local intergranular friction ratio (from numerical simulations).

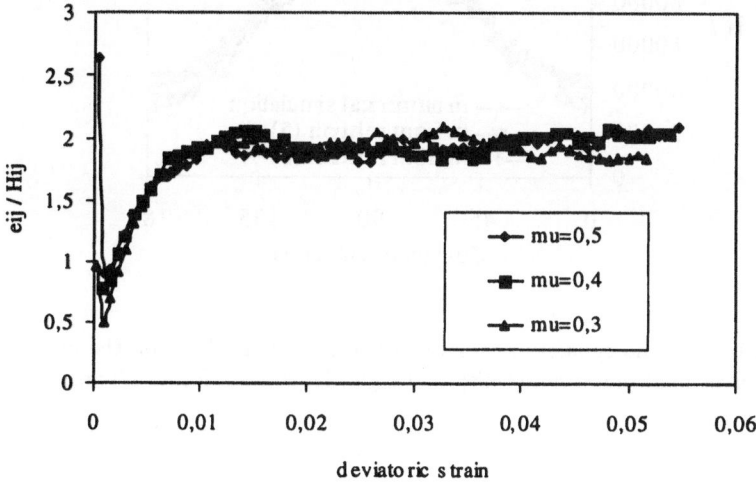

Fig. 3. Relation between internal parameter e_{ij} and the contact fabric tensor H_{ij} (from numerical simulations).

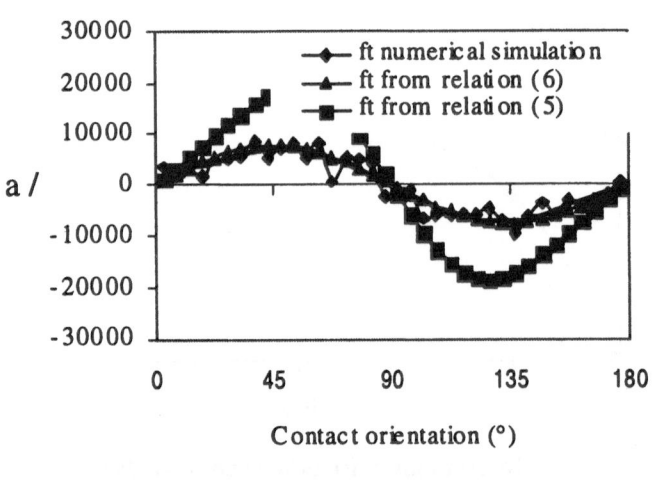

a /

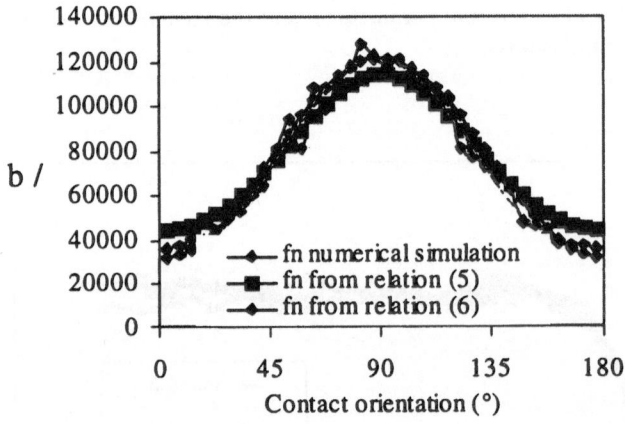

b /

Fig. 4. Analysis of the capacity of relations (5) and (6) to describe the distributions of contact forces in granular materials; a/: tangential contact force, b/: normal contact force

4   Kinematic averaging operator

Different kinematic averaging operators have been proposed in the literature. The operator derived from the "best fit method" [10] seems to be very simple. This operator can be defined by considering that the displacement field given by the displacement gradient tensor defined at the macro level a_{ij}, is the best estimate of the real local displacement measured at the centre of gravity of the particles (u_i^m). Then a_{ij} is considered to be given by the minimization of the square deviation between the two fields.

$$s = \sum_{k=1}^{N_0} E_i^k E_i^k = \sum_{k=1}^{N_0} \left[\delta a_{ij} l_j^k - (\delta u_i^m - \delta u_i^n) \right] \left[\delta a_{ij} l_j^k - (\delta u_i^m - \delta u_i^n) \right] \quad (12)$$

k characterizes the contact between particles n and m. N_0 is the number of contacts by unit volume.

Writing $\partial s / \partial a_{pq}$ leads to the following result for cylindrical or spherical particles

$$\delta \varepsilon_{ij} = \frac{1}{2N_0} \left\{ \left[\sum_{k=1}^{N_0} \frac{\delta l_i^k}{l^k} n_p^k \right] G_{pj} + \left[\sum_{k=1}^{N_0} \frac{\delta l_j^k}{l^k} n_p^k \right] G_{pi} \right\} \quad (13)$$

with $G_{ij} H_{jp} = \delta_{ip}$ and $H_{ij} = \langle n_i n_j \rangle$.

In the case of isotropic material we obtain

$$\delta \epsilon_{ij} = \frac{d}{2N_0} \sum_{k=1}^{N_0} \left(\frac{\delta l_i^k}{l^k} n_j^k + \frac{\delta l_j^k}{l^k} n_i^k \right) \quad (14)$$

d: dimension of the space.

These formulations can be used when considering different meanings for variable δl_i^k. Three possible expressions for δl_i^k have been analyzed [8]:

- the relative displacement at each contact,
- the displacement of the centers of particles in contact,
- the displacement of the centers of neighbouring particles.

These three hypotheses correspond to different local kinematic fields considered. The more general hypothesis corresponds to the analysis using the local displacement between neighbouring particles. A more restrictive hypothesis corresponds to the local displacement between particles in contact. It is clear that the local displacements of these kinds of particle are restrained by the non-penetrability condition. The last hypothesis is much more restrictive because, first, this kind of kinematics is restrained by the condition of non-penetrability, and moreover does not take into account the particular kinematics of the centers of particles linked to particle rotation [2]. Figure 5 shows that the use of the relative displacement at contacts does not allow the global strain tensor to be evaluated correctly. The values obtained from the averaging operator are much smaller than the measured boundary values. This figure shows that by using the relative displacements of the centers of particles in contact, the strain tensor is evalued better. However this evaluation is not very accurate and morover depends on the analyzed configuration. If, for a loading increment, the initial configuration is considered, the components of compressive strain are underestimated; if the final configuration is considered, the components of extensive strain are underestimated (Fig.6). The use of the relative displacements of the centers of neighbouring particles gives a very accurate evaluation of the strain tensor measured from the displacements of the boundaries. So these results show

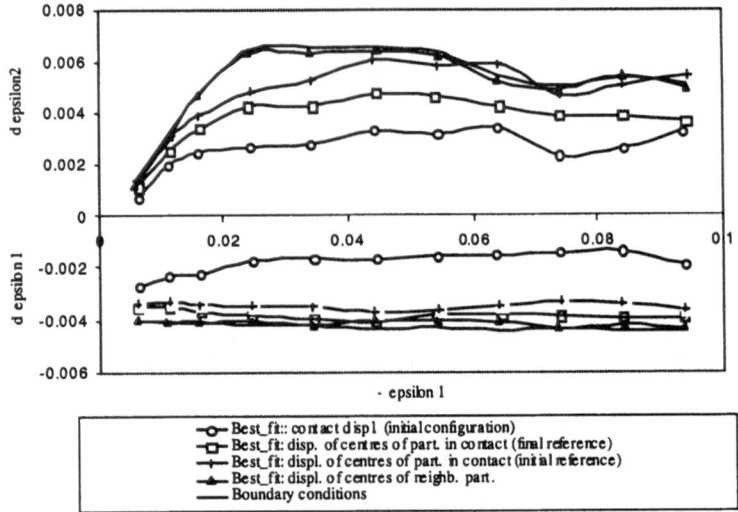

Fig. 5. Increments of strain defined from the boundary conditions and from the best fit approach.

clearly that the local kinematic variable which has to be considered cannot be defined only at contacts between particles. It can also be pointed out that the scale to consider at the local level corresponds to an array of particles and not a single contact. Other formulations have been proposed in the literature to derive the strain tensor from the local displacements, in particular formulations given by Kruyt [9] or by Bagi [1]. It can be noted that these formulations take into account two local variables one is the displacement of particles in contact, and the second is a local variable characterising the local arrays in the vicinity of the

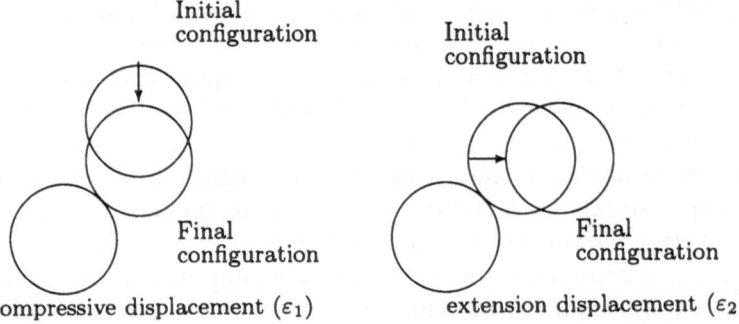

Fig. 6. Analysis of local kinematics: final configuration is better to evaluate ε_1 for compressive displacements, and initial configuration is better to evaluate ε_2 for extension displacements.

considered contact. So these approaches lead to the same conclusion: that the correct local level to consider for kinematic variables is an array of particles.

5   Kinematic localisation operator

The previous paragraph shows that the local kinematic variable which allows the strain tensor to be defined is the relative displacement of neighbouring particles. So the simpler localisation operator derived from the strain tensor has been analyzed, using different scale levels

$$\delta l_i^k = \delta \varepsilon_{ij} l_j^k \tag{15}$$

l_i^k : branch vector linking two neighbouring particles

Variable $\delta l_i^k \left(n^k \right)$ has been compared to the local values measured in the simulation by considering different definitions of neighbouring particles (only particles in contact or particles located at a given distance from each other). Figure 7 shows clearly that the local relative displacement of particles in contact cannot be described from relation (15). In particular the local condition of non-penetrability leads to particular conditions for the relative normal displacement of particles in contact. This local condition cannot be taken into account by

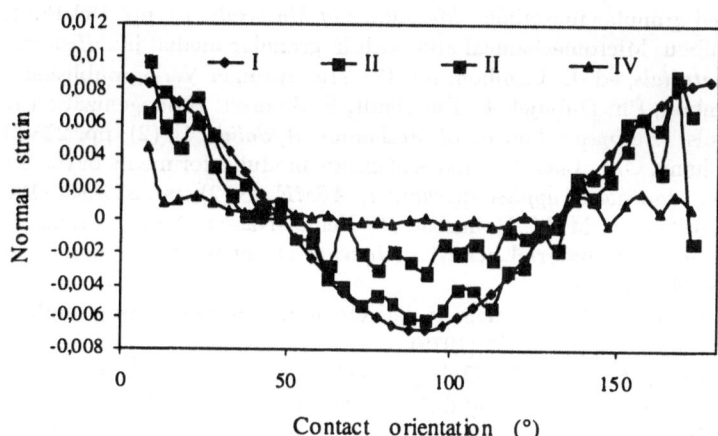

Fig. 7. Kinematic localisation operator considering different levels of scale; normal kinematics :

I: $\Delta u_n \left(\overrightarrow{n} \right) = n_i \Delta \varepsilon_{ij} n_j \left(r^A + r^B \right)$
II: $\Delta u_n \left(\overrightarrow{n} \right) = \left\langle \Delta U_n^B - \Delta U_n^A \right\rangle \quad d \left(A, B \right) < 2.27 \times 2 r^m$
III: $\Delta u_n \left(\overrightarrow{n} \right) = \left\langle \Delta U_n^B - \Delta U_n^A \right\rangle \quad d \left(A, B \right) < 1.136 \times 2 r^m$
IV: $\Delta u_n \left(\overrightarrow{n} \right) = \left\langle \Delta U_n^B - \Delta U_n^A \right\rangle \quad d \left(A, B \right) = 2 r^m$

relation (15). Figure 7 shows that relation is only valid if neighbouring particles are considered. This analysis leads to the same conclusion as in paragraph 4: that the correct local level to consider for kinematic variables is an array of particles.

6   Conclusions

The usual static averaging operator is well established. The static localisation operator will be much more accurate if it takes into account an internal variable characterizing the local contact law and another one for characterizing the internal fabric of the medium. The most relevant local kinematic variables to achieve a statistical homogenization procedure seem to be defined at the scale of the local arrays of particles ; the scale of the contact between particles does not seem to be the appropriate scale. Then one of the difficulties of statistical homogenization lies in the fact that the variables characterizing the contact between particles are not enough to achieve a complete homogenization procedure. Local variables characterizing the local array must also be taken into account. This necessity leads to much more complex approaches.

References

1. K. Bagi: Stress and strain in granular assemblies. *Mechanics of Materials*, 22, pp. 165-177 (1996).
2. J. P. Bardet: Observations on the effects of particle rotations on the failure of idealized granular materials. *Mechanics of Materials*, 18, pp. 159-182 (1994).
3. B. Cambou: Micromechanical approach in granular media, in: *Behaviour of granular materials*, ed. B. Cambou, pp. 171-216, Springer Verlag publisher (1998).
4. B. Cambou, Ph. Dubujet, F. Emeriault, F. Sidoroff: Homogenization for granular materials. *European Journal of Mechanics, A/Solids*, 14 (2), pp. 225-276,(1995).
5. C. S. Chang, C. L. Liao: Estimates of elastic modulus for media of randomly packed granule. *Journal of applied mechanics, ASME*, 48 (2), pp. 339-344,(1981).
6. J. Christofferson, M .M. Mehrabadi, S. Nemat-Nasser: A micromechanical description of granular material behavior. *Journal of applied mechanics, ASME*, 48 (2), pp. 339-344, (1981).
7. P. A. Cundall, O. D. L. Strack: A discrete numerical model for granular assemblies. *Geotechnique*, 29, pp. 47-65,(1979).
8. F. Dedecker, M. Chaze, Ph. Dubujet, B.Cambou: Specific features of strain in granular materials. *Mechanics of Cohesive-Frictional Materials*, 5, pp. 173-193,(2000).
9. N. P. Kruyt, L. Rothenburg: Micromechanical definition of the strain tensor for granular materials. *Journal of applied mechanics*, 118 , pp. 706-711,(1996).
10. C. L. Liao, T. P. Chang, D. H. Young, C. S. Chang: Stress-Strain relationships for granular materials based on the Hypothesis of Best Fit. *Int. J. Solids Structures*, 34, pp. 4087-4100, (1997).
11. E. Sanchez-Palencia, A. Zaoui: Homogenization techniques for composite media. *Springer-Verlag, Berlin*, (1987).
12. J. Weber: Recherche concernant les contraintes intergranulaires dans les milieux pulvérulents. *Bulletin de liaison des Ponts et Chaussées*, Paris, 20, pp. 1-20, (1966).

From discontinuous models towards a continuum description

M. Lätzel, S. Luding, H. J. Herrmann

Institute for Computer Applications, University of Stuttgart, Pfaffenwaldring 27, 70569 Stuttgart, Germany

Abstract. One of the essential questions of material sciences is how to bridge the gap between microscopic quantities, like contact forces and deformations, in a granular assembly, and macroscopic quantities like stress, strain or the velocity-gradient. A two-dimensional shear-cell is examined by means of discrete element simulations. Applying a volume averaging formalism, one obtains volume fractions, coordination numbers, and fabric properties. Furthermore, the stress tensor and the "elastic" (reversible) mean displacement gradient can be derived. From these macroscopic quantities, some material properties can be computed by different combinations of the tensorial invariants.

Because an essential ingredient of both simulations and experiments are rotations of the independent grains, we apply a Cosserat type description. Therefore we compute quantities like the couple stress and the curvature tensor as well as a combination of them, the "torque-resistance".

1 Introduction

Macroscopic continuum equations for the description of the behavior of granular media rely on constitutive equations for stress, strain, and other physical quantities describing the state of the system. One possible way to obtain an observable like, for example, the stress is to perform discrete particle simulations [1–3] and to average over the "microscopic" quantities in the simulation, in order to obtain the averaged macroscopic quantity. Besides the trivial definitions for averages over scalar and vectorial quantities like density, velocity, and particle-spin, one can find slightly different definitions for stress and strain averaging procedures in the literature [3–12] (see also the contributions by Cambou, Goddard, Kruyt and Lanier in this book).

The aim of this paper is to review recent results for tensorial, averaged continuum quantities, and to present new results connected to the rotational degrees of freedom. In section 2 the setup of the system is described and the simulation method is discussed. In section 3 our averaging method is introduced and in section 4 applied to some scalar quantities. Section 5 contains the definitions and averaging strategies for fabric, stress, and elastic deformation gradient and, in addition, some material properties are extracted from their combinations. The rotations of our particles are taken into account in section 6 and averages are presented for the corresponding macroscopic quantities.

2   Model system and simulation

2.1   The Couette shear-cell setup

In the simulations presented in this study, a two-dimensional Couette shear-cell is used, as sketched in Fig. 1. N particles are confined between an outer ring and an inner ring with radius R_o and R_i, respectively. The particles are of slightly different size in order to reduce ordering effects. The boundary conditions are based on an experimental realization [13–15]. For more details on other simulations, see [3, 15–18].

In the steady state shear situation where averages are taken, the outer wall is fixed and the inner wall rotates and thus introduces a slow shear deformation in the system. The simulations are started in a dilute state with an extended outer ring while the inner ring already rotates counter-clockwise with constant angular frequency $\Omega = 2\pi/T_i = 0.1\,\mathrm{s}^{-1}$ and period $T_i = 62.83\,\mathrm{s}$. The radius of the outer ring is reduced within about two seconds to reach its desired value R_o and thereafter it is kept fixed. If not explicitly mentioned, averages are performed after about three rotations at $t = 180\,\mathrm{s}$ (to get rid of the arbitrary initial configuration), and during about one rotation, until $t = 239\,\mathrm{s}$.

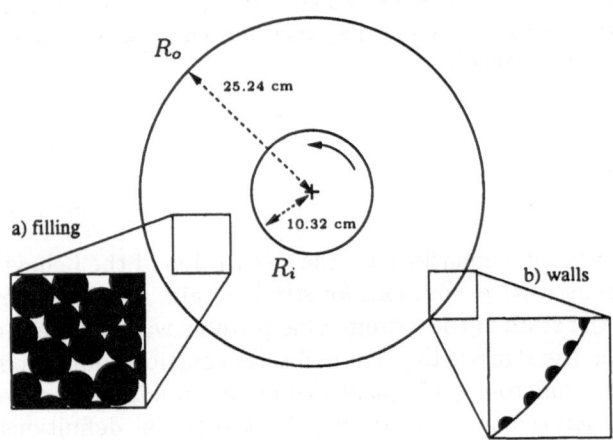

Fig. 1. A schematic plot of the model system

In the simulations different global volume fractions $\bar{\nu} = \sum_{p=1}^{N} V^p / V_{\mathrm{tot}}$ of the shear-cell are examined. The sum runs over all particles p with height h, diameter d_p and thus volume $V^p = \pi h (d_p/2)^2$ in the cell with $V_{\mathrm{tot}} = \pi h (R_o^2 - R_i^2)$. In this study three different $\bar{\nu}$ are examined. In the following the simulations will be referred to as A, B and C with $\bar{\nu} = 0.8084$, 0.8149 and 0.8194, respectively. For the different simulations the number of large and small particles is varied like N_{small}: 2511, 2545, 2555 and N_{large}: 400, 394, 399. For the calculation of the global volume fraction, the small particles glued to the wall are counted with half their volume only, and thus contribute with $\bar{\nu}_{\mathrm{wall}} = 0.0047$ to $\bar{\nu}$ [3]. The

Table 1. Microscopic material parameters of the model

property	values
diameter d_{small}, mass m_{small}	7.42 mm, 0.275 g
diameter d_{large}, mass m_{large}	8.99 mm, 0.490 g
reference diameter $\bar{d}$	8.00 mm
wall-particle diameter d_{wall},	2.50 mm
system/disk-height h	6 mm
normal spring constant k_n	352.1 N/m
normal viscous coefficient γ_n	0.19 kg/s
Coulomb friction coefficient μ	0.44
tangential spring constant k_t	267.1 N/m
bottom friction coefficient μ_b	2×10^{-5}
material density ϱ_0	1060 $\text{kg}\,\text{m}^{-3}$

properties of the particles, i.e. the parameters used for the force laws described in detail in the next subsection, are summarized in table 1.

2.2   The discrete element model

The elementary units of granular materials are mesoscopic grains. In our simulations [3] the grains are treated as rigid particles but deform locally at their contact points. We relate the interaction forces to the virtual overlap δ and to the tangential displacement of two particles during contact. The force laws used are material dependent and have to be validated by comparison with experimental measurements [19–21].

When the force $\boldsymbol{f}_i$ acting on particle i is known (contributions to $\boldsymbol{f}_i$ stem either from other particles, from boundaries or from external forces), the problem is reduced to the integration of Newton's equations of motion. Since we are performing two dimensional (2D) simulations, we have three equations for each particle, two for the linear and one for the rotational degree of freedom [3]. Particle-particle interactions are short range and active on contact only; attractive forces and the presence of other phases are neglected, i.e. we focus on "dry granular media". In the following the force laws accounting for excluded volume, dissipation, and friction are introduced.

Two particles i and j, with diameter d_i and d_j, respectively, interact only when they are in contact so that their overlap $\delta = \frac{1}{2}(d_i + d_j) - (\boldsymbol{r}_i - \boldsymbol{r}_j) \cdot \boldsymbol{n}$ is positive, with the position vector $\boldsymbol{r}_i$ of particle i and the unit vector $\boldsymbol{n} = (\boldsymbol{r}_i - \boldsymbol{r}_j)/|\boldsymbol{r}_i - \boldsymbol{r}_j|$ that points from j to i. The symbol '·' denotes the scalar product of vectors. The force $\boldsymbol{f}_i^c$ acting on particle i at its contact c with particle j is decomposed as $\boldsymbol{f}_i^c = \boldsymbol{f}_{\text{n,el}} + \boldsymbol{f}_{\text{n,diss}} + \boldsymbol{f}_{\text{t}}$. The first contribution to the force – accounting for the excluded volume of each particle – is an repulsive elastic force

$$\boldsymbol{f}_{\text{n,el}} = k_n \delta \boldsymbol{n} \,, \tag{1}$$

where k_n is proportional to the material's modulus of elasticity. Since we are interested in disks rather than spheres, we use a linear spring that follows Hooke's law, whereas in the case of elastic spheres, the Hertz contact law would be more appropriate.

The second contribution, a viscous dissipation, is given by the damping force in the normal direction

$$f_{n,diss} = \gamma_n \dot{\delta} n \ , \tag{2}$$

where γ_n is a phenomenological, viscous dissipation coefficient and $\dot{\delta} = -v_{ij} \cdot n = -(v_i - v_j) \cdot n$ is the relative velocity in the normal direction. The two normal components are simply added to $f_n = f_{n,el} + f_{n,diss}$.

The third contribution to the contact force – accounting for tangential friction – can be chosen in the simplest case, according to Coulomb, as $f_{t,c} \leq -\mu |f_n| t$, where μ is the friction coefficient and $t = v_{ij}^t / |v_{ij}^t|$ is the tangential unit-vector parallel to the tangential component of the relative velocity $v_{ij}^t = v_{ij} - (v_{ij} \cdot n)n$. Since $f_{t,c}$ is non-smooth and undetermined at $v_{ij}^t = 0$, we also introduced a tangential spring as a necessary ingredient to obtain a positive tangential restitution found from collision experiments with various materials [19]. When two particles get into contact at time t_0, one assumes a "virtual" spring between their contact points, where

$$\eta = \left(\int_{t_0}^{t} v_{ij}^t(t') dt' \right) \cdot t \tag{3}$$

is the total tangential displacement of this spring at time t, build up during contact duration $t - t_0$. Note that due to its definition η can either be positive or negative so that $\eta = \eta t$ can be anti-parallel to t. The restoring (static) frictional force, $f_{t,s} = -k_t \eta$, with the stiffness of the tangential spring k_t, can thus be oriented parallel or anti-parallel to t.

The two forces $f_{t,s}$ and $f_{t,c}$ are combined by taking the minimum value

$$f_t = -\min(k_t \eta, \mu |f_n|) t \ . \tag{4}$$

All contact force components and a bottom friction $f_b = -\mu_b m g v_i / |v_i|$ sum up to the total force

$$f_i = \sum_c (f_{n,el} + f_{n,diss} + f_t) + f_b \ . \tag{5}$$

3 From the micro- to a macro-description

In the previous section, the model system was introduced from the microscopic point of view. In this framework, the knowledge of the forces acting on each particle is sufficient to model the dynamics and the statics of the system. Tensorial quantities like the stress or the deformation gradient are not necessary for a discrete modelling. However, subject of current research is to establish a correspondence to continuum theories by computing tensorial quantities like the stress σ, the strain ε, as well as scalar material properties like, e.g., the bulk and shear moduli [6, 8, 9] (see also the chapters of Cambou, Goddard, Kruyt, Lanier and Radjai in this book). In the course of this process, we first discuss averaging strategies, before presenting some results.

3.1 Averaging strategy

Most of the measurable quantities in granular materials vary strongly both in time and on short distances. Thus, during the computation of the averages presented later on, we have to average over or to reduce the fluctuations. In order to suppress the fluctuations, we perform averages in both time and space. This is possible due to the chosen boundary condition. The system can run for a long time in a quasi-steady state and, due to the cylindrical symmetry, all points at a certain distance from the origin are equivalent to each other. Therefore, averages are taken over many snapshots in time with time steps Δt and on rings of material at a center-distance r with width Δr so that the averaging volume of one ring is $V_r = 2\pi h r \Delta r$. For the sake of simplicity (and since the procedure is not restricted to cylindrical symmetry), the averaging volume is denoted by $V = V_r$ in the following. The averaging over many snapshots is somehow equivalent to an ensemble average. However, we remark that different snapshots are not necessarily independent of each other as discussed in [3] and the duration of the simulation might be too short to explore a representative part of the phase space.

Finally, we should remark that the most drastic assumption used for our averaging procedure is the fact, that all quantities are smeared out over one particle. Since it is not our goal to solve for the stress field inside one particle, we assume that a measured quantity is constant inside the particle. This is almost true for the density, but not, e.g., for the stress. However, since we average over all positions with similar distance from the origin, i.e. averages are performed over particles with different positions relative to a ring, details of the position dependency inside the particles will be smeared out anyway.

3.2 Averaging formalism

The mean value of some quantity Q is defined as

$$Q = \frac{1}{V} \sum_{p \in V} w_V^p V^p Q^p \tag{6}$$

with the particle volume V^p, the particle quantity $Q^p = \sum_{c=1}^{C^p} Q^c$, and the quantity Q^c attributed to contact c of particle p which has C^p contacts. The weight w_V^p accounts for the particle's contribution to the average, and corresponds to the fraction of the particle volume that is covered by the averaging volume. Since an exact calculation of the area of a circular particle that lies in an arbitrary ring is rather complicated, we assume that the boundaries of V are locally straight, i.e. we cut the particle in slices, see [3] for details.

4   Results on macroscopic scalar quantities

In the following we apply the averaging formalism to obtain various macroscopic quantities. In table 2 the computed quantities as well as the pre-averaged particle quantities are shown. In this study $\otimes$ denotes the dyadic product and '·' is used for the order-reduction by one for each of the two tensors at left and right. In the data plots, a rescaled, dimensionless radius $\tilde{r} = (r - R_i)/\tilde{d}$ is used, which gives the distance from the inner wall in units of typical diameters.

Table 2. Quantities computed by using the averaging formalism see [3, 17, 18] for details and a derivation

quantity	Q^p	Q
volume fraction ν	1	$\dfrac{1}{V}\sum_{p\in V} w_V^p V^p$
mass flux density $\nu\boldsymbol{v}$	$\boldsymbol{v}^p$	$\dfrac{1}{V}\sum_{p\in V} w_V^p V^p \boldsymbol{v}^p$
fabric tensor $\mathbf{F}$ (contact number density)	$\displaystyle\sum_{c=1}^{C^p} \boldsymbol{n}^c \otimes \boldsymbol{n}^c$	$\dfrac{1}{V}\displaystyle\sum_{p\in V} w_V^p V^p \sum_{c=1}^{C^p} \boldsymbol{n}^c \otimes \boldsymbol{n}^c$
stress tensor $\boldsymbol{\sigma}$	$\dfrac{1}{V^p}\displaystyle\sum_{c=1}^{C^p} \boldsymbol{f}^c \otimes \boldsymbol{l}^{pc}$	$\dfrac{1}{V}\displaystyle\sum_{p\in V} w_V^p \sum_{c=1}^{C^p} \boldsymbol{f}^c \otimes \boldsymbol{l}^{pc}$
deformation gradient $\boldsymbol{\epsilon}$	$\dfrac{\pi h}{V^p}\displaystyle\sum_{c=1}^{C^p} \boldsymbol{\Delta}^{pc} \otimes \boldsymbol{l}^{pc} \cdot \mathbf{A}$	$\dfrac{\pi h}{V}\left(\displaystyle\sum_{p\in V} w_V^p \sum_{c=1}^{C^p} \boldsymbol{\Delta}^{pc} \otimes \boldsymbol{l}^{pc}\right) \cdot \mathbf{A}$ with $\mathbf{A} = \mathbf{F}^{-1}$

4.1   Volume fraction

As a first example for an averaged scalar quantity, the local volume fraction ν, see table 2, is computed. The volume fraction (related to the local density $\varrho(r) \approx \varrho^p \nu$ with the material density ϱ^p) is shown in Fig. 2 rescaled in units of ΩR_i. Starting from a nearly uniform volume fraction over the whole cell, after three rotations, a dilated zone forms near to the inner wheel as a consequence of the applied shear. This effect is less pronounced for higher initial global densities. In the outer region of the shear cell ($\tilde{r} > 10$), the structure of the packing remains frozen, i.e. not much reorganization takes place within the duration of the simulation.

4.2   Mass flux density

As a second scalar quantity, the tangential component of the mass flux density νv_ϕ is investigated. Dividing the mass flux density by the volume fraction, one

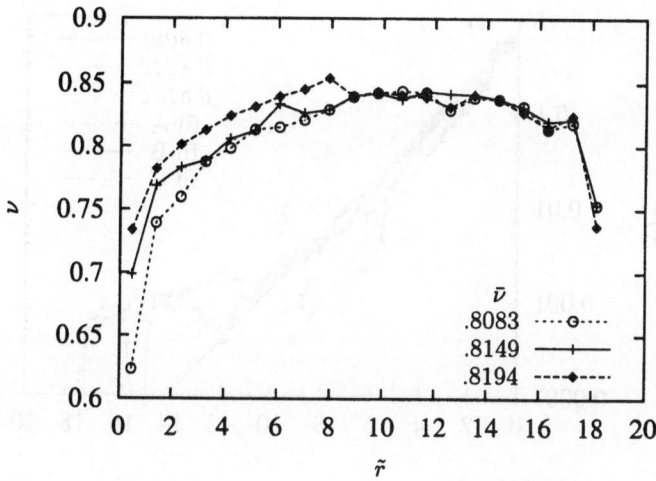

Fig. 2. Volume fraction ν, plotted against the dimensionless distance from the origin $\tilde{r}$, for different initial global densities $\bar{\nu}$.

gets the tangential velocity v_ϕ shown in Fig. 3. The simulation data are fitted by $v_\phi(r) = v_0 \exp(-\tilde{r}/s)$ with v_o: 0.670, 0.756, 0.788 and s: 1.662, 1.584, 1.191, thus showing an exponential profile corresponding to the shear band (forced close to the inner ring by our boundary conditions). The shear band, has a width of a few (~ 8) particle diameters, before the velocity v_ϕ reaches the noise level.

Given the velocity field, it is possible to compute the velocity gradient $\boldsymbol{\nabla v}$ by means of numerical differentiation. From $\boldsymbol{\nabla v}$ one can derive the deformation rate $D_{r\phi} = \frac{1}{2}\left[\frac{\partial v_\phi}{\partial r} - \frac{v_\phi}{r}\right]$ and the continuum rotation rate $W_{r\phi} = \frac{1}{2}\left[\frac{\partial v_\phi}{\partial r} + \frac{v_\phi}{r}\right]$, see [17] for details and [22] for a similar approach.

5   Macroscopic tensorial quantities

In this section, the averaged, macroscopic tensorial quantities in our model system are presented. The fabric tensor describes the contact network, the stress tensor describes the stress distribution due to the contact forces, and the elastic deformation gradient is a measure for the corresponding elastic, reversible deformations.

5.1   Fabric tensor

In assemblies of grains, the forces are transmitted from one particle to the next only at the contacts of the particles. Therefore the local geometry and direction of each contact is important [3, 23]. The fabric tensor in table 2 is

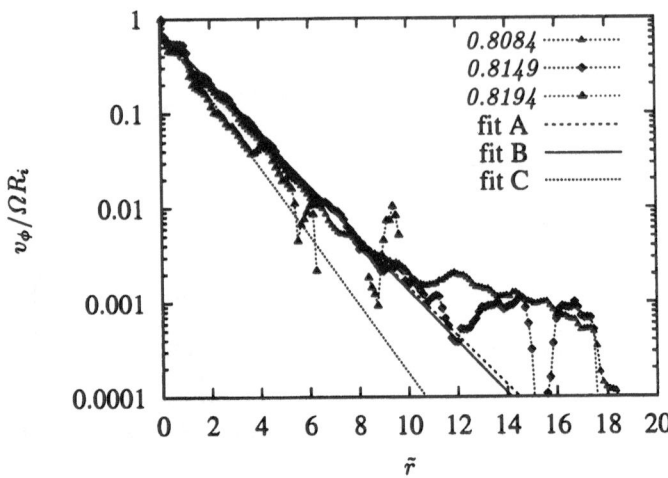

Fig. 3. Tangential velocity v_ϕ normalized by the velocity of the inner wheel ΩR_i, plotted against $\tilde{r}$. The lines are the fits to the simulation data for $\tilde{r} = 0.25$ to 8.1, using $v_\phi(r) = v_0 \exp(-\tilde{r}/s)$ with v_o: 0.670, 0.756, 0.788 and s: 1.662, 1.584, 1.191, for increasing density, as given in the inset, respectively

symmetric by definition and thus consists of up to three independent scalar quantities in two dimensions. The first of them, the trace (or volumetric part) $F_V = \text{tr}(\mathbf{F}) = (F_{\max} + F_{\min})$, is the contact number density, with the major and the minor eigenvalues $F_{\max}$ and $F_{\min}$, respectively. With other words, one obtains the relation $\text{tr}(\mathbf{F}) = \nu \mathcal{C}$ with reasonable accuracy, where $\mathcal{C}$ is the average number of contacts per particle.

The second scalar, the deviator $F_D = F_{\max} - F_{\min}$, accounts for the anisotropy of the contact network to first order, and the third, the angle ϕ_F, gives the orientation of the "major eigenvector", i.e. the eigenvector corresponding to $F_{\max}$, with respect to the radial outwards direction. In other words, the contact probability distribution is proportional to the function $F_V + F_D \cos(2(\phi - \phi_F))$ [3, 16], when averaged over many particles, an approximation which is not always reasonable [24].

The trace of the fabric tensor and the mean number of contacts increases with increasing distance from the inner ring, and is reduced in the vicinity of the walls due to ordering. With increasing local density, the trace of $\mathbf{F}$ is systematically increasing, while the deviatoric fraction F_D/F_V seems to decrease; this means that a denser system is more isotropic concerning the fabric. The major eigendirection is tilted counter-clockwise (for counter-clockwise rotation of the inner ring) by somewhat more than $\pi/4$ from the radial outwards direction, except for the innermost layer and for the strongly fluctuating outer region, where the small deviator does not allow for a proper definition of ϕ_F anyway.

5.2   Stress tensor

The stress tensor, see table 2, is proportional to the dyadic product of the force f^c acting at a contact c with its branch vector l^{pc}, which accounts for the distance over which the force is transmitted, see [3] for details. The trace of the stress tensor, i.e. the volumetric stress, is almost constant over the whole shear-cell besides fluctuations. In contrast, the non-diagonal elements of σ decay proportional to r^{-2} with increasing distance r from the inner ring. The deviatoric fraction σ_D/σ_V also decays like F_D/F_V, when moving outwards from the shear band at the inner ring.

5.3   Elastic deformation gradient

To achieve the material properties of a granular ensemble one is interested in the stress-strain relationship of the material. The strain ϵ can be obtained by time integration of the velocity gradient, see subsection 4.2, and subsequent symmetrization and linearization. We present an alternative technique, the application of "Voigt's hypothesis" which assumes that the deformation is uniform and that every particle displacement conforms to the corresponding mean displacement field, but fluctuates about [3, 25]. This relates the actual deformations to a "virtual" reference state where all contacts start to form, i.e. particles are just touching with $\delta = 0$, see [3] for a detailed derivation. The result is a non-symmetric tensor ϵ, which is *not* the strain, instead we refer to it as the elastic deformation gradient, see table 2.

The volumetric part of the elastic deformation gradient is largest in the shear zone and this effect is stronger the lower the global density. It is easier to compress the dilute material closer to the inner ring, as compared to the denser material in the outer part. The deviatoric fraction of ϵ, is also decaying with increasing distance from the center, similar to the deviatoric fractions of fabric and stress.

5.4   Material properties

At first we have a closer look on the orientations of the tensors plotted in Fig. 4 in the inner part of the shear cell. In the outer part, the deviatoric fraction is usually around 10 per-cent, i.e. so small that the orientations become too noisy to allow for a proper definition. We find that all orientation angles ϕ show the same qualitative behavior, however, the fabric is tilted more than the stress which, in turn, is tilted more than the deformation gradient. Thus, the three tensorial quantities examined so far are *not* co-linear.

We also compute the mean-field expectation values for σ and ϵ, in order to get a rough estimate for the orders of magnitude of the following results. Replacing f^c by its mean $\bar{f} = k_n \bar{\delta} n^c$, the radius a_p by $a = \bar{a}$, l^{pc} by an^c, and k_n/h by k'_n one gets $\bar{\sigma} = (k'_n \bar{\delta}/\pi a)\,\mathbf{F}$ for the stress. Performing some similar replacements for the elastic deformation gradient leads to $\bar{\epsilon} = (\pi/k'_n)\,\bar{\sigma} \cdot \mathbf{A} = (\bar{\delta}/a)\,\mathbf{I}$.

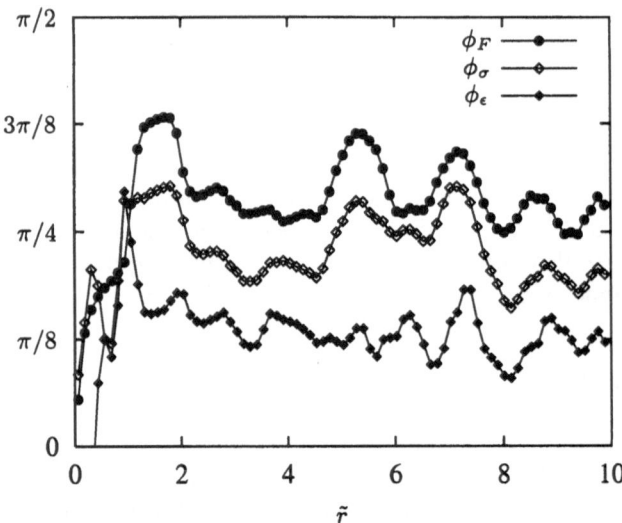

Fig. 4. Orientation of the tensors **F**, σ, and ε, plotted against the distance from the inner ring from simulation B. The data are shown only for about 10 layers of particles counted from the inner ring. Solid circles are fabric, open diamonds are stress, and solid diamonds are elastic deformation gradient data

The material stiffness, $\bar{E}$, can be defined as the ratio of the volumetric parts of stress and strain, so that one obtains the mean field prediction

$$\bar{E} = (k'_n/2\pi)\,\text{tr}(\mathbf{F})\ .\tag{7}$$

In Fig. 5 the rescaled stiffness of the granulate is plotted against the trace of the fabric for all simulations. Note that all data collapse almost on a line, but the mean-field value underestimates the simulation data by some per-cent. Additional simulation data (not shown here) for different k_n and even data from simulations with neither bottom- nor tangential friction collapse with the data for fixed k_n and different volume fractions, shown here. The few data points which deviate most are close to the boundaries, where the material is strongly layered. The deviation from the mean field prediction (*solid* line in Fig. 5) seems to disappear in the absence of shear – a fact which can be accepted since the mean field expression (7) does not account for the shear displacement.

In Fig. 6 the ratio of the deviatoric parts of stress and strain is plotted against the trace of the fabric. We did not use the traditional definition of the shear modulus [8], since our tensors are not co-linear as shown in Fig. 4. Like the material stiffness, both quantities are proportional, for points near or within the shear band. In the outer part of the shear-cell the particles are strongly inter-locked and thus resist much more against shear, and therefore G diverges. For increasing global density, the critical contact number density also grows, at a critical density.

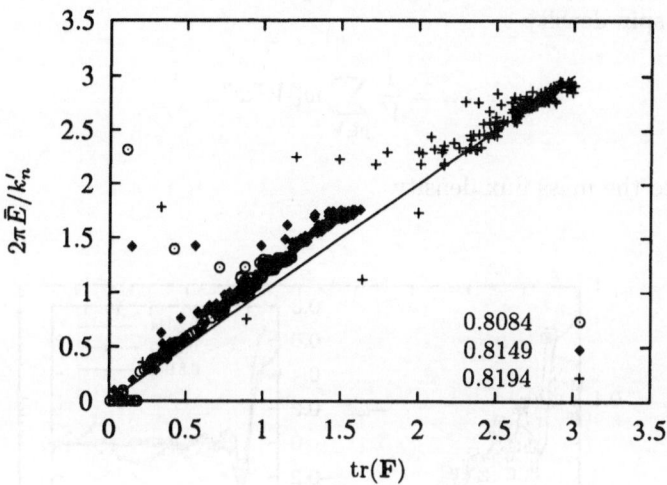

Fig. 5. Granulate stiffness $2\pi\bar{E}/k'_n = \mathrm{tr}(\sigma)/\mathrm{tr}(\epsilon)$, plotted against $\mathrm{tr}(\mathbf{F})$ from all simulations. Every point corresponds to one ring of 150, i.e. $\Delta r \approx (1/8)d_{\mathrm{small}}$

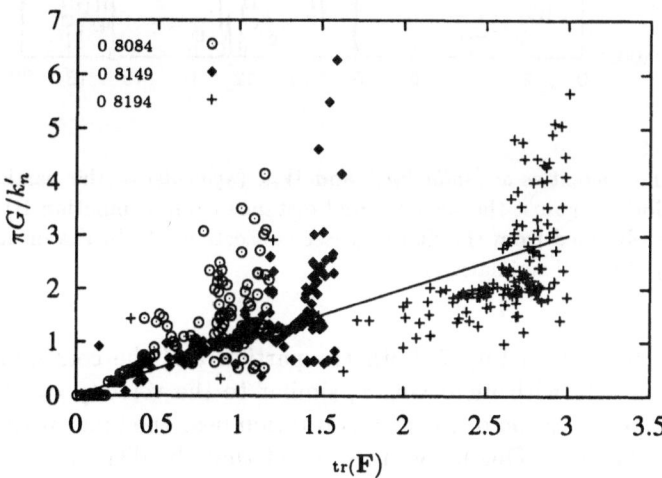

Fig. 6. Scaled granulate shear resistance $\pi G/k'_n = \mathrm{dev}(\sigma)/\mathrm{dev}(\epsilon)$ plotted against $\mathrm{tr}(\mathbf{F})$ from all simulations. The line indicates the identity curve

6   Rotational degrees of freedom

The particles in the model are able to rotate so that also quantities concerning the rotational degree of freedom are of interest [22]. In Fig. 7 the macroscopic particle rotations ω, the continuum rotation $W_{r\phi}$ and the excess rotation $\omega^* = \omega - W_{r\phi}$, are displayed. The spin is averaged using $Q^p = \omega^p$ in (6) so that one obtains the spin density

$$\nu\omega = \frac{1}{V} \sum_{p \in V} w_V^p V^p \omega^p \;, \tag{8}$$

in analogy to the mass flux density.

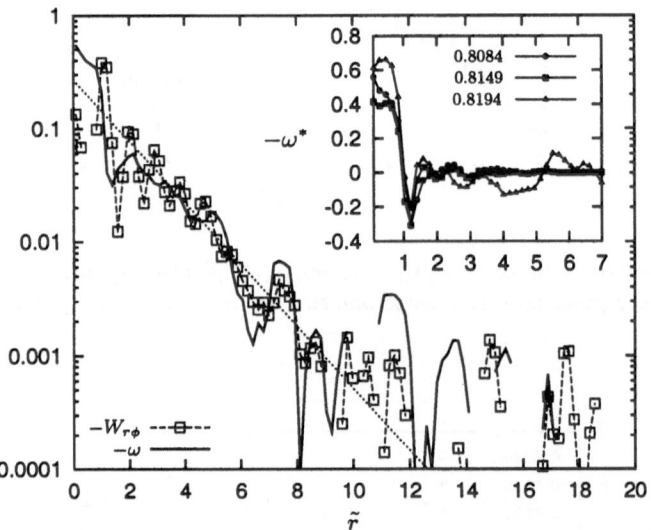

Fig. 7. Angular velocities ω (*solid line*) and $W_{r\phi}$ (*symbols*) of the particles and the continuum, plotted against the scaled radial distance (from simulation B). The *dotted* line is $W_{r\phi}$ as obtained from the fit to v_ϕ, see subsection 4.2. In the inset, the excess spin is displayed for all simulations

As it can be seen in Fig. 7, both the particle and the continuum rotation decay exponentially with increasing $\tilde{r}$, similar to the velocity v_ϕ. The inset of Fig. 7 shows an oscillation of the excess rotation near the inner wheel, from one disk layer to the next. This is due to the fact that the disks in adjacent layers are able roll over each other in the shear zone.

Because of the evident importance of the rotational degree of freedom, this we use the theoretical framework of a Cosserat continuum [22, 26, 27]. In addition to the stress and the displacement gradient one can define the couple stress $\mathbf{M}$ and the curvature $\boldsymbol{\kappa}$. The couple stress tensor is defined here, in analogy to the stress, as

$$\mathbf{M} = \frac{1}{V} \sum_{p \in V} w_V^p \sum_{c=1}^{C^p} (\boldsymbol{l}^{pc} \times \boldsymbol{f}^c) \otimes \boldsymbol{l}^{pc} \;, \tag{9}$$

where the force is replaced by the torque due to the tangential component of the force, and the '×' denotes the vector-product. In a two dimensional system, only the two components M_{zr} and $M_{z\phi}$ of the tensor are non-zero. The values of M_{zr} as a function of $\tilde{r}$ are shown in Fig. 8(a). Note that $\mathbf{M} = \mathbf{0}$, when the sum of the torques acting on one particle vanishes in static equilibrium. In our steady state shear situation $\mathbf{M}$ fluctuates around zero, except for a large value in the shear band, close to the inner wall. In analogy to ϵ we define

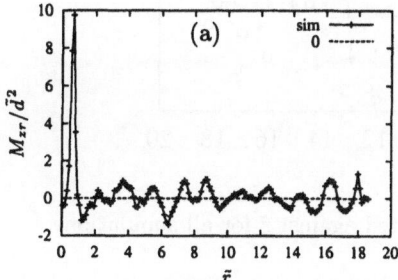 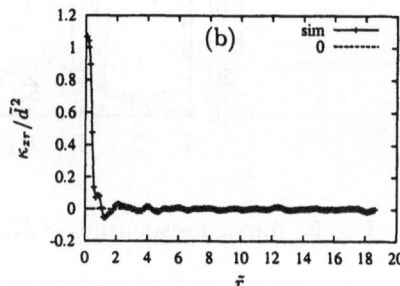

Fig. 8. (a) Plot of the couple stress $M_{zr}/\tilde{d}^2$ against $\tilde{r}$, (b) plot of the curvature $\kappa_{zr}/\tilde{d}^2$ against $\tilde{r}$ from simulation B

$$\kappa = \frac{\pi h}{V} \left(\sum_{p \in V} w_V^p \sum_{c=1}^{c^p} (l^{pc} \times \mathbf{\Delta}^{pc}) \otimes l^{pc} \right) \cdot \mathbf{A} , \qquad (10)$$

where the local contact displacement $\mathbf{\Delta}^{pc}$ is replaced by the corresponding angular vector $l^{pc} \times \mathbf{\Delta}^{pc}$. The values of the curvature κ_{zr} are plotted in Fig. 8(b) against $\tilde{r}$ with similar qualitative behavior as M_{zr}. The other components $M_{z\phi}$ and $\kappa_{z\phi}$ lead to no new insights and are omitted here.

Since we are interested in the role the rotational degree of freedom plays for the constitutive equations, we define the "torque resistance" μ_c as the ratio of the magnitudes of the couple stress and the curvature components. This quantity describes how strongly the material resists against applied torques. In Fig. 9 the torque resistance is plotted for the three simulations. In the dilute regions near the inner wheel, where the particles are able to rotate more easily, μ_c is smaller than in the dense outer part, where the particles are interlocked and thus frustrated. This behavior is consistent with the results for increasing global densities, i.e. the torque resistance increases with density. Note that the strongest fluctuations are due to the division by small κ_{zr} values and have no physical meaning in our interpretation.

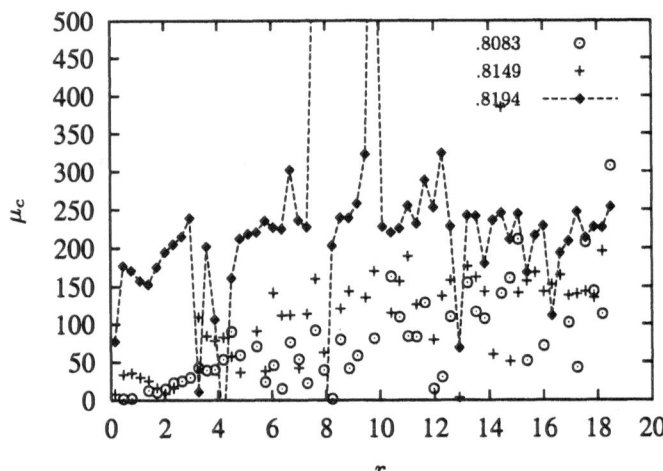

Fig. 9. Torque resistance $\mu_c = M_{zr}/\kappa_{zr}$ plotted against $\tilde{r}$ for all simulations.

7 Summary and conclusion

Discrete element simulations of a 2D Couette shear cell were presented and used as the basis for a micro-macro averaging procedure. The boundary conditions were chosen to allow for averaging over large volumes (rings with width Δr, where Δr can be much *smaller* than the particle diameter) and over a steady state shear and thus over long times. A shear band is localized close to the inner, rotating cylindrical wall. The configurations changed rather rapidly in the shear band, whereas the system is almost frozen in the outer part.

The averaging strategy used assumes the quantities to be homogeneously smeared out over the whole particle which is cut in slices by the averaging volumes. This so called slicing method shows discretization effects in the range of averaging volume widths Δr from about one to one fifth of a particle diameter, whereas the results become independent of the size of the averaging volume for smaller Δr. For Δr much larger than the particle-size, the microscopic details – including rotations – are not longer resolved properly.

The material density, i.e. the volume fraction, the coordination number, the fabric tensor, the stress tensor and the elastic, reversible deformation gradient were obtained by the averaging procedure. The fabric is linearly proportional to the product of volume fraction and coordination number. In the shear band, dilation together with a reduction of the number of contacts is observed. The mean volumetric stress is constant in radial direction while the mean deformation gradient decays with the distance from the inner wall. The ratio of the volumetric parts of stress and strain gives the material stiffness of the granulate, which is small in the shear band and larger outside, due to dilation.

In the shear band, large deviators, i.e. anisotropy, of all tensorial quantities are found, however, decreasing algebraically with increasing distance from the inner wall. The isotropy of the tensors grows only slightly with increasing density and all tensors are tilted counter-clockwise from the radial direction by an angle of the order of $\pi/4$. The system organizes itself such that more contacts are created to act against the shear and also the shear resistance increases with the contact density. An essential result is that the macroscopic tensors are *not* co-linear, i.e. their orientations are different. The orientation of the fabric is tilted most, that of the deformation gradient is tilted least and thus, the material cannot be described by a simple elastic model involving only the two Lamé constants (or bulk modulus and Poisson's ratio) as the only parameters. The stress and the deformation gradient are seemingly interconnected via the fabric tensor.

Finally, the particle rotation was measured in analogy to the particle velocity. Subtraction of the continuum rotation from the particle rotation leads to the excess eigen-rotation of the particles with respect to the mean rotation, in the spirit of a micro-polar or Cosserat continuum theory. In analogy to the stress and elastic deformation gradient, we defined couple stress and curvature. The quotient of the respective non-zero components gives a "torque-resistance" which increases with increasing local density and stress.

References

1. H. J. Herrmann, J.-P. Hovi, and S. Luding, editors. *Physics of dry granular media - NATO ASI Series E 350*. Kluwer Academic Publishers, Dordrecht, 1998.
2. P. A. Cundall and O. D. L. Strack. A discrete numerical model for granular assemblies. *Géotechnique*, 29(1):47–65, 1979.
3. M. Lätzel, S. Luding, and H. J. Herrmann. Macroscopic material properties from quasi-static, microscopic simulations of a two-dimensional shear-cell. *Granular Matter*, 2(3):123–135, 2000. cond-mat/0003180.
4. L. Rothenburg and A. P. S. Selvadurai. A micromechanical definition of the cauchy stress tensor for particulate media. In A. P. S. Selvadurai, editor, *Mechanics of Structured Media*, pages 469–486. Elsevier, 1981.
5. P. A. Cundall, A. Drescher, and O. D. L. Strack. Numerical experiments on granular assemblies; measurements and observations. In *IUTAM Conference on Deformation and Failure of Granular Materials*, pages 355–370, Delft, 1982.
6. J. D. Goddard. Microstructural origins of continuum stress fields - a brief history and some unresolved issues. In D. DeKee and P. N. Kaloni, editors, *Recent Developments in Structered Continua. Pitman Research Notes in Mathematics No. 143*, page 179, New York, 1986. Longman, J. Wiley.
7. R. J. Bathurst and L. Rothenburg. Micromechanical aspects of isotropic granular assemblies with linear contact interactions. *J. Appl. Mech.*, 55:17, 1988.
8. N. P. Kruyt and L. Rothenburg. Micromechanical definition of strain tensor for granular materials. *ASME Journal of Applied Mechanics*, 118:706–711, 1996.
9. C.-L. Liao, T.-P. Chang, D.-H. Young, and C. S. Chang. Stress-strain relationship for granular materials based on the hypothesis of best fit. *Int. J. Solids Structures*, 34:4087–4100, 1997.

10. E. Kuhl, G. A. D´Addetta, H. J. Herrmann, and E. Ramm. A comparison of discrete granular material models with continuous microplane formulations. *Granular Matter*, 2:113–121, 2000.
11. F. Dedecker, M. Chaze, Ph. Dubujet, and B. Cambou. Specific features of strain in granular materials. *Mech. Coh. Fric. Mat.*, 5:174–193, 2000.
12. F. Calvetti, G. Combe, and J. Lanier. Experimental micromechanical analysis of a 2d granular material: relation between structure evolution and loading path. *Mech. Coh. Fric. Mat.*, 2:121–163, 1997.
13. D. Howell and R. P. Behringer. Fluctuations in a 2d granular Couette experiment: A critical transition. *Phys. Rev. Lett.*, 82:5241, 1999.
14. C. T. Veje, D. W. Howell, and R. P. Behringer. Kinematics of a 2D granular Couette experiment. *Phys. Rev. E*, 59:739, 1999.
15. C. T. Veje, D. W. Howell, R. P. Behringer, S. Schöllmann, S. Luding, and H. J. Herrmann. Fluctuations and flow for granular shearing. In H. J. Herrmann, J.-P. Hovi, and S. Luding, editors, *Physics of Dry Granular Media*, page 237, Dordrecht, 1998. Kluwer Academic Publishers.
16. S. Schöllmann. Simulation of a two-dimensional shear cell. *Phys. Rev. E*, 59(1):889–899, 1999.
17. S. Luding, M. Lätzel, and H. J. Herrmann. From discrete element simulations towards a continuum description of particulate solids. In H. Kalman, A. Levy, and M. Hubert, editors, *The Third Israeli Conference for Conveying and Handling of Particulate Solids*, pages 3.125–3.130, Tel-Aviv, 2000. The Forum for Bulk Solids Handling.
18. S. Luding, M. Lätzel, W. Volk, S. Diebels, and H. J. Herrmann. From discrete element simulations to a continuum model. In *European Congress on Computational Methods in Applied Sciences and Engineering, ECCOMAS 2000*, Barcelona, 2000. in press.
19. S. F. Foerster, M. Y. Louge, H. Chang, and K. Allia. Measurements of the collision properties of small spheres. *Phys. Fluids*, 6(3):1108–1115, 1994.
20. L. Labous, A. D. Rosato, and R. Dave. Measurements of collision properties of spheres using high-speed video analysis. *Phys. Rev. E*, 56:5715, 1997.
21. E. Falcon, C. Laroche, S. Fauve, and C. Coste. Behavior of one inelastic ball bouncing repeatedly off the ground. *Eur. Phys. J. B*, 3:45–57, 1998.
22. A. Zervos, I. Vardoulakis, M. Jean, and P. Lerat. Numerical investigation of granular interfaces kinematics. *Mech. Cohes.-Frict. Matter*, 5:305–324, 2000.
23. J. D. Goddard. Continuum modeling of granular assemblies. In H. J. Herrmann, J.-P. Hovi, and S. Luding, editors, *Physics of Dry Granular Media*, pages 1–24, Dordrecht, 1998. Kluwer Academic Publishers.
24. M. M. Mehrabadi, S. Nemat-Nasser, H. M. Shodja, and G. Subhash. Some basic theoretical and experimental results on micromechanics of granular flow. In *Micromechanics of granular media*, Amsterdam, 1988. Elsevier.
25. C.-L. Liao and T.-C. Chang. A generalized constitutive relation for a randomly packed particle assembly. *Computers and Geotechnics*, 20(3/4):345–363, 1997.
26. E. Cosserat and F. Cosserat. *Theorie des Corps Deformables*. Herman et fils, Paris, 1909.
27. A. C. Eringen. Theory of micropolar elasticity. In H. Liebowitz, editor, *Fracture*, volume 2, pages 621–729, New York and London, 1968. Academic Press.

From solids to granulates - Discrete element simulations of fracture and fragmentation processes in geomaterials.*

G. A. D'Addetta ,[1] F. Kun,[2,3] E. Ramm,[1] H. J. Herrmann[2]

[1] Institute of Structural Mechanics, University of Stuttgart, Pfaffenwaldring 7, D-70569 Stuttgart, Germany
[2] Institute of Computer Applications I, University of Stuttgart, Pfaffenwaldring 27, D-70569 Stuttgart, Germany
[3] Department of Theoretical Physics, Kossuth Lajos University, P.O. Box: 5, H-4010 Debrecen, Hungary

Abstract. We present a two-dimensional model of heterogeneous cohesive frictional solids where the material structure is idealized by a discrete granular particle assembly. Our discrete model is composed of convex polygons which are linked together by simple beams accounting for cohesive effects. Varying its parameters the model naturally interpolates between a continuous solid state and a discontinuous dry granular state of the material. In order to demonstrate the wide applicability of the model simulations ranging from the quasi-static uniaxial loading and shearing of a solid to the dynamic fragmentation due to explosion, impact and collision of solids, will be presented. A comparison with experimental observations is carried out in order to verify the qualitative application of the model. To provide a geometrical description of the damage state, different characteristic quantities will be introduced and their relevance will be examined.

1 Introduction

From a physical point of view geomaterials, like concrete, ceramics or marl can be considered as cemented granulates forming a heterogeneous macroscopic solid. The failure mechanisms of this materials are characterized by complex failure modes under various loading situations and a highly anisotropic bias due to their inhomogeneous microstructure. The growth and coalescence of microcracks lead to the formation of macroscopic crack patterns and, finally, to a fragmentation into separate particle clusters, forming a solid–granulates mix.

Depending on the observation scale this class of materials shows different physical and geometrical properties. Taking into account the point of view and the chosen scale the structural fluctuations play a more or less dominant role and thus determine the condition under which a material can be regarded as homogeneous. This is accompanied by the fact, that with a decreasing resolution length

---

* Joint paper of the presentations by G. A. D'Addetta *'Application of a discrete model to the fracture process of cohesive frictional materials'* and F. Kun *'A discrete element method for fragmentation processes'*.

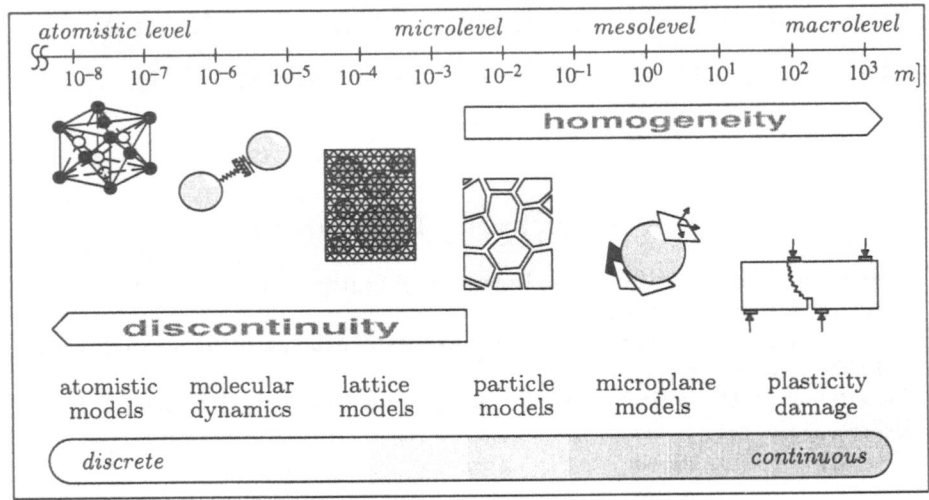

Fig. 1. Observation scales and applicable models for geomaterials like concrete.

on a specific level the internal material structure is less identifiable up to the point when the material is considered as continuous. The observed size scales in geomaterials like concrete are typically subdivided into hierarchical levels, like the atomic, micro-, meso- and macrolevel [29]. The basic question regarding the analytical description of a certain material focuses on how the constitutive behaviour can be described on different size scales.

The range of applicability of different simulation models is directly related to the observation scale, as can be seen in Fig. 1, where our choice of the scale definitions used throughout the paper is noted. For example in the context of concrete modelling the application area ranges from the simulation of a crystal structure and calcium silicate hydrates over a concrete particle stack and laboratory scale measurements on test specimens to structural scale simulations of large structures like earth dams or cooling towers. Basically a class of continuum models can be applied at each size scale, if the local quantities, e.g. damage, are smeared over a certain region. Usually a verification of this models is done via a comparison with experiments of finite size. Emerging heterogeneities, anisotropies and discontinuities are then cast in the form of macroscopic continuum variables. In order to allow for a realistic localization and to overcome mathematical difficulties, as the loss of ellipticity, an enhancement of the classical continuum models is necessary. To account for the neighboured influence within a material, like crack interaction or stress redistribution effects, a physically motivated internal length scale has to be introduced into the model. Usual enhancements of continuum formulations are based on the introduction of localization limiters like non–local, gradient–enhanced, viscous or COSSERAT formulations. From the point a localization phenomenon, like a crack, occurs due to a specific loading the material cannot be treated as continuous any longer. Due to the creation and continuous motion of the evolving crack surfaces the fracture and fragmentation of the solid is difficult to handle numerically. Therefore, most continuous simulation models cannot account for the discrete nature of material failure in a natural way. As an alternative, discrete models like particle, lattice or granular dynamics models have been developed. Although prohibitive in large scale computations this class of models is able to predict and simulate the fracture behaviour of small scale applications of geomaterials.

In this paper we present a combined beam–particle model as an example for an improved discrete element simulation scheme. In our approach heterogeneous materials are considered as a cohesive granular frame represented by polygonal microstructure elements which are bonded together by beams, in order to allow for failure induced anisotropy in an unsmeared fashion. The material is not thought as granular material ab initio, but gradually develops towards this constitution by a gradual cracking of the beams. Due to the characteristics of the disorder present in the microscopic structure and the size of the particles, an internal length scale is incorporated intrinsically into the model.

The paper is organized as follows: Section 2 provides an outline of the theoretical background of the model. In section 3 various applications of the model will be presented ranging from the elastic deformation through fracturing to the complete fragmentation of solids. The failure evolution of basic loading scenarios in the framework of solid mechanics, like uniaxial compression, tension and shear will be used in order to verify the presented numerical scheme and the effects taking place at the microstructural level. Different failure related characteristic quantities for the description of the failure within the material will be investigated and their relation to continuum–based formulations will be pointed out. This is followed by a study of the catastrophic fragmentation of solids, like the impact of a projectile on a rectangular bar, the explosion of a circular solid and the collision of disc–shaped macroparticles. In this context the transition from damage to fragmentation is particularly studied.

Since the fracture evolution occurs very rapidly most of the available experimental results are obtained from the analysis of the debris in the final, relaxed state of the processes. It will be demonstrated that this simulation method allows us to monitor quantities which are hard to measure or are not measurable at all in experiments. Hence, this model provides a deeper understanding of the fracture and fragmentation processes well beyond the description of the final state.

2   Description of the model

Our model of cohesive frictional materials is an extension of those models which are used to study the behaviour of granular materials applying randomly shaped convex polygons to symbolize grains [13, 27]. The construction of the model is composed of three major steps. Namely, the implementation of the granular structure of the solid, the determination of the interaction of grains, and finally capturing the breaking of the solid. This section gives a detailed overview of the three steps of the model construction.

2.1 Granularity

In order to take into account the complex structure of the granular solid we use arbitrarily shaped convex polygons, *i.e.* we divide the solid into grains by a VORONOI cellular structure. The VORONOI construction is a random tessellation of the plane into convex polygons. In general, this is obtained by putting a random set of points onto the plane and then assigning each point that part of the plane which is nearer to it than to any other point. In our case, to get an initial configuration of the polygons we construct a so-called vectorizable random lattice, which is a VORONOI construction with slightly reduced disorder (see Ref. [17]). It is performed by first putting a regular grid onto the plane and

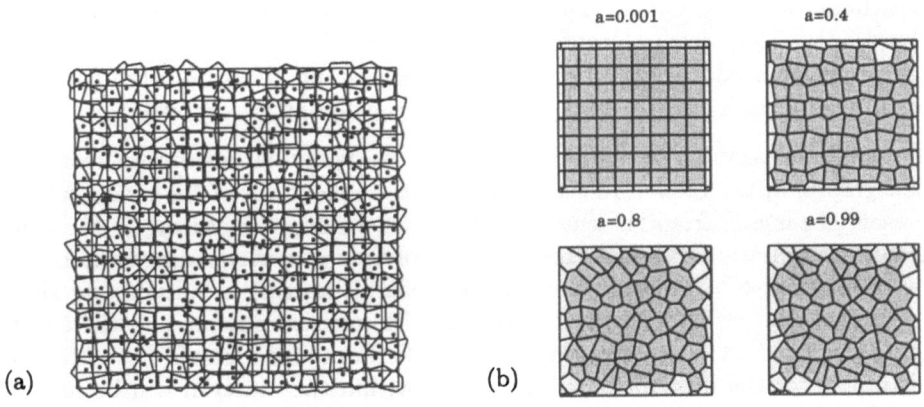

(a)   (b)

Fig. 2. (a) Construction of the vectorizable random lattice. The dots indicate the points used for the tessellation. These points were thrown independently and randomly onto the squares of the regular grid. (b) Random lattices generated at different values of the parameter a. The smooth surface of the samples was obtained by cutting straight boundaries after the VORONOI construction.

throwing points randomly and independently in a square of the side length a centered on the plaquettes of the regular grid. Using these points for the VORONOI construction, the randomness of the tessellation can be controlled by tuning the value of the parameter a between 0 and the lattice spacing of the grid 1. Fig. 2 presents an example of the vectorizable random lattice including the underlying grid, the random points used for the tessellation and the final polygonal structure. The advantage of the vectorizable random lattice compared to the ordinary Poissonian VORONOI tessellation is that the number of neighbours of each polygon is limited which makes the computer code faster and allows us to simulate larger systems.

The convex polygons of this VORONOI construction are supposed to model the grains of the material, see also Refs. [13, 27]. This way the structure of the solid is built on a microscopic scale. Each element is thought of as a large collection of atoms, however, in the simulation these polygons are the smallest particles interacting elastically with each other. The polygons have three continuous degrees of freedom in two dimensions: the two coordinates of the centre of mass and the rotation angle.

2.2 Elastic behaviour of the solid

In the framework of the discrete element approach the elastic behaviour of the solid is captured by defining proper interactions on the microscopic level, between the polygons. In the case of spherical particles the Hertz contact law provides the force between two particles as a function of the overlap distance. For randomly shaped particles it is impossible to derive such a simple interaction law, hence, we introduce an approximate method.

The polygons are considered to be rigid bodies. They are not breakable and not deformable but they can overlap when they are pressed against each other. The overlap represents up to some extent the local deformation of the grains. Usually the overlapping polygons have two intersection points which define the contact line as shown in Fig. 3. The total force $\mathbf{F}_p^{ij}$ acting between polygons i and j can be decomposed into normal $F^{N,ij}$ and tangential $F^{T,ij}$ components with respect to the contact line, *i.e.*

$$\mathbf{F}_p^{ij} = F^{N,ij} \cdot \mathbf{n} + F^{T,ij} \cdot \mathbf{t}, \tag{1}$$

where $\mathbf{n}$ denotes the unit vector pointing in the direction perpendicular to the contact line, and $\mathbf{t}$ is the tangential unit vector.

In order to simulate the elastic contact force between touching grains we introduce a repulsive force between the overlapping polygons. This force is proportional to the overlapping area A divided by a characteristic length L_c of the interacting polygon pair. Our choice of L_c is given by $1/L_c = 1/2(1/r^i + 1/r^j)$, where r^i, r^j are the radii of circles of the same area as the polygons. This normalization is necessary in order to reflect the fact that the spring constant is proportional to the elastic modulus divided by a characteristic length. In the case of a linear spring this characteristic length is simply the equilibrium length (initial length) of the spring. The direction of the force is chosen to be perpendicular to the contact line of the polygons. Further, damping and friction of the touching polygons according to COULOMB's friction law are also implemented. Therefore, the complete form of the normal force $F^{N,ij}$ contains an elastic and a damping contribution, while the tangential component $F^{T,ij}$ is responsible for the friction:

$$F^{N,ij} = -\frac{E_p A_p}{L_c} - m_{eff}^{ij} \cdot \gamma^N \cdot v_{rel}^N, \tag{2}$$

$$F^{T,ij} = min(-m_{eff}^{ij}\gamma^T|v_{rel}^T|, \mu|F^{N,ij}|), \tag{3}$$

where E_p is the particle YOUNG's modulus and A_p is the overlapping area, see also Fig. 3. γ^N denotes the damping coefficient and γ^T and μ are the friction

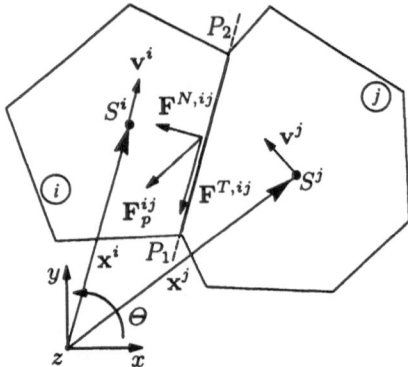

Fig. 3. To calculate the elastic contact force $\mathbf{F}_p^{ij}$ between two particles i and j one has to obtain the overlap area (white area). The intersection points P_1, P_2 define the contact line $P_1 P_2$. The force is applied at the centre of the contact line and the direction of the force is perpendicular to $P_1 P_2$.

coefficients. The relative velocity at the contact and the effective mass can be computed according to

$$\mathbf{v}_{rel} = \mathbf{v}^j - \mathbf{v}^i \qquad m_{eff}^{ij} = \frac{m^i \cdot m^j}{m^i + m^j}, \qquad (4)$$

where $\mathbf{v}^i$ and m^i refer to the velocity and mass of a polygon i, respectively.

In order to bond the particles together it is necessary to introduce a cohesion force between neighbouring polygons. For this purpose we introduce beams, which were extensively used recently in crack growth models [8, 9]. The centres of mass of neighbouring polygons are connected by elastic beams, which exert an attractive, restoring force between the grains, and can break in order to model the fragmentation of the solid. Because of the randomness contained in the VORONOI–tessellation the lattice of beams is also random. An example of a random lattice of beams coupled to the VORONOI polygons can be seen in Fig. 4. A beam between sites i and j is thought of having a certain cross section A^{ij} giving to it not only longitudinal but also shear and bending elasticity. This cross section is the length of the common side of the neighbouring polygons in the initial configuration. The length of the beam l^{ij} is defined by the distance of the centres of mass. The elastic behaviour of the beams is governed by two material dependent constants. For a beam between sites i and j the normal, shear and bending flexibilities are given by

$$a^{ij} = \frac{l^{ij}}{E_b A^{ij}} \qquad b^{ij} = \frac{l^{ij}}{G_b A^{ij}} \qquad c^{ij} = \frac{l^{ij^3}}{E_b I^{ij}}, \qquad (5)$$

where E_b and G_b are the YOUNG's and shear moduli of the beam, A^{ij} is the area of the beam section, and I^{ij} is the moment of inertia of the beam for flexion. A fixed value of E_b was used for all beams and b^{ij} was chosen to be $b^{ij} = 2a^{ij}$,

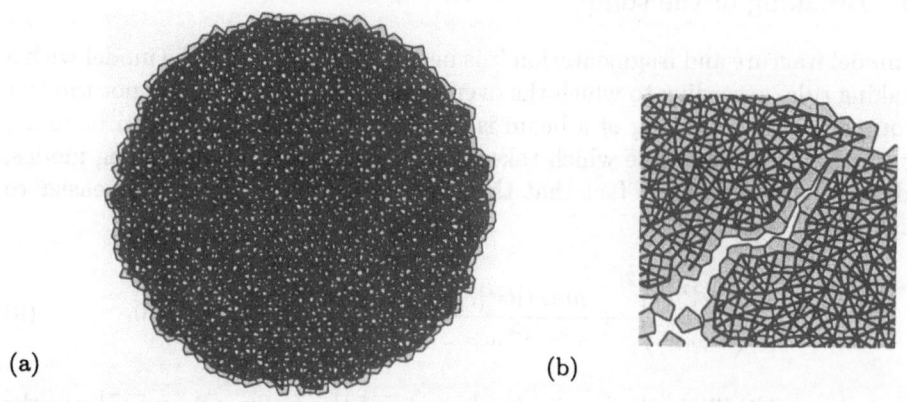

Fig. 4. (a) Elastic beams connecting the VORONOI polygons in a disc-shaped sample. Due to the randomness in the VORONOI tessellation the lattice of beams is also random. (b) At the broken beams along the side of the polygons cracks are formed inside the solid.

representing a POISSON's ratio of the beams of $\nu_b = 0.0$. The length, the cross section and the moment of inertia of each beam are determined by the random initial configuration of the polygons giving rise to the disorder in the beam lattice. The beam YOUNG's modulus E_b and the particle YOUNG's modulus E_p are, in principal, independent.

In the local frame of the beam three continuous degrees of freedom are assigned to both lattice sites (centres of mass) connected by the beam, which are for site i, the two components of the displacement vector (u_x^i, u_y^i) and a bending angle Θ^i. For the beam between sites i and j one has the longitudinal force acting at site i:

$$F_{b,x}^i = \alpha^{ij}(u_x^j - u_x^i), \qquad (6)$$

the shear force

$$F_{b,y}^i = \beta^{ij}(u_y^j - u_y^i) - \frac{\beta^{ij}l^{ij}}{2}(\Theta^i + \Theta^j), \qquad (7)$$

and the flexural torque at site i

$$M_{b,z}^i = \frac{\beta^{ij}l^{ij}}{2}(u_y^j - u_y^i + l^{ij}\Theta^j) + \delta^{ij}l^{ij^2}(\Theta^j - \Theta^i), \qquad (8)$$

where $\alpha^{ij} = 1/a^{ij}$, $\beta^{ij} = 1/(b^{ij} + 1/12c^{ij})$, and $\delta^{ij} = \beta^{ij}(b^{ij}/c^{ij} + 1/3)$. Consequently, this calculation formulae resemble the TIMOSHENKO beam theory. It can be shown that this beam model is a discretization of the simplified COSSERAT equations of continuum elasticity which can be used to describe the elastic behaviour of the granular solids, instead of the LAMÉ equations [23].

2.3 Breaking of the solid

To model fracture and fragmentation it is necessary to complete the model with a breaking rule, according to which the over-stressed beams break. For not too fast deformations the breaking of a beam is only caused by stretching and bending. We impose a breaking rule which takes into account these two breaking modes, and which can reflect the fact that the longer and thinner beams are easier to break [8]:

$$p_b^{ij} = \left(\frac{\epsilon_b^{ij}}{\epsilon_{b,max}} \right)^2 + \frac{max(|\Theta^i|, |\Theta^j|)}{\Theta_{max}} \geq 1 \qquad ; \ \epsilon_b^{ij} \geq 0, \qquad (9)$$

where $\epsilon_b^{ij} = \Delta l^{ij}/l^{ij}$ is the longitudinal strain of the beam, Θ^i and Θ^j are the rotation angles at the two ends of the beam and $\epsilon_{b,max}$ and Θ_{max} are threshold values for the two breaking modes. In the simulations we used the same threshold values $\epsilon_{b,max}$ and Θ_{max} for all the beams.

The first term of Eq. (9) takes into account the role of stretching and the second term the role of bending. Varying the threshold values the relative importance of the two modes in the beam breaking can be controlled. The primary microscopic fracture process within geomaterials is caused by a tensile failure of the cohesive bonds between the grain boundaries. The beams and their fracture are thought to represent this cohesive effect within this combined beam–particle model. Therefore, the breaking of beams in the simulations is allowed solely under stretching.

The time evolution of the system is followed by solving numerically the equation of motion of the individual polygons (molecular dynamics) for the translational and rotational degrees of freedom:

$$m^i \ddot{\mathbf{x}}^i = \sum_{j=1}^{N} \mathbf{F}^{ij},$$
$$i = 1, \ldots, N \qquad (10)$$
$$I_p^i \ddot{\Theta}^i = \sum_{j=1}^{N} M_z^{ij},$$

where N is the total number of polygons in the sample, and I_p^i denotes the moment of inertia of polygon i with respect to its centre of mass. The force $\mathbf{F}^{ij}$ and the torque M_z^{ij} contains the contribution of the polygon-polygon contacts in Eqs. (2), (3) and that of the beams in Eqs. (6), (7), (8). In the simulation code a GEAR Predictor-Corrector scheme of fifth order is used for the solution of the NEWTON equations in Eq. (10).

During the simulation the left hand side of Eq. (9) is evaluated at each iteration time step for all the existing stretched beams. The breaking of beams means that those beams for which the condition of Eq. (9) holds are removed from the calculation (see also Fig. 4 (b)), *i.e.* their elastic constants are set to zero. It should be noticed that an update of the beam connections is carried out after each time step after solving the equation of motion of all particles and taking into account the updated coordinates of the centres of mass. Removed beams are never restored during the simulation. The surfaces of the grains, on which beams are broken, represent cracks. The energy of the broken beams is released in creating these new crack surfaces inside the solid.

2.4   Stress calculation

A well established quantity within the framework of constitutive modelling of granular media is the average stress tensor. The basic idea pointed out in various publications in the recent years is the achievement of a homogenized solution of the stress state within a sample of contacting particles by analysis of the internal or external work within a specific sample volume V. We extend this idea to the combined beam–particle model by enhancing the general particle contact related formulation by a beam connection related part. The following derivation is based on the external work done at the sample under consideration of the application of the divergence theorem to the weak form of the equilibrium equations, similar to the procedure in [6]. The average stress tensor can be cast in matrix form:

$$\overline{\sigma} = \frac{1}{V} \sum_c \mathbf{F}_c \otimes \mathbf{x}_c \qquad \text{with} \qquad \mathbf{F}_c = \begin{bmatrix} F_{p,x}^{ij} + F_{b,x}^{ij} \\ F_{p,y}^{ij} + F_{b,y}^{ij} \end{bmatrix} \tag{11}$$

The complete force vector $\mathbf{F}_c$ neglects the moment components and contains only the x- and y-components of the particle contact force in Eq. (1) and those of the beams in Eqs. (6) and (7). $\mathbf{x}_c$ denotes the position vector of the boundary particles, where the loading is applied to the sample, and contains the coordinates of the corresponding centres of gravity.

The summation in Eq. (11) is taken over all contacts c within the particle set. In this context the term 'contact' is used for the description of both, particle contacts and beam connections. It should be noted that in the two–dimensional framework considered here the sample volume V reduces to the area enclosed by the centres of gravity of the boundary particles. The stress calculation procedure is applied solely to the quasi–static loading simulations with dense granular systems in section 3.1 within this paper. Therefore, the assumption of neglecting the particle rotations within the calculation of the stress tensor, similar as applied for the relative velocity in Eq. (4), seems quite comprehensible [2].

3   Simulation results

In order to illustrate the application range of the combined beam–particle model numerical simulations taking into account samples with different geometries and particle quantities have been carried out. Examples ranging from the elastic deformation over the fracturing to the complete fragmentation of a solid are examined. The simulations can be categorized into two groups: Quasi–static uniaxial loading and shearing of rectangular solids or dynamic loading situations, like impact, explosion or collision simulations of solids.

3.1   Quasi–static loading scenarios

The following simulation results are thought as academic examples highlighting the basic features of the combined beam–particle model in the framework of quasi–static application in the field of solid and damage mechanics. The emphasis lies on the qualitative application of the model to realistic experimental setups in order to reproduce typical failure phenomena in heterogeneous materials and to investigate the microstructural failure in terms of statistical considerations [4]. It should be mentioned that no parameter identification with respect to the output of experiments has been done for the following simulations, yet. Therefore, within the oncoming diagrams units are not indicated in order to focus only on the qualitative aspects of the simulations. A collection of the most important parameters is given in Table 1 at the end of this section.

Compression Simulations A rectangular specimen consisting of 1000 particles (25 × 40) with a height to width ratio of 1.6 was uniaxially loaded in vertical direction under constant strain rate conditions by moving inward the adjacent loading plates, as shown in Fig. 5.

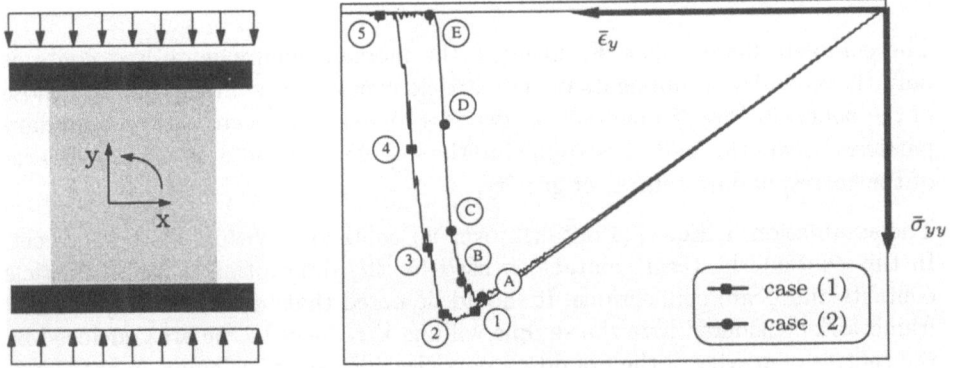

Fig. 5. Loading setup and stress-strain diagrams of compression simulations.

Within this compression study two different boundary conditions have been examined: with and without lateral confinement of the inferior and upper boundaries of the specimen, noted as case (1) and (2) within the following text. In order to provide the lateral confinement for case (1) the boundary particles have been kept fixed in x- and Θ-direction, whereas no lateral confinement of these boundary particles is provided for case (2). The stress-strain diagrams for both cases are plotted in Fig. 5 and reveal a linear-elastic regime with a constant 'macroscopic' elasticity modulus up to the peak, followed by a rather sharp drop indicating the 'macroscopic' disintegration of the material. In this framework the beam breaking can be considered as 'microscopic' failure of the material and the dominating energy dissipation mechanism. The complete failure is a highly

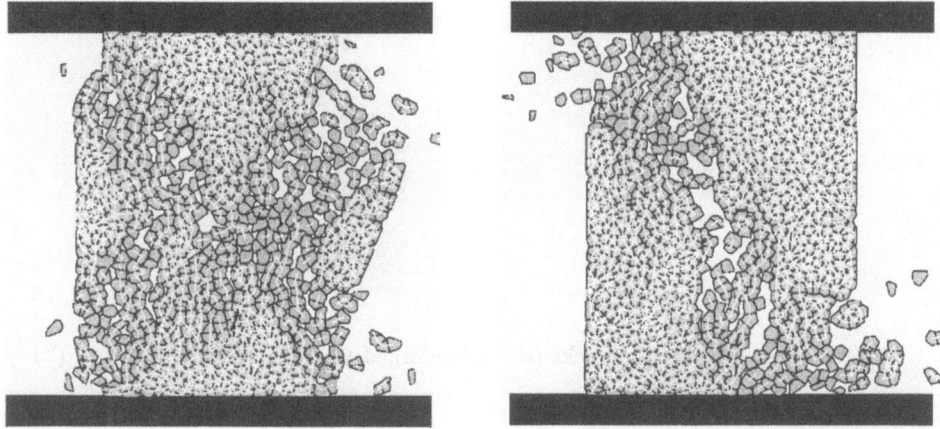

Fig. 6. Fractured state of specimens with (1) and without lateral confinement (2).

dynamic process driven by progressive beam bursts and determines the sharp drop within the softening region in Fig. 5. The average longitudinal strain $\bar{\epsilon}_y$ in the diagram is computed by averaging the length decrease over the horizontal cross–section, while the corresponding normal stress $\bar{\sigma}_{yy}$ in y-direction is calculated according to the average stress principle in Eq. (11).

As shown in the graphical output of the simulation program in Fig. 6 for case (1), following a localization of deformations on diagonal bands within the specimen, the failure planes start to orient on the two main diagonals connecting the edges of the specimen. The loss of lateral stiffness due to progressive beam breaking in the horizontal direction leads to a bulging of the specimen accompanied by the development of column–type structures on the two diagonals of the specimen. This hourglass failure mode agrees qualitatively well with the failure formation of uniaxially compressed concrete cylinders with a persistent effect of friction. The development of a biaxial compressive state hinders cracks to appear within a triangular region neighbouring the boundary confinements.

Similar as observed in experiments with concrete in [30] a buckling of the arising load transferring 'particle bridges' leads to the final failure of the tested specimens and defines the dominant microstructural effect for cracking in compressive situations. The array of parallel splitting cracks enclosing the vertical particle bridges becomes unstable at a certain load level and buckles [4], as schematically depicted in Fig. 7 (a), thus leading to a diagonal failure zone. For comparative reasons an enlarged view of a diagonal shear zone taken from the output of the simulation program is given in Fig. 7 (b) and indicates the agreement. Experiments with different sands and rod–like particles in [20] confirm the described failure mechanism and thus the results of our numerical calculations and the experiments with concrete in [30]. It is noteworthy to point out that the same basic failure mechanism may appear in a non–cohesive material like the sand tested in [20] as well as in strongly cohesive materials like the con-

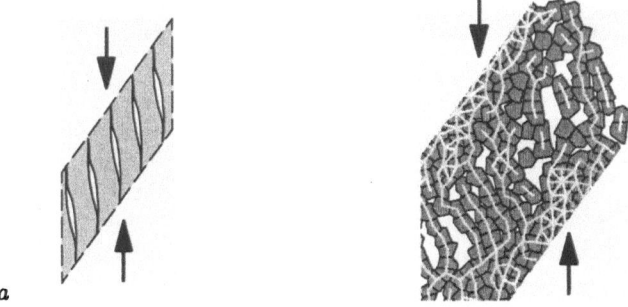

Fig. 7. Array of splitting cracks (a) and enlarged view of simulation output (b).

crete tested in [30]. Further, the effect of the rotational resistance, intrinsically included within our model, is emphasized as a basic microdeformation mechanism in [20]. An example for the effect of the rotational resistance at contacts is shown in Fig. 8 for the lateral boundary zone of a compressed sample. The evolution of tensile cracks parallel to the loading direction (①, ②) lead to a rotation of the particles within a particle column (③, ④) and thus to a bending of the column. The failure results in a buckling of this particle chain due to the bending mechanism (⑤). Interestingly, with the combined beam particle model the same effect as reported in [10], where a discrete element simulation with spheres including a rolling resistance term has been applied to the examination of shear band development, is reproduced.

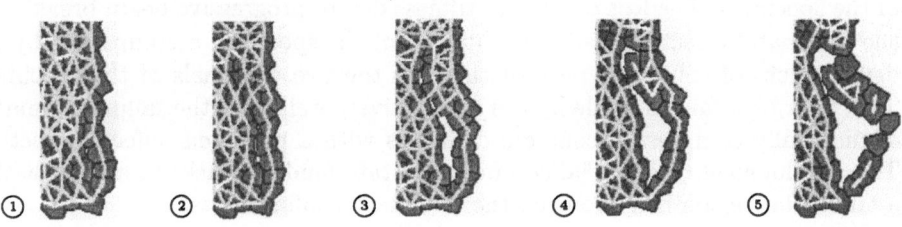

Fig. 8. Buckling of a boundary particle column within a compression simulation.

Based on this investigation it can be concluded, that in compressive simulations tensile splitting on the microlevel leads to a shear-crack failure on the macrolevel and therewith resembles correctly the reported crack mechanisms, which are difficult to observe in experiments.

In contrast to the failure mode for the confined specimen, in case (2) the macroscopic failure behaviour is determined by two non–crossing shear lines emerging from the right side, moving to the bottom and then growing upwards to the opposite edge. Again, similar to experiments with concrete prisms [29] loaded between brushes internally inclined fracture planes develop and by shearing and

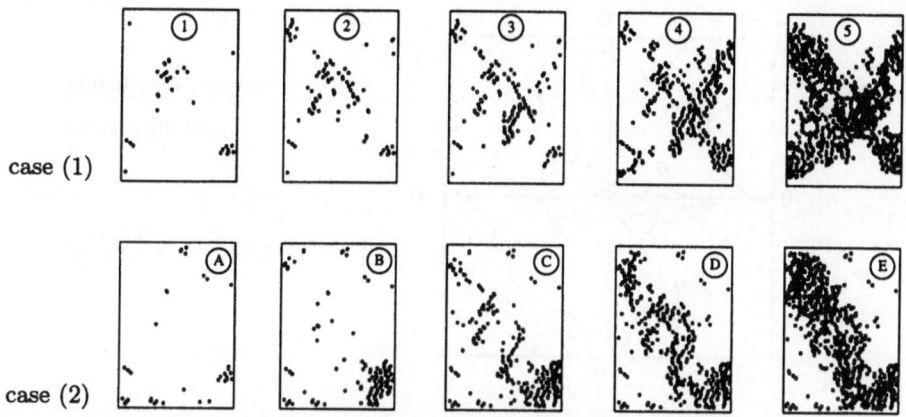

case (1)

case (2)

Fig. 9. Failure evolution for cases (1) and (2).

buckling of the corresponding load transfer bridges determine the global fracture behaviour. Midpoints of the broken beams marked in Fig. 9 give further evidence on the cracking within the specimens [16] and thus give a better insight of the developing failure planes. The corresponding picture shows the history of the failure evolution according to their appearance in the stress-strain diagram in Fig. 5. Simulations with varying slenderness ratios, not reported further, led to a good qualitative agreement with the experimental fracture observations regarding the inherent geometric size effect, as described in [29].

In the context of a geometric description of the state of damage different failure related characteristic quantities can be examined. The most obvious one is the amount of broken beams within a sample and the corresponding relation between the time or strain scale and the density of broken beams. This measure is an integral quantity since it does not depend on the position of the corresponding beams within the sample. Numerical calculations with different particle numbers under consideration of a constant height to width ratio confirm a strong size dependent effect. With an increasing particle and subsequent beam amount a lower value of the density of broken beams at peak load and at the end of the softening path is found. The size dependent behaviour is in agreement with the theoretical calculations for pure lattices shown in [5] and clarifies that the broken beam density is no objective measure of the state of damage within the specimen.

The polar plot of the damage fabric represents a more objective failure related characteristic quantity in the context of a geometric damage description. Hence, this is a convenient way to monitor the evolution of fractured beams with respect to the directionality. For case (1) the polar graphs are drawn in Fig. 10 according to the points defined in the diagram of Fig. 5. The outer graph shows the initial beam distribution of the examined sample and could be considered as limiting fracture distribution of all beams. All distributions are normalized by the factor

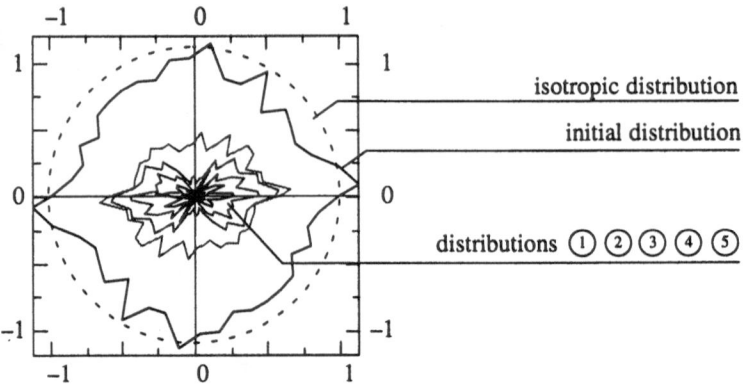

Fig. 10. Damage distribution of evolving failure for case (1).

$1/2\pi$. It should be noted, that the polar plot of this data is symmetric with respect to the origin, due to a double counting of each beam as a result of the numerical update algorithm. A comparison of the initial distribution with the dotted circular curve shows the good agreement with the completely isotropic distribution. The tendency to an anisotropic failure of the sample for the confined specimen (case (1)) within the descending branch of the stress-strain diagram in Fig. 5 is apparent. Interestingly up to point ④ almost half of the horizontal beams ($-45° \leq \alpha \leq 45°$) have been broken, while nearly no breaking is observed in the vertical direction. The previously noted angles are defined against the horizontal x-axis of the sample. As described before the remaining vertical beam chains are still able to transfer the load and therefore a complete failure results first by a breakage of this transfer mechanism up to point ⑤ after the brittle unloading of the material.

The defect correlation length can be considered as a further failure related characteristic quantity, which will be studied for case (2) within the oncoming paragraphs. Due to its coupling to the particle size it represents the inherent internal length scale in the model. The distribution of distances of two consecutively broken beams $d_{i,i+1}$ is depicted in the histogram in Fig. 11 (a). Curve Ⓐ shows

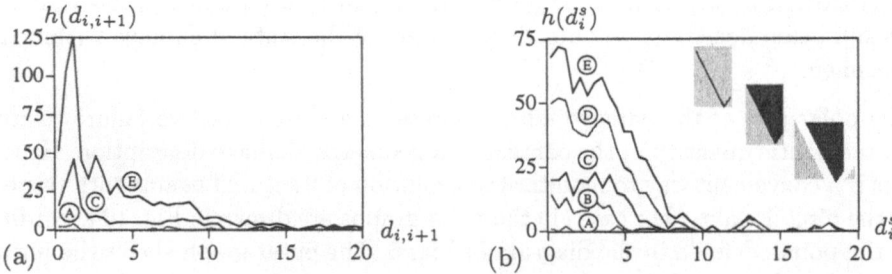

Fig. 11. Distance from next crack (a) and from shear fault (b).

a distributed behaviour of the distances of successive defects in the range of half the system height. At this stage the next broken beam can be found anywhere within the sample with equal probability. From stage ©️ up to stage Ⓔ a clear preference of the small distances can be observed and thus a correlation of the defects is obtained. According to a recent publication [12] this behaviour could be interpreted as a switch–over from a crack nucleation to a crack propagation mode. The theoretical ansatz presented in [5] based on the separation of the histogram into an exponential and a constant part introduces a defect correlation length. Depending on the existence of this length for a certain data representation, the transition point between the two described modes and the corresponding time stage can be determined. In our case, similar as for the simulations with the models in [12] and [5] no distinct correlation length can be determined as defined by such an ansatz, but the general trend is obvious. Interestingly, as shown in [11] earlier experimental oriented publications reveal a similar effect by the application of this ideas to acoustic emission tests of rocks. The distribution of the normal distances to the principal shear fault d_i^s, as shown in Fig. 11 (b), points out the finite width of the evolving shear zone. A plateau for distances between 0 and 5 can be clearly recognized, whereby the average beam length is 1.1 and the average particle size is 1. With increasing vertical loading the specimen separates into three parts that are pushed apart as schematically depicted within the diagram in Fig. 11 (b). This fracture behaviour not only leads to a nearly constant increase of distances between 0 and 5, but also results in an increase of distances in the area between 5 and 8, representing the border zone of the shear fault. This underlines the fact that the shear zone width remains constant from the beginning of the localization at stage Ⓑ to the end of the faulting at stage Ⓔ. With the help of the histogram in Fig. 11 (b) the thickness of the shear band can be determined. Neglecting the effect of the described block separation mode a thickness of 10 times the average particle size is found. The histograms in Fig. 12 show the distances of sequential failure event pairs in direction of the shear fault $d_{i,i+1}^t$ and perpendicular to it $d_{i,i+1}^n$. Interestingly, the probability to get very small distances tangential to the shear fault direction is very pronounced compared to the perpendicular direction.

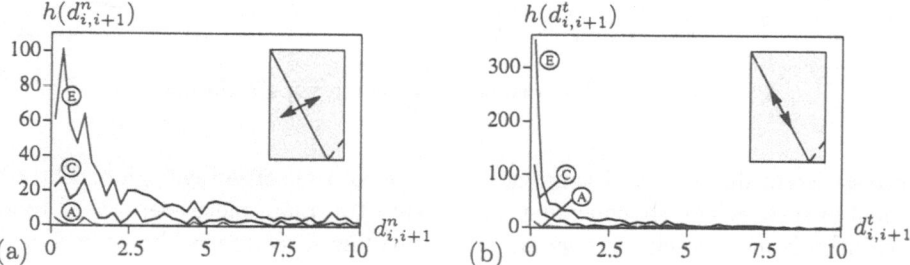

Fig. 12. Distance from next crack normal (a) and tangential (b) to shear fault direction.

In summary, the different failure related characteristic quantities have shown to help in understanding the general failure mechanism from a microscopic point of view. The crack density describes a non–objective integral quantity for the cracks within a specimen. The polar plots of the damage distribution contain information about the predominant failure directions and the correlation length of the defects characterizes the spatial distribution of the cracks.

A comparison with other discrete simulation schemes shows clearly the advantages of this model regarding the determination of the failure evolution. Pure static lattice simulations of compressive loading scenarios, as presented in [26] or [24] cannot lead to convincing results concerning the macroscopic failure behaviour, due to the lack of complexity within the formulation of the compressive beam/truss failure behaviour. The dynamic combined beam–particle model, however, is able to represent this complexity by its inherent particle contact feature, realistically modelling the motion within a compressed material sample. At least from that point on when a beam breaks and an open crack surface appears the behaviour is completely controlled by the particle dynamics and so influences the general failure mechanism significantly.

Tension simulation A similar particle sample as used for the compression simulations with a height to width ratio of 1.6 was subjected to uniaxial tension. On this account the acceleration and velocity of the particles at the inferior and superior boundary were pointing outwards of the specimen.

In this case the failure behaviour is completely determined by the beam lattice and its corresponding stiffness and statistical values. Again, the stress-strain relation results in a linear response up to the peak level [4]. In contrast to the compression simulation almost no fluctuations within the stress-strain diagram, as a result of an energy release due to beam cracking, are observed before peak. The failure evolution is depicted in Fig. 13 starting from the peak point of the

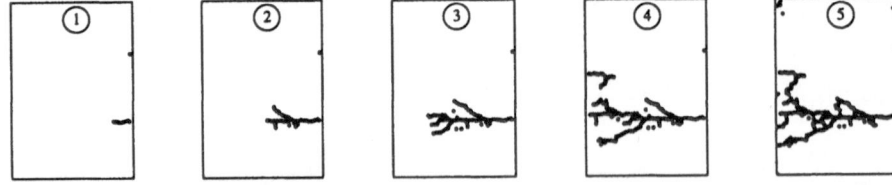

Fig. 13. Failure evolution of the tension simulation.

stress–strain diagram and ranging to the end of the softening region. The fragmented state where the specimen breaks into two main pieces separated by one main horizontal crack, orthogonal to the loading direction is shown in Fig. 14, right. Additionally, some short arrested cracks starting from the main crack appear, indicating the effect of a bending moment resulting from an increasing load eccentricity due to crack formation. It should be noticed that the macroscopic failure starts by the cracking of one weak beam with a low effective stiffness at the boundary. In comparison to the compressive simulations the failure zone is highly localized. Again, the observed failure pattern agrees with experimental results found in the literature [29]. Although, in contrast to this results, no uniformly distributed microcracking, often observed in the pre–peak region of tensile concrete experiments, could be reproduced. Due to the rapid effect of material degradation by the dynamically driven progressive beam breaking in combination with a narrow beam stiffness distribution, the sample tends to localize in a small band of adjacent particles.

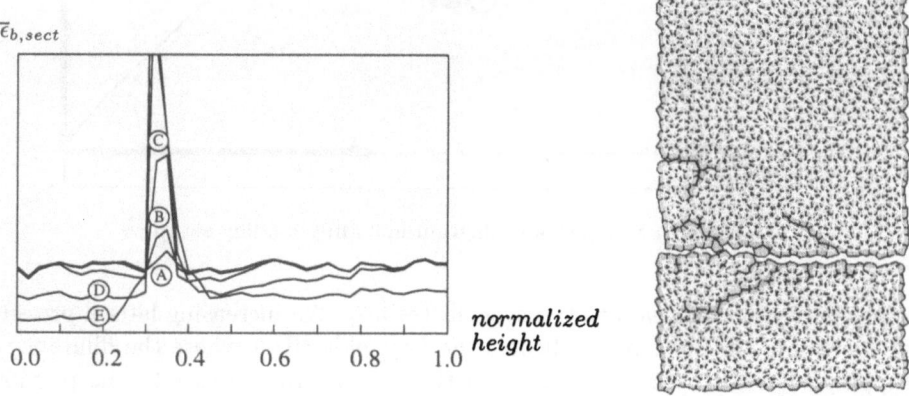

Fig. 14. Averaged beam strains and simulation output at stage ④ of simulation.

By taking the average value of the beam strain components in the vertical direction $\bar{\epsilon}_{b,sect}$ in every zone between two adjacent particle layers over the height a deeper insight into the localization of deformations can be achieved. Curves Ⓐ to Ⓔ in Fig. 14, left, characterize the situation before the peak stress is reached up to the end of the softening branch. Of course, this representation gives only a qualitative picture when compared to experiments like those presented in [3], where parallel strain gauges were used along the loading axis of an elongated beam to monitor the effect of microcracking. In the 'elastic' regime, before cracking occurs (Ⓐ) an almost constant distribution of the averaged strains over the height indicates the negligible effect of the inherent heterogeneity. The strain rate in this regime can be proven to be constant by the distribution of the averaged strain increments $\dot{\bar{\epsilon}}_{b,sect}$ over the specimen height. The localization pattern is clearly visible by an increase of the average strain (Ⓑ), followed by a decrease of $\bar{\epsilon}_{b,sect}$ first near the localization region (Ⓒ) and later on, on the complete upper and lower part of the specimen (Ⓓ,Ⓔ). This can be interpreted as a partial unloading of the material within the specimen thickness emerging in the vicinity of the crack and spreading to the boundaries.

Simple shear simulation A quadratic sample consisting of 900 (30 × 30) particles was sheared by a constant shear velocity v_s at the top, while the bottom was completely fixed. The height was kept constant in order to resemble the setup of a simple shear test as schematically depicted within the stress–strain diagram in Fig. 15. Within this figure the vertical average stress at the upper boundary $\bar{\sigma}_{yy}$

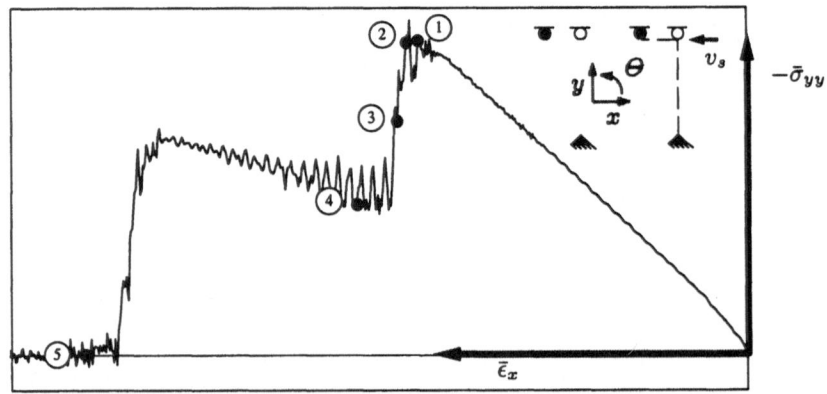

Fig. 15. Stress-strain diagram including loading setup.

is plotted versus the average transversal strain $\bar{\epsilon}_x$. An increasing lateral pressure at the top of the specimen indicates the Reynolds effect where the dilatancy is suppressed due to the fixed height of the sample. In contrast to the previous simulations no sudden failure is obtained. However, a drop to half of the peak value may be interpreted as the start of a shear band localization, as can be deduced from stages ① and ② within the illustration of the failure evolution in Fig. 16. Afterwards the vertical stress increases again as an outcome of a further suppression of the dilatation. Hence, the slope of the curve is smaller due to the fact that the influence of the particle contact stiffness is growing against the influence of the beam stiffness. Under consideration that the particle stiffness is 10 times lower than the beam stiffness (see Table 1) this behaviour is quite comprehensible. After beam breaking occurs within this diagonal band, the frictional particle contact plays the dominant role with regard to the global failure behaviour. At stage ④ the reaction force distribution is no longer uniformly distributed at the top boundary of the specimen. Larger pressure is needed at the

Fig. 16. Failure evolution of the simple shear simulation.

right corner to keep the initial height of the specimen compared to a low value at the left corner. Therefore, the beams within the right area are completely compressed. The complete failure of the specimen is accomplished by fracturing of these vertical beam chains as shown for the compressive simulation in Fig. 7. This can be concluded if the damage distributions at stages ④ and ⑤ in Fig. 17 (a) are compared as well.

Basically within the first part of the simulation beams oriented in the direction of the principal tensile stress break. The formation of a shear band from the lower left to the upper right corner becomes obvious, if the graphical output of the simulation program at stage ④ in Fig. 17 (b) is regarded. A tendency towards this direction can be recognized by examination of the inner curves of the damage distribution in Fig. 17 (a). In a later stage of the simulation represented by the outer curve, the trend shows no longer a pronounced diagonal direction, but rather a slight tendency towards the horizontal direction.

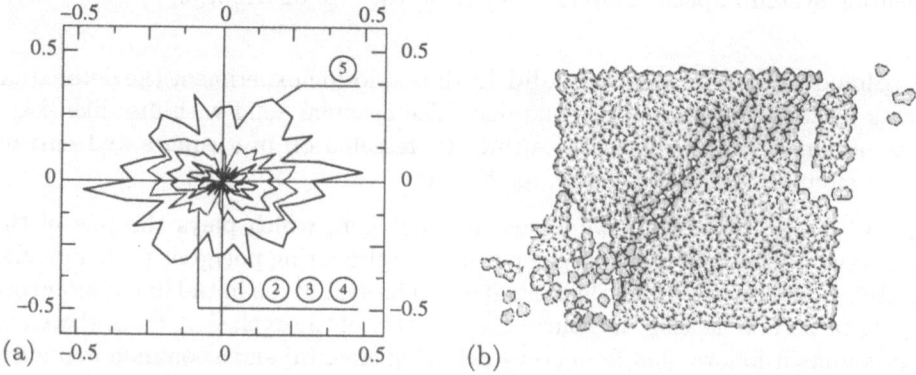

Fig. 17. Damage distribution (a) and graphical output at stage ④ of simulation (b).

Experimental data for shear failure are often compared to results using discontinuous models through notched four-point-shear beam experiments [22, 29] or related shear box experiments [25]. Nevertheless the results have shown that the general trend of shear simulations could be qualitatively accessed by the application of the combined beam–particle model.

3.2   Dynamic fragmentation of solids

Fragmentation, *i.e.* the breaking of particulate materials into smaller pieces is a ubiquitous process that underlies many natural phenomena and industrial processes. The length scales involved in this process range from the collisional evolution of asteroids to the degradation of materials comprising small agglomerates employed in industrial processes. On the intermediate scale there are several industrial and geophysical examples concerning the usage of explosives in mining and oil shale industry, fragments from weathering, coal heaps, rock fragments from chemical and nuclear explosions. The most striking observation about fragmentation is that the measured fragment size distributions exhibit power law behaviour with exponents between 1.9 and 2.6 concentrating around 2.4. Power law behaviour of small fragment masses seems to be a common characteristic of brittle fracture [1, 7, 8, 19, 21, 28]. Beside the size distribution of the debris, there is also particular interest in the energy required to achieve a certain size reduction. Collision experiments revealed that the mass of the largest fragment normalized by the total mass shows power law behaviour as a function of the specific energy, *i.e.* imparted energy normalized by the total mass [1, 7, 19].

In order to investigate catastrophic fragmentation of solids we performed simulations under three different experimental conditions: we studied fragmentation due to an explosion which takes place inside the solid [13], break-up of a solid block by shooting a projectile into it [13], and finally the collision of macroscopic bodies was considered [14, 15]. In the following we present a detailed analysis of the wave propagation, crack nucleation, crack growth and time evolution of fragmenting system. Special emphasis is put on the size distribution of fragments.

Explosion of a disc shaped solid In the explosion experiment the detonation takes place in the centre of a solid disc. The granular solid with disc-like shape was obtained starting from the VORONOI–tessellation of a square and cutting out a circular disc in the centre, see Fig. 4.

In the centre of the solid we choose one polygon, which plays the role of the explosive. Initial velocities are given to the neighbouring polygons perpendicular to their common sides with the central one. The sum of the initial linear momenta has to be zero, reflecting the spherical symmetry of the explosion. From these two constraints it follows that for a polygon having mass m^i and a common boundary of length A^{ij} with the explosive centre, the initial velocity is proportional to $\frac{A^{ij}}{m^i}$. The sum of the initial kinetic energies defines the energy E_o of the explosion. (For the parameter values and the initial conditions of the simulation see Table 1 at the end of this section.) As a result of these initial conditions a circularly symmetric outgoing compression wave is generated in the solid. In our context this means that there is a well–defined shell where the average longitudinal strain of the beams $< \epsilon_b^{ij} > = < \Delta l^{ij}/l^{ij} >$ is negative. This compression wave is not homogeneous in the sense that not all the beams in this region are compressed . If the angle of a beam with respect to the radial direction is close to $\pi/2$ a beam can be slightly elongated within the compression wave. Since the overall shape of the solid has the same symmetry as the compression wave it is possible to avoid geometrical asymmetries, which would arise for example in the explosion of a rectangular sample due to the corners. In the disordered solid the initial compression wave gives rise to a complicated stress distribution, in which the over-stressed beams break according to the breaking rule Eq. (9). The simulation is stopped if there is no beam breaking during 300 successive time steps. Free boundary conditions were used in all simulations.

Due to the beam breaking the solid eventually breaks apart, *i.e.* at the end of the process it consists of well separated groups of polygons. These groups of polygons, connected by the remaining beams, define the fragments. In the simulation of the explosion we are mainly interested in the time evolution of the fragmentation process and the mass distribution of fragments at the end of the process. In the time evolution of the explosion two regimes can be distinguished. The initial regime is controlled by the compression wave and the disorder of the solid. The amplitude of the shock wave is proportional to the ratio of the average initial velocity of the polygons to the longitudinal sound speed of the solid. The width and the speed of the wave are mainly determined by the grain size and the YOUNG's moduli.

Since the beams are not allowed to break under compression the compression wave can go outward almost unperturbed and an elongation wave is formed behind it. Due to the elongation wave a highly damaged region is created in the vicinity of the explosive centre where practically all the beams are broken and all the fragments are single grains. This highly damaged region is called the mirror spot. Since the breaking of the beams, *i.e.* the formation of cracks in the solid, dissipates energy after some time the growth of the damage stops. The size of this mirror spot is determined by the initial energy of the explosion, by the dissipation rate and by the breaking thresholds (see Fig. 18). During and after the formation of the mirror spot when the outgoing compression and elongation waves go through the solid, the weakest (*i.e.* the longest and thinnest) beams break in an uncorrelated fashion creating isolated cracks in the system. The uncorrelated beam breaking is dominated by the quenched disorder of the solid structure. This first uncorrelated regime of the explosion process lasts till the compression wave reaches the free boundary of the solid. From the free boundary the compression wave is reflected back with opposite phase generating an incoming elongation wave. The constructive interference of the incoming and outgoing elongation waves gives rise to a highly stretched zone close to the boundary. The

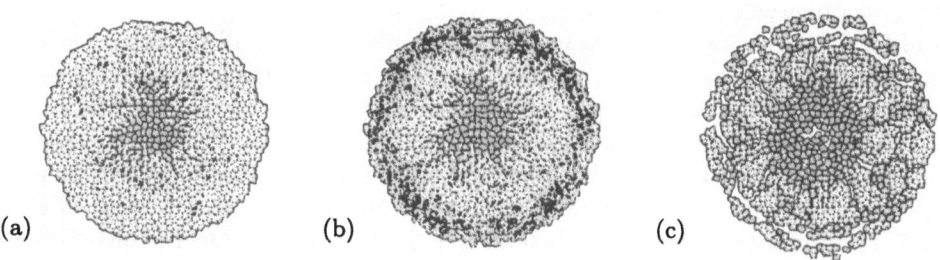

(a)   (b)   (c)

Fig. 18. Explosion of a disc-shaped solid. Snapshots of the evolving system are presented when the initial compression wave reaches the boundary of the solid ($t = 0.0001s$ (a)), the constructive interference of the incoming and outgoing elongation waves breaks the boundary layer ($t = 0.001s$ (b)) and the final breaking scenario ($t = 0.004s$ (c)). Black coloured beams indicate that the strain is close to the stretching threshold $\epsilon_{b,max}$.

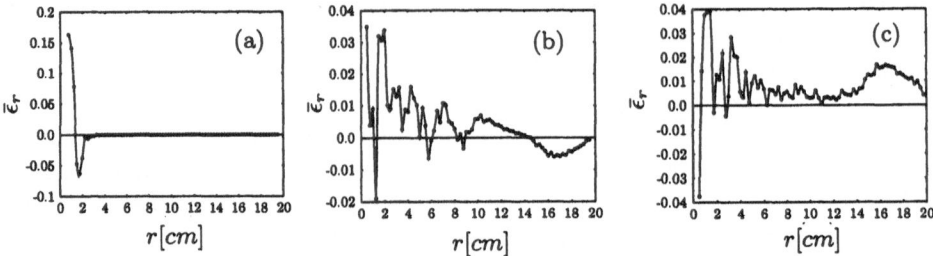

Fig. 19. Propagation of the elastic wave in the disc–shaped solid. (*left*) after $t = 10^{-5} sec$ of the initial hit, (*middle*) after $t = 3 \cdot 10^{-4} sec$ the compression wave is approaching the boundary and (*right*) after $t = 5 \cdot 10^{-4}$ the constructive interference of the incoming and the outgoing elongation waves.

beams having small angle with respect to the radial direction have the largest elongation. In this zone a large number of beams break causing usually the complete break-off of a boundary layer along the surface of the solid. The thickness of this detached layer is roughly half the width of the incoming elongation wave (see Fig. 18). The fragments of this boundary layer fly away in the radial direction with a high velocity carrying with them a large portion of the total energy in the form of their kinetic energy. After that the system starts to expand. This overall expansion initiates cracks going from inside to outside and from outside to inside. The branching of single cracks and the interaction of different cracks give rise to the final fragmentation of the solid (Fig. 18). This second part of the evolution of the explosion process is dominated by the correlation of the cracks.

The propagation of the elastic waves, when the beam breaking is switched off, is presented in Fig. 19 under consideration of the average radial strain $\bar{\epsilon}_r$. One can observe the peak of the initially imposed shock, the propagation of the compression and elongation waves and the formation of the highly stretched zones at the boundary.

We performed simulations alternatively fixing the two breaking parameters $\epsilon_{b,max}$ and Θ_{max} and changing the value of the other one, keeping all the other parameters of the simulations fixed. In both cases the mass distribution of fragments was obtained. The fragment mass histograms are presented in Fig. 20. The lower cutoff of the histograms is determined by the size of the unbreakable polygons (smallest fragments), while the upper cutoff is given by the finite size of the system (largest fragment). It can be observed that increasing the value of the varied parameter the distributions tend to a 'limiting curve', but the histograms follow a power law for practically all the parameter pairs for at least one order of magnitude in mass:

$$F(m) \sim \alpha m^{-\beta}. \tag{12}$$

The effective exponents β were obtained from the estimated slopes of the curves. Apart from the case of extremely small breaking parameters the exponent β only

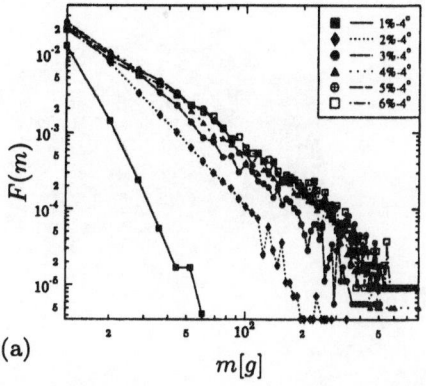
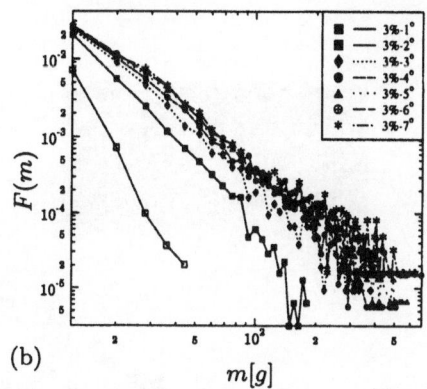

(a) $m[g]$

(b) $m[g]$

Fig. 20. (a) The fragment mass histograms varying the stretching threshold $\epsilon_{b,max} = 1\% - 6\%$. The bending threshold is fixed $\Theta_{max} = 4°$. The contribution of the single polygons are ignored. For increasing $\epsilon_{b,max}$ the curves tend to a limit, which is determined by the fixed bending mode. (b) The fragment mass histograms varying the bending threshold $\Theta_{max} = 1° - 7°$. The stretching threshold is fixed $\epsilon_{b,max} = 3\%$. For increasing Θ_{max} the curves tend to a limit, which is determined by the fixed stretching mode

slightly varies around $\beta = 2.0$, indicating a more or less universal behaviour within the accuracy, with which β was determined (± 0.05). These values of the exponent β obtained numerically are in reasonable agreement with most of the experimental results [13].

Impact of a projectile with a solid block Besides the explosion a catastrophic fragmentation of solids can also be generated by an impact with a projectile [18].

We applied our model to study the fragmentation of a rectangular solid block due to an impact. One polygon at the lower middle part of the block is given a high velocity directed inside the block simulating an elastic collision with a projectile. The boundary conditions and the stopping condition were the same as in the explosion experiment. The breaking thresholds were chosen $\epsilon_{b,max} = 3\%$ and $\Theta_{max} = 4°$. The evolution of the fragmenting solid block is presented in Fig. 21. As in the case of the explosion, the initially generated compression wave plays a significant role. Since the energy of the collision is concentrated around the impact site of the projectile the damage is the largest in that region. The completely destroyed zone, where all the beams are broken stretches inside the solid in the forward direction resulting in the break-up of the solid. When the shock wave reaches the boundary at the side of the solid opposite to the collision point it gives rise to the break-off of a boundary layer. The fragments of this layer fly away in the forward direction with a high velocity. Some small fragments from the vicinity of the collision point are scattered backward. The damage in the direction perpendicular to the projectile is not strong, the broken boundary

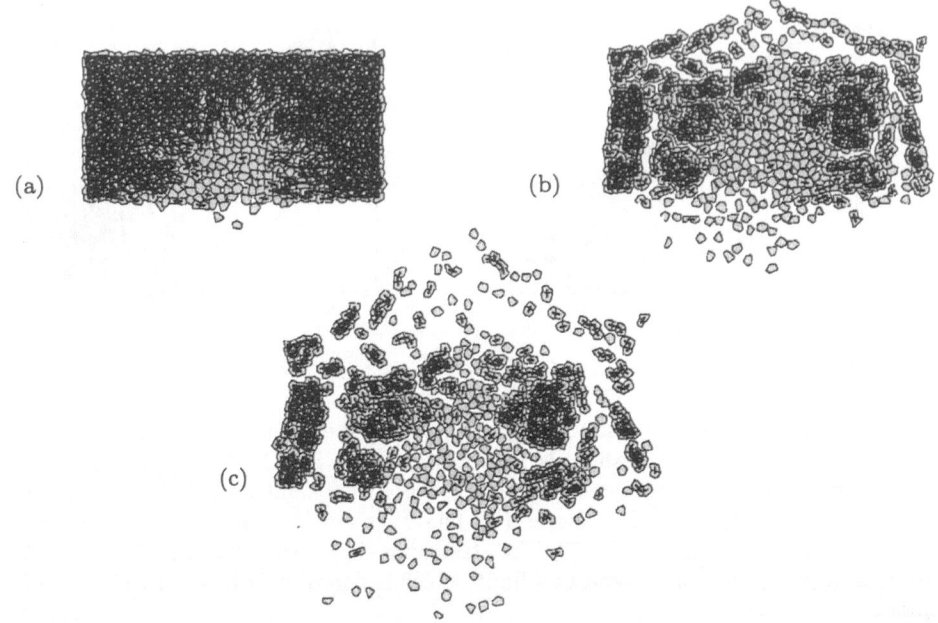

Fig. 21. Fragmentation of a block composed of elastic grains. One grain at the lower middle part of the block is given a high velocity directed inside the block. Here the velocity was $400m/s$. The size of the block was chosen to be $40cm \times 20cm$. Snapshots of the evolving system are presented at $t = 0.0004s$ (a), $t = 0.0015s$ (b) and $t = 0.003s$ (c).

layer is thicker and the speed of the fragments is smaller. Results of laboratory experiments on high velocity impacts can be found in Refs. [1, 7, 19, 21, 28]. In

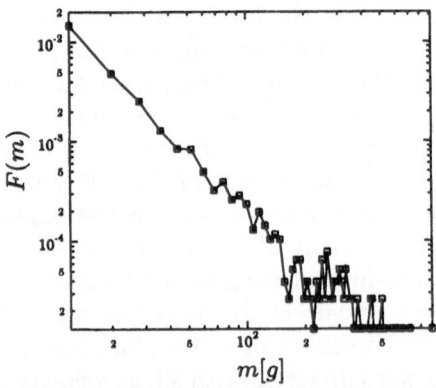

Fig. 22. Fragment mass histogram of the impact simulation.

Ref. [19] a picture series obtained by a high speed camera is presented showing the time evolution of an impact experiment. Our results are qualitatively in good agreement with the experimental observations.

The resulting fragment mass histogram $F(m)$ is presented in Fig. 22. Similarly to the explosion experiment, $F(m)$ shows power law behaviour for approximately one order of magnitude in mass. The value of the effective exponent is $\beta = 1.98 \pm 0.05$, which agrees well with most of the experimental results [14].

Collision of macroscopic bodies In nature the collision of macroscopic solids is a common fragmentation mechanism the consequences of which can be widely observed. It is well known that in the flow of granular materials a large part of the kinetic energy of the grains is dissipated in the vicinity of their contact zone during the collisions. Beside the viscous and plastic effects, the dissipation by damaging is also an important source of energy loss in the flow. Collision of particles occurs also in the solar system in planetary rings. In this case the energy dissipation due to impact damage might also influence the large scale structure formation in the rings. On larger length scale in the solar system, the so-called collisional evolution of asteroids due to subsequent collisions, and the formation of rubble piles in the asteroid belt is still a challenging problem [7]. Among industrial applications the breakup of agglomerates in chemical processes can be mentioned. Due to experimental difficulties, the computer simulation of microscopic models is an indispensable tool in the study of these impact

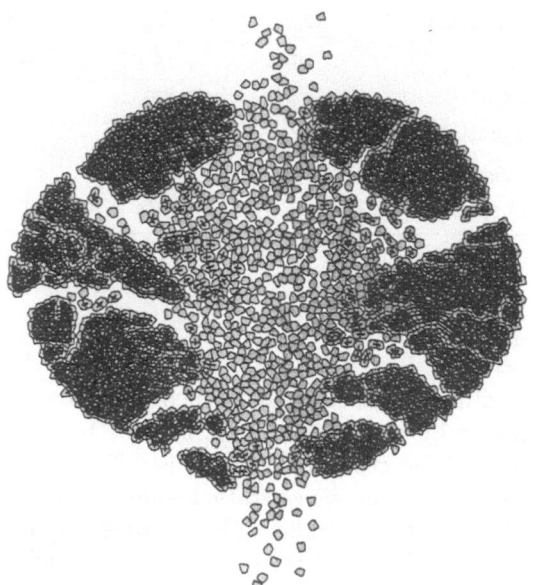

Fig. 23. Final scenario of a collision of two discs of the same size.

phenomena. To study fragmentation of macroscopic bodies due to collision we performed simulations of the collision of two–dimensional discs of the same size with zero impact parameter varying the energy of the collision in a broad interval. The typical final breaking scenario of a collision process can be seen in Fig. 23.

Analyzing the energetics of the fragmentation process and the resulting size distribution of fragments we identified two distinct final states of the collision process, *i.e.* damaged and *fragmented* states with a sharp transition in between. With a detailed study of the behaviour of the fragment mass distribution in the vicinity of the transition point and its dependence on the finite particle size we gave numerical evidence that the transition point behaves as a critical point and the *damage-fragmentation* transition occurs as a continuous phase transition. The control parameter of the transition was chosen to be the dimensionless ratio η of the energy of impact and the binding energy of the sample, and the order parameter was associated to the mass of the largest fragment divided by the total mass [15].

Table 1. The parameter values used in the simulations (*Sim* 1 = quasi–static simulations, *Sim* 2 = dynamic fragmentation simulations.

Parameter	Symbol	Unit	Sim 1	Sim 2
Density	ρ	g/cm^3	5	5
Particle YOUNG's modulus	E_p	dyn/cm^2	10^{10}	10^{10}
Beam YOUNG's modulus	E_b	dyn/cm^2	10^{11}	$5 \cdot 10^9$
Beam elongation threshold	$\epsilon_{b,max}$	%	3	3
Beam bending threshold	Θ_{max}	deg	3	4
Time step	dt	s	10^{-6}	10^{-6}
Diameter of the disc	d	cm	-	40
Energy of the explosion	E_o	erg	-	$5 \cdot 10^9$
Average initial speed	v_o	m/s	-	200
Estimated sound speed	c	m/s	-	900

4   Conclusions

We presented a two-dimensional discrete model of solids connecting unbreakable, undeformable elements by breakable, elastic beams. To demonstrate the capabilities of the model in the study of fracture and fragmentation processes we performed simulations of various quasi-static loading scenarios, furthermore, the model was applied to investigate catastrophic fragmentation of solids due to explosion, shooting a projectile, and due to collision. The results of the simulations were found to be in reasonable agreement with the experimental observations which shows that our discrete element simulation model is a powerful tool for the study of the degradation of geomaterials.

Still our study makes a certain number of technical simplifications which might be important for a full quantitative grasp of fracture and fragmentation phenomena. Most important seems to us the restriction to two dimensions, which should be overcome in future investigations. The existence of elementary, non-breakable polygons restricts breaking on lower scales and hinders us from observing the formation of powder of a shattering [8]. Further, a sound quantification of the model parameters concerning realistic adaptation to geomaterials should be contemplated for the future.

An advantage of our model with respect to other discrete models is that we can follow the trajectory of each fragment, which is often of big practical importance and that we know how much energy each fragment carries away. The polygonal structure of our solid allows us to realistically model granular or polycrystalline matter, and considering as well cell repulsion as beam connectivity gives us a rich spectrum of possibilities ranging from breaking through bending to the effect of dilatancy. If one or the other mechanism is turned off we have the extreme cases of an elastic homogeneous solid and a compact dry granular packing. In the framework of our discrete model it is rather straightforward to impose and study any kind of structural inhomogeneity. It was also demonstrated that during the simulation with this discrete model of solids it is possible to monitor such quantities which are hard to measure or not measurable. Hence, this way of treatment can provide a deeper understanding of the processes studied.

Acknowledgement

The authors are grateful for the financial support of the German Science Foundation (DFG) within the research group *Modellierung kohäsiver Reibungsmaterialien* under grant no.VE 163/4-1.4. F. Kun is indebted for the support of the Alexander von Humboldt Foundation and also for the Bolyai Janos fellowships of the Hungarian Academy of Sciences.

References

1. N. Arbiter, C. C. Harris, G. A. Stamboltzis: Soc. Min. Eng. **244**, 119 (1969)
2. R. J. Bathurst, L. Rothenburg: J. Appl. Mech. **55**, 17 (1988)
3. L. Cedolin, S. dei Poli, I. Iori: J. Engng. Mech. **113**, 327 (1987)
4. G. A. D'Addetta, F. Kun, E. Ramm: Int. J. Fract., submitted for publication (2000)
5. A. Delaplace, G. Pijaudier-Cabot, S. Roux: J. Mech. Phys. Solids **44**, 99 (1996)
6. A. Drescher, G. De Josseling De Jong: J. Mech. Phys. Solids **20**, 337 (1972)
7. A. Fujiwara, A. Tsukamoto: Icarus **44**, 142 (1980)
8. H.-J. Herrmann, S. Roux (eds.): *Statistical Models for the Fracture of Disordered Media* (North Holland, Amsterdam 1990)
9. H.-J. Herrmann, A. Hansen, S. Roux: Phys. Rev. B **39**, 637 (1989)
10. K. Iwashita, M. Oda: J. Engng. Mech. **124**, 285 (1998)
11. D. Krajcinovic: *Damage Mechanics* (North-Holland, Amsterdam 1996)
12. D. Krajcinovic, M. Vujosevic: Int. J. Solids Structures **35**, 4147 (1998)
13. F. Kun, H.-J. Herrmann: Comp. Meth. Appl. Mech. Engng. **138**, 3 (1996)

14. F. Kun, H.-J. Herrmann: Int. J. Mod. Phys. C **7**, 837 (1996)
15. F. Kun, H.-J. Herrmann: Phys. Rev. E **59**, 2623 (1999)
16. F. Kun, G. A. D'Addetta, H.-J. Herrmann, E. Ramm: Comp. Ass. Mech. Engng. Sci. **6**, 385 (1999)
17. K. B. Lauritsen, H. Puhl, H. J. Tillemans: Int. J. Mod. Phys. C **5**, 909 (1994)
18. S. A. Magnier, F. V. Donzé: Mech. Coh. Frict. Mat. **3**, 257 (1998)
19. T. Matsui, T. Waza, K. Kani, S. Suzuki: J. Geophys. Res. **87 B13**, 10968 (1982)
20. M. Oda, H. Kazama: Géotechnique **48**, 465 (1998)
21. L. Oddershede, P. Dimon, J. Bohr: Phys. Rev. Lett. **71**, 3107 (1993)
22. A. V. Potapov, M. A. Hopkins, C. S. Campbell: Int. J. Mod. Phy. C **6**, 399 (1995)
23. S. Roux: 'Continuum and discrete description of elasticity and other rheological behavior'. In Ref. [8], pp. 87-113
24. E. Schlangen, J. G. M. Van Mier: Cem. Concr. Res. **14**, 105 (1992)
25. E. Schlangen, E. J. Garboczi: Engng. Fract. Mech. **57**, 319 (1997)
26. H. Schorn: 'Numerical Simulation of Composite Materials as Concrete'. In: *Fracture Toughness and Fracture Energy of Concrete*, ed. by F. Wittmann (Elsevier, Amsterdam, The Netherlands 1986) pp. 177–188
27. H. J. Tillemans, H.-J. Herrmann: Physica A **217**, 261 (1995)
28. D. L. Turcotte: J. Geophys. Res. **91 B2**, 1921 (1986)
29. J. G. M. Van Mier: *Fracture Processes of Concrete* (CRC Press, Boca Raton 1997)
30. R. A. Vonk: *Softening of Concrete Loaded in Compression* (PhD Thesis, Technical University of Eindhoven, The Netherlands 1992)

Microscopic modelling of granular materials taking into account particle rotations

W. Ehlers, S. Diebels, T. Michelitsch

Institute of Applied Mechanics, Chair II, University of Stuttgart, Pfaffenwaldring 7, 70569 Stuttgart, Germany

Abstract. Granular material is characterized by the appearance of localization phenomena, as for instance the formation of shear bands under the influence of gravity. By means of a micromechanically motivated discrete element method (DEM), two-dimensional simulations of monodisperse circular disks are performed, where both translational and rotational degrees of freedom of the particles are taken into account by the consideration of *Newton*ian equations of motion for the translations and by *Euler*ian equations of motion for the rotations of the single particles. It turns out that even for the simplest contact laws, e. g. a combination of *Coulomb* and *Newton* type friction for the tangential contact of monodisperse particles and a repulsive damped spring normal contact force, shear bands are obtained. In the regime of small relative tangential velocities, the viscous part of the frictional contact law becomes effective. Then, "slow" relative tangential velocities are surpressed corresponding to an enforcing of rolling modes characterized by zero relative tangential velocities in the contact points and leading to an instability that corresponds to shear banding. The DEM simulations furthermore suggest that the size distribution of the assembly modifies the shape of the shear band but is not necessary for its formation. These propositions seem to be in agreement with the experimental observations reported in the paper by *Viggiani et al.* [23] included in this volume.

1 Introduction

The difficulty to describe the overall physical properties of granular material is caused by its internal irregularities. Granular material represents a state of matter that can neither be fully described as a standard solid nor as a regular fluid. Solid granular material as for instance concrete or rock is characterized by a huge amount of cohesive friction, while non-cohesive materials such as sand behave like a solid under pressure but like a fluid when fluidizing effects occur. Furthermore, this behaviour may depend on the time scale of observation.

So far, no general constitutive law exists that characterizes all forms of appearance of granular material. The macroscopic response of granular media on external loading is governed by very complex non-linear hysteretic stress-strain curves in the elasto-plastic or the elasto-viscoplastic regime which are strongly sensitive to the loading history, indicating irreversible changes of sites of the particles [14, 21]. Furthermore, a significant volume increase (dilatancy) can occur during shear loading [14, 22]. A cohesive granular material can occur as a solid on a short time scale, but after a long time of observation, the fluid character may reveal. For non-cohesive granular material such as sand or powders, a typi-cal feature is the absence of the elastic domain even for small shear-loading, but the immediate occurrence of plastic flow. These observations indicate that even arbitrarily small shear stresses immediately induce irreversible particle motions. In the present simulation, there is some evidence that the coupling of transla-tional and rotational degrees of freedom of the particles plays an important role for this effect. The complexity of the constitutive behaviour of granular material is caused by the fact that the macroscopic response of a granular assemblage is an interplay of the distributions of particle sizes, shapes and spatial distribution, the particle density, the distribution of particle contacts (and related to it, the fabric tensor) as well as the constitutive law that holds for the single particle itself. This diversity of influences governing the constitutive behaviour of granu-lar material makes it very difficult to directly relate the macroscopic behaviour to the microscopic properties of the particles. Notwithstanding these difficulties, considerable work has been done in the last few decades to establish constitutive laws that hold at least in limited ranges [1, 9, 11, 14].

In principle, there are two different strategies to describe granular matter. One strategy is the microscopic approach where numerical computations are carried out by use of the discrete element method (DEM) developed in 1979 by *Cundall & Strack* [7]. The DEM is physically motivated and employs the *New-tonian* equations of motion for each particle. Meanwhile, the DEM has become a standard method to simulate 2-d or 3-d assemblies, and it has turned out that it provides reasonable benefit to investigate processes *inside* a granular system [17, 21]. Despite this approach is physically motivated, one does not get around to introduce phenomenological assumptions on the contact forces between the particles. The disadvantage of the microscopic approach is the huge number of degrees of freedom (equations of motion) which have to be integrated numeri-cally. On the other hand, the macroscopic approach removes this disadvantage by introducing only a few field variables that contain the physical information as a result of a homogenization process consisting in smearing out the microscopic degrees of freedom over a representative elementary volume (REV). However, the complexity of macroscopic models stems from the difficulty to accurately describe the material response. Nevertheless, granular material has been suc-cessfully described on the macroscale by use of the Theory of Porous Media (TPM) established by *Bowen* [4, 5], *de Boer & Ehlers* [3] and *Ehlers* [10, 11]. The TPM approach is either based on a real or on a virtual averaging pro-cess relating the microscopic to the macroscopic scale. Furthermore, the TPM describes the behaviour of porous materials such as granular soil or polymeric or metallic foams on a scale much larger than the characteristic diameter of the pores. In addition, there is a micropolar extension of the TPM [8, 9, 12–14] which represents a *Cosserat* approach [6, 15] to porous materials. Apart from the translational degrees of freedom, this approach takes into account the rotational motion of the material points in the sense of additional degrees of freedom.

An important characteristic property of granular material is the appearance of localization phenomena such as shear bands under certain loading conditions as for instance the influence of gravity or external shear loading. Macroscopic model computations [12–14] indicate that shear band formation is caused by microscopic failure which takes place locally depending on the external loading conditions and on the spatial fluctuations of the material characteristics. This localized failure then leads to instabilities as for instance shear banding, slope slidings and avalanching. However, the determination of an internal length scale to yield the shear band width is still an open problem. There are conjectures that the shear band width has the order of magnitude of ten (average) particle diameters [18, 23], independent of the size of the sample. This magnitude seems to be related to the correlation length of the granular material. But so far, according to our knowledge, a rigorous theory yielding the characteristic length of shear bands is still absent.

The goal of this paper is to show that the microscopic DEM approach leads to good results in modelling shear band failure and furthermore gives the macroscopic TPM its micromechanical foundation.

2   Kinematics

Using the DEM approach, the granular material is modeled as an assemblage of N *rigid* particles. Each particle i has the constant mass[1]

$$m = \int_M dm. \tag{1}$$

The spatial position $\mathbf{x}$ of any material point P within an arbitrary particle is given by the position $\mathbf{x}_M$ of the center of mass M of this particle and by the local body-fixed position vector $\bar{\mathbf{x}}$, compare Fig. 1:

$$\mathbf{x}(t) = \mathbf{x}_M(t) + \bar{\mathbf{x}}(t). \tag{2}$$

Therein, the position of the center of mass is defined by

$$\mathbf{x}_M(t) = \frac{1}{m} \int_M \mathbf{x}\, dm. \tag{3}$$

Furthermore, the rigid body assumption requires that $|\bar{\mathbf{x}}| = \text{const.}$ Following this, the motion $\bar{\mathbf{x}}(t)$ of a material point with respect to the center of mass of the particle consists of a pure rotation given by

$$\bar{\mathbf{x}}(t) = \mathbf{R}(t)\, \bar{\mathbf{x}}(t_0). \tag{4}$$

Therein, $\mathbf{R}$ is a proper orthogonal tensor exhibiting

$$\mathbf{R}^T \mathbf{R} = \mathbf{I} \quad \text{and} \quad \det \mathbf{R} = 1. \tag{5}$$

---

[1] Particle indices are surpressed whenever possible.

The superscript $(\cdot)^T$ denotes the transposition of the tensorial object. The rotation tensor $\mathbf{R}$ is related to the axis of rotation $\mathbf{e}_\varphi$ and to the angle of rotation φ by the *Euler-Rodriguez* formula

$$\mathbf{R} = \mathbf{e}_\varphi \otimes \mathbf{e}_\varphi + (\mathbf{I} - \mathbf{e}_\varphi \otimes \mathbf{e}_\varphi)\cos\varphi + (\mathbf{e}_\varphi \times \mathbf{I})\sin\varphi. \tag{6}$$

Therein, the tensor product between vectors and tensors is given by

$$\mathbf{e}_\varphi \times \mathbf{I} = [\overset{3}{\mathbf{E}}(\mathbf{e}_\varphi \otimes \mathbf{I})]^{\underline{2}}, \tag{7}$$

where $\overset{3}{\mathbf{E}}$ is the *Ricci* permutation tensor (fundamental tensor of third order) and $(\cdot)^{\underline{2}}$ defines a contraction towards a 2-nd rank tensor [2].

Taking the time derivative of (4) yields

$$\dot{\bar{\mathbf{x}}} = \mathbf{\Omega}\bar{\mathbf{x}}, \tag{8}$$

where $\mathbf{\Omega} = \dot{\mathbf{R}}\mathbf{R}^T$ represents the so-called gyration tensor, which is related to the angular velocity $\boldsymbol{\omega}$ of the rigid body by

$$\mathbf{\Omega}(t) = -\overset{3}{\mathbf{E}}\boldsymbol{\omega}(t). \tag{9}$$

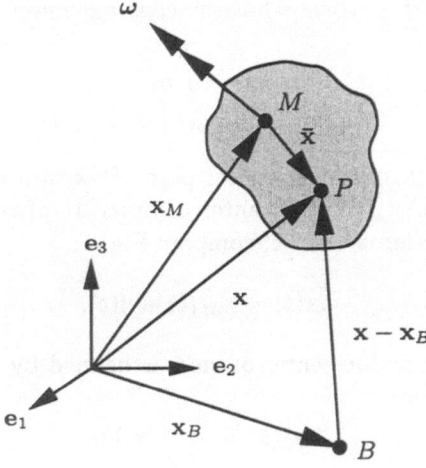

Fig. 1. Schematic representation of a single rigid particle consisting of material points P rotating with the angular velocity $\boldsymbol{\omega}$ around its center of mass M.

3   Equations of motion

In this section, the *Newtonian* equations of motion are set up for rigid particles of arbitrary shape. A brief outline of the derivation starting from the axioms of conservation of momentum and moment of momentum is presented in the following. To that end, the motion of the particles is characterized by the generalized motion $q = (x_M, \varphi)^T$, where $\varphi = \varphi e_\varphi$ represents the pseudo vector of the particle rotation.

The balance of momentum

$$\dot{p} = k \tag{10}$$

of any particle under consideration relates the change of momentum p of the particle to the forces k acting upon it. The momentum p is defined by

$$p = \int_M \dot{x} \, dm, \tag{11}$$

while the forces k are split into tractions t acting on the surface ∂V and into body forces f acting in volume V. Thus,

$$k = \int_{\partial V} t \, da + \int_V f \, dv. \tag{12}$$

Taking into account (1)–(3), the balance of momentum (10) can be rewritten with respect to the center of mass:

$$m \ddot{x}_M = k. \tag{13}$$

Furthermore, the balance of moment of momentum requires that the change of angular momentum h_B with respect to any fixed but arbitrary point B is given by the moments m_B of the forces k acting on the body with respect to the same point B:

$$\dot{h}_B = m_B. \tag{14}$$

Therein, the angular momentum is governed by the moment of momentum

$$h_B = \int_M (x - x_B) \times \dot{x} \, dm, \tag{15}$$

while the moments acting on the body result from

$$m_B = \int_{\partial V} (x - x_B) \times t \, da + \int_V (x - x_B) \times f \, dv. \tag{16}$$

Given (15), the angular momentum with respect to the center of mass is obtained as

$$h_M = \int_M \bar{x} \times (\omega \times \bar{x}) \, dm =: \bar{\Theta}_M \omega. \tag{17}$$

Therein, the tensor of inertia $\bar{\Theta}_M$ referred to the particle center of mass is given by

$$\bar{\Theta}_M = \int_M [(\bar{x} \cdot \bar{x}) I - (\bar{x} \otimes \bar{x})] \, dm = \bar{\Theta}_M^T. \tag{18}$$

In the case of rigid particles, the time dependence of $\bar{\Theta}_M$ is only caused by the particle rotation around the center of mass with angular velocity $\boldsymbol{\omega}$. With respect to a corotational frame fixed in M, the tensor of inertia is constant. As a consequence, the corotational *Green-Naghdi* derivative

$$\overset{\circ}{\bar{\Theta}}_M = \dot{\bar{\Theta}}_M - \Omega\,\bar{\Theta}_M - \bar{\Theta}_M\,\Omega^T \tag{19}$$

vanishes. Thus,

$$\dot{\bar{\Theta}}_M = 2\,(\Omega\,\bar{\Theta}_M)_{\text{sym}}. \tag{20}$$

Combination of (14), (17) and (20) finally yields the balance of moment of momentum in the form of the *Euler*ian gyroscopic equations of rotational motion

$$\dot{\mathbf{h}}_M = \bar{\Theta}_M\,\dot{\boldsymbol{\omega}} + 2\,(\Omega\,\bar{\Theta}_M)_{\text{sym}}\,\boldsymbol{\omega} = \mathbf{m}_M. \tag{21}$$

Hence, (13) and (21) represent the complete set of the *Newton*ian and *Euler*ian equations of motion of rigid particles allowing for the determination of the generalized motion $q = (\mathbf{x}_M, \boldsymbol{\varphi})^T$, if the external forces acting on the particles are known. In a general setting, these forces are related to the motion of the particles by constitutive laws. In the case of granular material, where particles interact via contacts, the corresponding constitutive equations are governed by contact laws.

4 Contact laws

In this section, a set of constitutive equations is discussed which allows for the description of the essential effects occurring in the deformation of granular materials. In the present case, the rheological model of the contact laws is assumed to consist of springs, dashpots and a frictional element. This combination allows for the description of the elasto-viscoplastic behaviour observed in granular materials. The resulting contact law is sufficient to study the formation of shear bands in model computations and it does not seem reasonable to use more complicated laws until the basic mechanism leading to localization, shear bands and failure in granular media is not better understood on the microscale. Furthermore, the qualitative description of instability phenomena such as shear banding is determined by only a few parameters; a well-known situation for self-organized systems, compare e. g. *Haken & Wunderlin* [16]. On the other hand, several effects like cohesion or tangential stiffness may strongly influence the quantitative behaviour. Therefore, in order to get insight into the governing physical processes, the presented model is restricted to simple contact laws and to two-dimensional assemblies consisting of monodisperse circular particles. In this case, the *Newton*ian equations (13) reduce to two scalar equations with respect to the directions $\mathbf{e}_1$ and $\mathbf{e}_2$ of the plane of translational motion, while the *Euler*ian equations (21) yield a single equation in the direction of $\mathbf{e}_3$ perpendicular to this plane, compare Fig. 2. As a consequence of this setting, $\dot{\bar{\Theta}}_M$ from (20) vanishes for rotations around $\mathbf{e}_3$.

The generalized coordinates of a particle now comprehend three degrees of freedom per particle, two translational coordinates x_1, x_2 in the directions of e_1 and e_2 indicating the position $\mathbf{x}_M$ of the center of mass and one angle of rotation φ around the axis e_3 perpendicular to the plane of translational motion. Thus,

$$q = (x_1, x_2, \varphi)^T. \qquad (22)$$

In order to achieve a compact writing, it is also convenient to introduce a generalized force vector via

$$k = (k_1, k_2, m_M)^T. \qquad (23)$$

Therein, k_1 and k_2 represent the forces acting on the particle in the plane of motion, while m_M is the moment of these forces with respect to the center of mass. Furthermore, the generalized mass matrix

$$M = \mathrm{diag}\{m, m, \Theta_M\} \qquad (24)$$

is introduced, where the moment of inertia for a homogeneous circular disk with radius r and mass m is given by $\Theta_M = m\,r^2/2$. The two *Newtonian* equations (13) and the *Eulerian* gyroscopic equation (21) can now be written in the compact form

$$M\ddot{q} = k. \qquad (25)$$

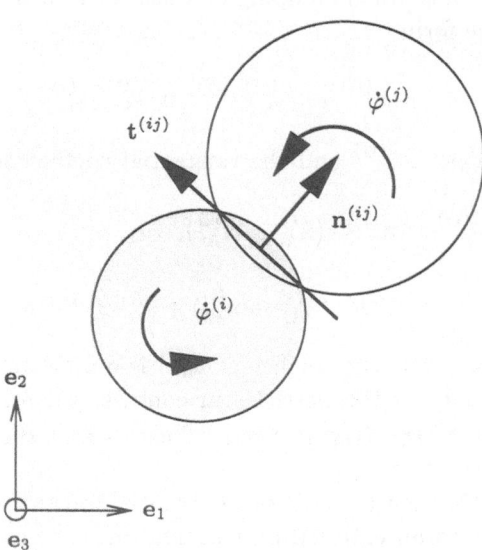

Fig. 2. Contact between two particles i and j. The idealized contact point is located on the connection between the particle centers.

The contact laws relate kinematic quantities to the forces acting in the contact area. The situation is schematically shown in Fig. 2. Even if the particles are assumed to be rigid for computational reasons, an overlap of the particles is allowed in the contact zone. The overlap takes into account the elastic properties of the particles. However, two particles i and j are in contact if their overlap defined as

$$\Delta_n^{(ij)} = r^{(i)} + r^{(j)} - |\mathbf{x}_M^{(j)} - \mathbf{x}_M^{(i)}| \tag{26}$$

is positive. In (26), $r^{(i)}$ and $r^{(j)}$ indicate the disk radii of particles i and j. In order to formulate the contact law which is required to solve (25), it is convenient to introduce the orthonormal basis $\{\mathbf{n}^{(ij)}, \mathbf{t}^{(ij)}, \mathbf{e}_3\}$ attached to the contact point between particles i and j, compare Fig. 2:

$$\mathbf{n}^{(ij)} = \frac{\mathbf{x}_M^{(j)} - \mathbf{x}_M^{(i)}}{|\mathbf{x}_M^{(j)} - \mathbf{x}_M^{(i)}|}, \tag{27}$$

$$\mathbf{t}^{(ij)} = \mathbf{e}_3 \times \mathbf{n}^{(ij)}.$$

Therein, $\mathbf{n}^{(ij)}$ denotes the outward oriented unit normal vector of particle i in the contact point with particle j, $\mathbf{t}^{(ij)}$ characterizes the corresponding tangential vector, and $\mathbf{e}_3$ indicates the unit vector in the direction perpendicular to the plane of the motion.

The relative velocity $\mathbf{v}^{(ij)}$ between the particles in the contact point is required to include *Newton*ian damping or viscosity into the contact laws, respectively. It takes the form

$$\mathbf{v}^{(ij)} = v_n^{(ij)} \mathbf{n}^{(ij)} + v_t^{(ij)} \mathbf{t}^{(ij)}. \tag{28}$$

The normal coefficient $v_n^{(ij)}$ and the tangential coefficient $v_t^{(ij)}$ are determined by

$$v_n^{(ij)} = \mathbf{n}^{(ij)} \cdot (\dot{\mathbf{x}}_M^{(j)} - \dot{\mathbf{x}}_M^{(i)}),$$

$$v_t^{(ij)} = \mathbf{t}^{(ij)} \cdot (\dot{\mathbf{x}}_M^{(j)} - \dot{\mathbf{x}}_M^{(i)}) + r^{(i)} \omega^{(i)} + r^{(j)} \omega^{(j)}, \tag{29}$$

where $\omega = \dot{\varphi}$. Note that the coefficient $v_n^{(ij)}$ is only determined by the relative translational velocity of the particles in contact, whereas the tangential coefficient $v_t^{(ij)}$ contains the translational velocities and the angular velocities as well.

To formulate the contact law, the generalized forces $\mathbf{k}$ entering equation (25) are decomposed into an external part $\mathbf{k}_e$ and into a second part $\mathbf{k}_c$ due to the contact forces, namely

$$\mathbf{k} = \mathbf{k}_e + \mathbf{k}_c. \tag{30}$$

In the presented case, $\mathbf{k}_e = (0, -g, 0)^T$ is the gravity always acting in the negative $\mathbf{e}_2$-direction of the plane model system. The generalized contact forces $\mathbf{k}_c$ affecting the motion of particle i have the form

$$\mathbf{k}_c = (f_1^{(i)}, f_2^{(i)}, m_M^{(i)})^T, \tag{31}$$

where the forces $\mathbf{f}^{(i)} = f_1^{(i)}\mathbf{e}_1 + f_2^{(i)}\mathbf{e}_2$ result from all contact forces $\mathbf{f}^{(ij)}$ acting on the particle i,

$$\mathbf{f}^{(i)} = \sum_j \mathbf{f}^{(ij)},$$

$$\mathbf{f}^{(ij)} = \lambda_{ij}(f_n^{(ij)}\mathbf{n}^{(ij)} + f_t^{(ij)}\mathbf{t}^{(ij)}),$$

(32)

while the moments $m_M^{(i)}$ are the sum of all moments acting on the particle i:

$$m_M^{(i)} = \sum_j m_M^{(ij)}.$$

(33)

It follows from

$$\mathbf{m}_M^{(ij)} = \lambda_{ij}\, r^{(i)} \mathbf{n}^{(ij)} \times \left(f_n^{(ij)}\, \mathbf{n}^{(ij)} + f_t^{(ij)}\, \mathbf{t}^{(ij)} \right)$$

(34)

that the moments acting on a particle i are only related to the tangential contact forces. Therefore, (34) results in

$$m_M^{(ij)} = \mathbf{m}_M^{(ij)} \cdot \mathbf{e}_3 = \lambda_{ij}\, f_t^{(ij)} r^{(i)}.$$

(35)

The sums in (32) and (33) are performed over all particles. The factor λ_{ij} fulfills $\lambda_{ij} = 1$ if particles i and j are in contact and $\lambda_{ij} = 0$ otherwise. Furthermore, $\lambda_{ii} = 0$.

The repulsive normal contact force is governed by a damped spring model related to a viscoelastic normal contact. Thus,

$$f_n^{(ij)} = -E^{(ij)}\Delta_n^{(ij)} + \eta_n^{(ij)} v_n^{(ij)},$$

(36)

where $E^{(ij)}$ and $\eta_n^{(ij)}$ are an effective elastic constant and a viscosity coefficient, respectively. For the tangential contact between two particles i and j, a combination of a *Coulomb*-type friction and a viscous force is assumed by

$$f_t^{(ij)} = \min \left\{ \eta_t\, |v_t^{(ij)}|,\ \mu|f_n^{(ij)}| \right\} \frac{v_t^{(ij)}}{|v_t^{(ij)}|},$$

(37)

where η_t is a second viscosity parameter and μ is the friction coefficient, respectively. It is important to note that a combination of the velocity-dependent viscous force and the *Coulomb*-type frictional force is related to the viscoplastic properties of the model under study. The tangential part of the contact law (37) is constructed in such a way that the system chooses the smaller term of both possibilities, *Coulomb*-type friction or viscous *Newtonian* friction. As a consequence, the forces in a static system are all normal.

Corresponding to (37), the viscous regime is determined by the size of normal contact forces. The higher $|f_n^{(ij)}|$, the larger may be $|v_t^{(ij)}|$ where viscous friction is still effective. From a certain threshold of normal loading (being mostly applied by the weight of the above aligned particles), this mechanism becomes that effective that the rolling modes become unstable and the particles start sliding on each other. This effect causes a macroscopic motion that occurs as a shear band formation. The above introduced contact laws are applied in the examples presented in Section 6.

5   Numerical aspects

The $3 \times N$ equations of plane motion for an assemblage of N cylindrical particles are solved numerically using a simple explicit *Verlet* integration scheme also known as explicit *Newmark* scheme. The generalized coordinates q^{n+1} at the new time step t^{n+1} are computed from the known values q^n by a *Taylor* series expansion up to the second order:

$$q^{n+1} = q^n + \dot{q}^n \Delta t + \frac{1}{2} \ddot{q}^n (\Delta t)^2, \qquad \Delta t = t^{n+1} - t^n. \qquad (38)$$

The generalized acceleration is determined by the generalized forces according to (25), i. e.

$$\ddot{q}^n = M^{-1} k^n. \qquad (39)$$

Due to the fact that the forces k^n are known, (38) may directly be solved for the new values of the generalized coordinates q^{n+1}. As a disadvantage, this explicit algorithm requires small time steps Δt to be stable. On the other hand, the idea can be extended straight forward to predictor-corrector schemes, which allow for greater time steps and increased accuracy.

To accelerate the computations, it is indispensable to implement neighbourhood lists for each particle in order to minimize the contact search. A neighbourhood list labels those particles in the immediate vicinity of the particle of interest. To be included into the neighbourhood list of a certain particle i, the positions of the surrounding particles j must be located within a circle of radius R surrounding i, compare Fig. 3, and must fulfill

$$|x_M^{(j)} - x_M^{(i)}| \leq R. \qquad (40)$$

For a certain particle i, the contact search and the calculation of the resulting contact forces are only performed for those particles j included in its neighbourhood list at each time step.

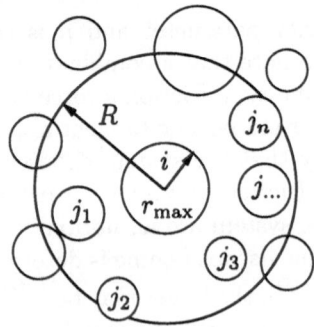

Fig. 3. Schematic neighbourhood list of particle i, the shaded particles belong to the list and are possible contact partners.

The neighbourhood list of a particle has to be updated before a contact event can take place with another particle up to now not included in the list. For dense granular materials, the time interval τ after that the neighbourhood list has to be updated is much greater than the time step Δt of the computation and it is determined by the highest possible relative velocity v_{max} in the specimen under study. Assuming moderate accelerations, the critical time to update the list is estimated as $\tau \approx (R - 2 r_{max}) / v_{max}$ where r_{max} denotes the largest particle radius. An estimation of the computational costs shows that an optimal radius R for the adjacency circle is approximately $R = 3 r_{max}$. The hereby achieved reduction of computation time allows for a significant increase in the number of particles taken into account in the simulations. Even if the numerical cost is of $O(N^2)$ to update the neighbourhood list, the procedure is efficient for slowly moving particles due to $\tau \gg \Delta t$.

6   Simulation examples and results

In this section, the above presented model is applied to an assemblage of monodisperse cylindrical particles. At the beginning, the particles fill a square box of 10×10 cm in dense packing. At time $t = 0$, a rigid wall at the left hand side of the box starts moving to the left with a constant velocity $v_{wall} = 4.5$ cm/s. A schematic view of the boundary-value problem is given in Fig. 4. Comparable experiments on cylindrical rods were performed by *Rossi* [19]. This type of model material is known as a *Schneebelli* medium.

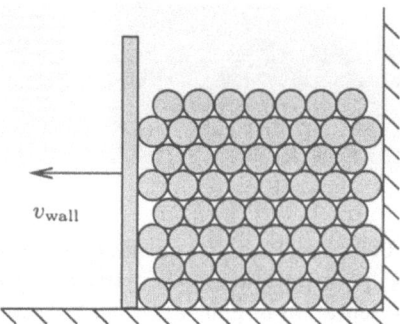

Fig. 4. Schematic view of the boundary-value problem.

The boundary-value problem is computed for an assemblage of 1,600 particles and for a second one consisting of 10,000 particles. Fig. 5 and Fig. 6 show the distribution of the translational kinetic energy in the specimens and the particle rotations at different positions u_{wall} of the moving wall. The grey scales indicate increasing energy with increasing brightness and negative (clockwise) rotations in black and positive (counterclockwise) rotations in light grey, respectively.

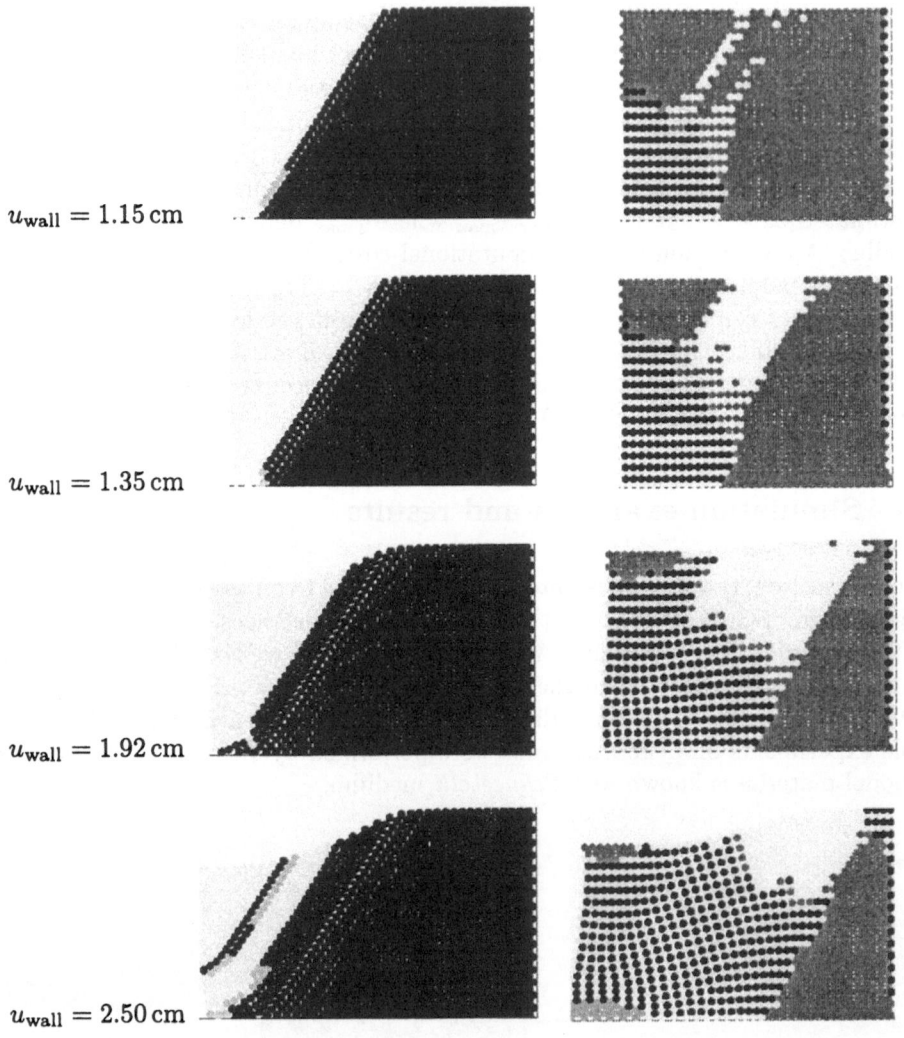

$u_{\text{wall}} = 1.15\,\text{cm}$

$u_{\text{wall}} = 1.35\,\text{cm}$

$u_{\text{wall}} = 1.92\,\text{cm}$

$u_{\text{wall}} = 2.50\,\text{cm}$

Fig. 5. Assemblage of 1,600 particles; *left*: translational kinetic energy increasing from black (zero) to light grey; *right*: rotation of the particles with clockwise orientation in black, counterclockwise orientation in light grey and zero values in grey.

In both examples, it can be seen that the kinetic energy localizes in a wedge close to the moving wall. While a large part of the specimen is undeformed, the wedge starts sliding and, during the beginning of the motion of the wall, it behaves nearly as a rigid body. The sliding wedge only deforms for large displacements of the moving wall. Note that the transition from the sliding wedge to the undeformed part of the medium is localized in a small shear band. This region of high shear rates is characterized by dilation and by strong particle rotations. While the localization of the particle rotations is still diffuse in the example with 1,600 particles, it becomes a sharp band in the case of 10,000 particles. In both cases, the computed thickness of the resulting band is strongly related to the size of the particles, since it approximately covers 10 particle diameters when the band is fully evolved. Furthermore, rolling modes occur at the front and the bottom of the specimen. By averaging the microscopic discrete particle model towards a macroscopic continuum, these rolling modes vanish over the REV, while the localized rotation gives rise to a *Cosserat* or a micropolar continuum, respectively.

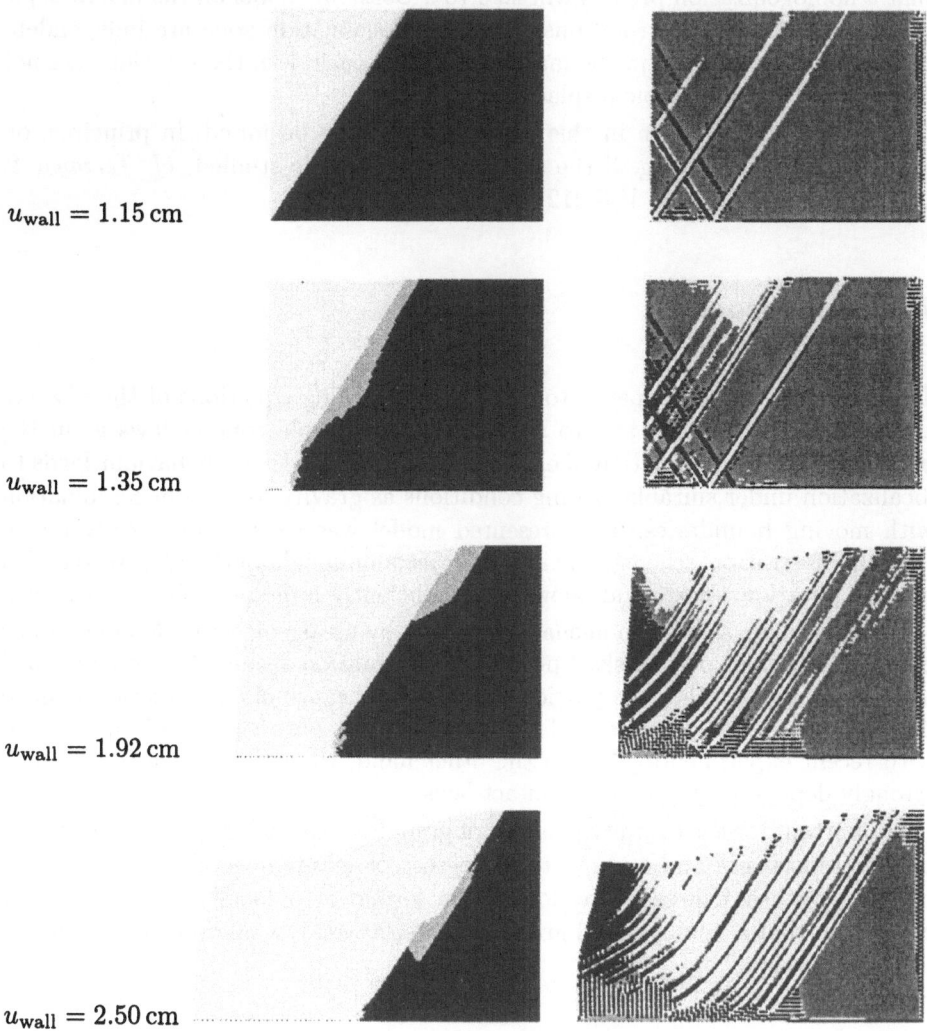

Fig. 6. Assemblage of 10,000 particles; *left*: translational kinetic energy increasing from black (zero) to light grey; *right*: rotation of the particles with clockwise orientation in black, counterclockwise orientation in light grey and zero values in grey.

In the second example with 10,000 particles, a second shear band becomes visible in the representation of the rotations. Therefore, the sliding of parts of the sample does not start continously but in blocks of a certain size. As can be seen from the last pictures included in Fig. 5 and Fig. 6, all particles start to rotate in the same direction, when the whole specimen is fluidized for large displacements of the left hand wall. This is possible because the model does not take into account tangential stiffness of the contacts. Due to the fact that all particles within the shear band rotate in the same direction, it is again recognized that a homogenization process will lead to a *Cosserat* model on the macroscopic scale, since the average rotations within the localization zone are independent degrees of freedom even in the macroscopic approach, i. e. the rotations are not related to the macroscopic displacement field.

The behaviour shown in this examples can also be found, in principle, on a macroscopic scale, e. g. if the stability of a slope is studied, *cf.* Terzaghi & Jelinek [20] or *Ehlers & Volk* [12–14].

7   Conclusions

The main goal of the paper is to recall the governing equations of the discrete element method (DEM) and to show that even simple contact laws as in the presented case a combination of elastic, viscous and frictional behaviour leads to localization under suitable loading conditions as gravity loading in combination with moving boundaries. The presented model was solved numerically by an explicit integration scheme. Taking into account neighbourhood lists reduced the computational costs and allowed for sufficiently large numbers of particles.

The basic physical phenomena were shown by means of 2-d DEM simulations of circular monodisperse disks. Specially, the formation of shear bands is included into the model. Neither the particle shapes nor the size of the particles seem to be stringent for the *formation* of shear bands. This observation is in agreement with recent experiments [23]. On the other hand, the quantitative results may strongly depend on the chosen contact laws.

Important for the overall dynamics of granular material appears the coupling of the translatoric and the rotatoric degrees of freedom of the particles. The interplay of these degrees of freedom for the formation of localization leads to the conclusion that a homogenized granular material yields a micropolar continuum of *Cosserat* type.

Acknowledgements

This work was performed in the framework of the research group "Cohesive-Frictional Materials". Financial support of the *Deutsche Forschungsgemeinschaft* (DFG) is greatfully acknowledged.

References

1. M. Becker and H. Lippmann: Plane plastic flow of granular model material. *Arch. Mech.* **29**, 829-846 (1977).
2. R. de Boer: *Vektor- und Tensorrechnung für Ingenieure.* Springer-Verlag, Berlin 1982.
3. R. de Boer and W. Ehlers: *Theorie der Mehrkomponentenkontinua mit Anwendung auf bodenmechanische Probleme*, Teil I. Forschungsberichte aus dem Fachbereich Bauwesen **40**, Universität-GH-Essen 1986.
4. R. M. Bowen: Theory of mixtures. In A. C. Eringen (ed.), *Continuum Physics*, Vol. III, pp. 1-127, Academic Press, New York 1976.
5. R. M. Bowen: Incompressible porous media models by use of the theory of mixtures. *Int. J. Engng. Sci.* **20**, 1129-1148 (1980).
6. E. Cosserat and F. Cosserat: *Théorie de Corps Déformable*. A. Hermann et fils, Paris 1909.
7. P. A. Cundall and O. D. L. Strack: A discrete numerical model for granular assemblies. *Géotechnique* **29**, 47-65 (1979).
8. S. Diebels: Constitutive modelling of micropolar porous media. In J.-F. Thimus *et. al.* (eds.), *Poromechanics - A Tribute to Maurice A. Biot*, pp. 71-76. A. A. Balkema, Rotterdam 1998.
9. S. Diebels: A macroscopic description of the quasi-static behavior of granular materials based on the Theory of Porous Media. *Granular Matter* **2**, 143-152 (2000).
10. W. Ehlers: *Poröse Medien – ein kontinuumsmechanisches Modell auf der Basis der Mischungstheorie.* Forschungsberichte aus dem Fachbereich Bauwesen **47**, Universität-GH-Essen 1989.
11. W. Ehlers: Constitutive equations for granular materials in geomechanical context. In K. Hutter (ed.), *Continuum Mechanics in Environmental Sciences and Geophysics*, CISM Courses and Lectures No. 337, pp. 313-402, Springer-Verlag, Wien 1993.
12. W. Ehlers and W. Volk: On shear band localization phenomena of liquid-saturated granular elasto-plastic porous solids accounting for fluid viscosity and micropolar solid rotations. *Mech. Cohesive-frictional Mater.* **2**, 301-320 (1997).
13. W. Ehlers and W. Volk: On shear band localization phenomena induced by elasto-plastic consolidation of fluid-saturated soils. In D. R. J. Owen, E. Oñate and E. Hinton (eds.), *Computational Plasticity – Fundamentals and Applications*, pp. 1656-1664, CIMNE, Barcelona 1997.
14. W. Ehlers and W. Volk: On theoretical and numerical methods in the theory of porous media based on polar and non-polar solid materials. *Int. J. Solids Structures* **35**, 4597-4616 (1998).
15. A. C. Eringen and C. B. Kafadar: Polar field theories. In A. C. Eringen (ed.), *Continuum Physics*, Vol. IV, Academic Press, pp. 1-73, New York 1976.
16. H. Haken and A. Wunderlin: *Die Selbststrukturierung der Materie.* Vieweg, Wiesbaden 1991.
17. M. Lätzel, S. Luding, and H. J. Herrmann. Macroscopic material properties from quasi-static, microscopic simulations of a two-dimensional shear-cell. *Granular Matter*, 2(3):123–135, 2000. cond-mat/0003180.
18. T. Marcher and P. A. Vermeer: Macromodelling of softening in non-cohesive soils. This issue, pp. 89-110.
19. P. Rossi: *Kinematische Modelluntersuchung von ebenen Grundbauproblemen.* Diplomarbeit, Institut für Geotechnik, Universität Stuttgart 1983.

274 W. Ehlers *et al.*

20. K. von Terzaghi and R. Jelinek: *Theoretische Bodenmechanik*. Springer-Verlag, Berlin 1954.
21. C. Thornton: Numerical simulations of deviatoric shear deformation of granular media. *Géotechique* **50**, 43-53 (2000).
22. C. Thornton and S. J. Antony: Quasi shear deformation of a soft particle system. *Powder Technology* **109**, 179-191 (2000).
23. G. Viggiani, M. Küntz and J. Desrues: Does shear banding in sand depend on grain size distribution? This issue, pp. 111-127.

Microstructured materials: local constitutive equation with internal lenght, theoretical and numerical studies

R. Chambon,[1] T. Matsuchima,[2] D. Caillerie[1]

[1] Laboratoire 3S Grenoble U J.F., I.N.P.G., C.N.R.S. U.M.R. 5521, B.P. 53X, 38041 Grenoble Cedex France
[2] University of Tsukuba Tsukuba Japan

Abstract. Starting from the theory of microstrucured materials, a local second gradient theory is developed by adding to the theory kinematical constraints. Using these constraints to eliminate some variables is one way to deal with such models. In some simple cases, it is possible to get analytical solutions of boundary value problems which can be used as benchmark for numerical studies. A finite element method able to take into account geometrical non linearities is done. In order to avoid difficulties arising usually with C1 continuity requirements, a field of Lagrange multipliers corresponding to the constraints introduced so far, is used. Results obtained with this finite element method are quite satisfactory.

1 Introduction

It is now recognized that the granular nature of some materials like soils and other geomaterials can not be ignored in many phenomena like for instance localization and consequently rupture. This paper is an attempt to explore the possibilities of enhanced models, clearly related with the microstructure of materials, to explain and model problems exhibiting a scale effect. The paper has two main parts. The first one is devoted to the theory, the development of analytical solutions and discussions about these solutions. The second part is based on an other viewpoint which allows to develop a Finite Element Method with Lagrange multipliers to solve geometrically non linear boundary value problems.

Let us give the principles of our notations. A component is denoted by the name of the tensor (or vector) accompanied with tensorial indices. All tensorial indices are in lower position as there is no need in the following of a distinction between covariant and contravariant components. Upper indices have other meanings and particularly are used to indicate the corresponding time or the number of iteration in iterative procedures. The summation convention with respect to repeated tensorial indices has been adopted. In order to avoid confusions squares are systematically denoted with parenthesis.

2   A general theory for continua with microstructure

2.1   Kinematic description of a continuum with microstructure

Kinematics of a classical continuum is defined by a displacement field denoted u_i function of the coordinates denoted x_i. In media with microstructure, a field of second order tensors denoted v_{ij} which models the strains and the rotations of the grains themselves is added to the previous description. It is called here micro kinematic gradient. In classical models the gradient of the displacement is used to define the internal virtual work as a linear form of the virtual displacement gradient. In the present case, it is consistent to consider the virtual work as a linear form with respect to the virtual displacement gradient, the virtual micro kinematic gradient and its gradient. Let us denote: F_{ij} the macro displacement gradient, $F_{ij} = \frac{\partial u_i}{\partial x_j}$, D_{ij} the macro strain $D_{ij} = \frac{1}{2}(F_{ij} + F_{ji})$, R_{ij} the macro rotation $R_{ij} = \frac{1}{2}(F_{ij} - F_{ji})$. v_{ij} denotes the micro kinematic gradient (this does not imply that this field is deriving from a displacement field), d_{ij} is the micro strain $d_{ij} = \frac{1}{2}(v_{ij} + v_{ji})$ and r_{ij} is the micro rotation $r_{ij} = \frac{1}{2}(v_{ij} - v_{ji})$. Finally h_{ijk} is the (micro) second gradient $h_{ijk} = \frac{\partial v_{ij}}{\partial x_k}$.

2.2   The internal virtual work

In order to define the internal virtual work the virtual variables corresponding to the previous kinematic variables, are defined and dual static variables are defined too. Due to the principle of material frame indifference the virtual work has to depend only on the virtual macro strain, the virtual relative deformation gradient (i.e: the difference between the virtual macro displacement gradient and the virtual micro kinematic gradient) and the virtual (micro) second gradient [4]. Denoting with a * virtual quantities, the internal virtual work for a given body Ω reads:

$$W^{\star i} = \int_\Omega (\sigma_{ij} D^\star_{ij} + \tau_{ij}(v^\star_{ij} - F^\star_{ij}) + \chi_{ijk} h^\star_{ijk})\, dv \qquad (1)$$

where σ_{ij} is called here the macro stress. τ_{ij} is an additive stress associated with the microstructure. It is not necessarily symmetric and is called microstress. χ_{ijk} which is related with $h^\star_{ijk}$ is called the double stress. It is worth noticing here that $D^\star_{ij}$ and $F^\star_{ij}$ are depending on a virtual displacement field $u^\star_i$ and $h^\star_{ijk}$ is depending on $v^\star_{ij}$.

2.3   The external virtual work

We assume that only classical body forces (denoted G_i) are applied, this means precisely that there is no body double force (see [4] for more general assumptions). On the boundary, we assume that the classical traction forces t_i and double surface tractions T_{ij} are acting. Finally, denoting as usual $\partial\Omega$ the boundary of Ω, the external virtual work reads:

$$W^{\star e} = \int_\Omega G_i u^\star_i\, dv + \int_{\partial\Omega} (t_i u^\star_i + T_{ij} v^\star_{ij})\, ds \qquad (2)$$

G_i is assumed to be known in every point of Ω, t_i and similarly T_{ij} are assumed to be known at least on a part of $\partial\Omega$.

2.4   The balance equations and the boundary conditions

By equating the external virtual work (equation 2) and the internal virtual work (equation 1) for all kinematic admissible virtual fields, (fields which are sufficiently smooth and which has a null value on the part of the boundary where the corresponding real field is prescribed), using one integration by part and the divergence formula yields the balance equations:

$$\frac{\partial(\sigma_{ij} - \tau_{ij})}{\partial x_j} + G_i = 0 \tag{3}$$

$$\frac{\partial \chi_{ijk}}{\partial x_k} - \tau_{ij} = 0 \tag{4}$$

and the static boundary conditions:

$$(\sigma_{ij} - \tau_{ij})n_j = t_i \tag{5}$$

$$\chi_{ijk}n_k = T_{ij} \tag{6}$$

where n_j is the external normal to the boundary $\partial\Omega$. In order to get a complete problem, it is necessary to prescribe on the boundary $\partial\Omega$ either static boundary conditions t_i and T_{ij} or the corresponding kinematic boundary conditions, u_i and v_{ij}.

3   Microstructured continuum with kinematic constraint: Second gradient models

3.1   Equations of a second gradient model

Now we add a kinematic constraint: the micro strain is equal to the macro strain.

$$F_{ij} = v_{ij} = \frac{\partial u_i}{\partial x_j} \tag{7}$$

This main assumption can be used in two different manners. A first one useful in finite element modeling is detailed in section 5. Another one is using equation 7, eliminating v_{ij} in every equation, and consequently assuming that $v_{ij}^* = \frac{\partial u_i^*}{\partial x_j}$. It is done in the following. Let us notice here that we have only one real unknown field u_i and similarly only one virtual displacement field, which is rather different from the virtual work principle applied in section 2. It is worth noticing a second point. As u_i^* and $\frac{\partial u_i^*}{\partial x_j}$ are not independent because the value of u_i and its tangential derivatives (along the boundary) can not vary independently, t_i and T_{ij} can not be taken independently. Let us denote D the normal derivative of any quantity

q, $(Dq = \frac{\partial q}{\partial x_k} n_k)$ and D_j the tangential derivatives $(D_j q = \frac{\partial q}{\partial x_j} - \frac{\partial q}{\partial x_k} n_k n_j)$. It is more convenient to rewrite the external virtual work with p_i and P_i like in the following virtual work principle equation.

$$\int_\Omega (\sigma_{ij} D_{ij}^\star + \chi_{ijk} \frac{\partial^2 u_i^\star}{\partial x_j \partial x_k}) \, dv = \int_\Omega G_i u_i^\star dv + \int_{\partial\Omega} (p_i u_i^\star + P_i Du_i^\star) ds \quad (8)$$

In this case p_i and P_i can be chosen independently. Application of virtual work principle equation 8 and two integrations by part give the balance equation and the boundary conditions. The balance equations read

$$\frac{\partial \sigma_{ij}}{\partial x_j} - \frac{\partial^2 \chi_{ijk}}{\partial x_j \partial x_k} + G_i = 0 \quad (9)$$

The boundary conditions are here less simple due to the relation between u_i and $v_{ij} = \frac{\partial u_i}{\partial x_j}$ and consequently between the corresponding virtual quantities on the part of the boundary where the forces and double forces are prescribed. Finally if boundaries are regular (which means existence and uniqueness of the normal for every point of the boundary $\partial\Omega$ of the studied domain), after one more integration by part we get:

$$\sigma_{ij} n_j - n_k n_j D\chi_{ijk} - \frac{D\chi_{ijk}}{Dx_k} n_j - \frac{D\chi_{ijk}}{Dx_j} n_k + \frac{Dn_l}{Dx_l} \chi_{ijk} n_j n_k - \frac{Dn_j}{Dx_k} \chi_{ijk} = p_i \quad (10)$$

$$\chi_{ijk} n_j n_k = P_i \quad (11)$$

where p_i and P_i are prescribed.

For granular material we have some insight into microscopic phenomena. In this case grains can be assumed as rigid and it is reasonable to neglect d_{ij}. Calvetti et. al. [1] show clearly that in most cases (except for some exotic loading paths which the data are not so clear even if the trends are similar) the macro rotation is the same as the average of the rotation of the grains (here as 2D experiments are performed, the grains are small rods). This justifies the use of a particular Cosserat model, a Cosserat model with a kinematic constraint similar to the one done here [3]. This particular case supports the opinion that the present assumption can be reasonable to assume that cohesive geomaterials (i.e. cohesive soils like clays, rocks and concrete) can be modeled by a second gradient continuum.

3.2   Local elasto-plastic second gradient models

For the theory developed so far, the constitutive equation can be seen as equations giving the stress and the double stress as a function of the local kinematics history. Going forward allows to develop elasto plastic model where the free energy and the mechanical dissipation are clearly defined and consequently to generalize to second gradient without any problem the classical thermodynamic statements of the elasto plasticity [3]. In this section we develop a simplified elasto-plastic second gradient model. It is assumed that the elastic part of the model is isotropic and linear. In this case as proved by Mindlin [7], we have

$$\dot{\sigma}_{ij} = \mathcal{K}^1_{ijkl} \dot{D}^e_{kl} \tag{12}$$

$$\dot{\chi}_{ijk} = \mathcal{K}^2_{ijklmn} \dot{h}^e_{lmn} \tag{13}$$

Where $\mathcal{K}^1_{ijkl}$ depends as usual on two different parameters and $\mathcal{K}^2_{ijklmn}$ depends on five different parameters.

We assume that the yield function depends only on σ_{ij} and that the plastic strain is only a classical one. Thus the model can be written

$$\dot{\chi}_{ijk} = \mathcal{K}^2_{ijklmn} \dot{h}_{lmn} \tag{14}$$

for loading and unloading, and

$$\dot{\sigma}_{ij} = \mathcal{K}^1_{ijkl} \dot{D}_{kl} \tag{15}$$

for unloading only and

$$\dot{\sigma}_{ij} = \mathcal{K}^1_{ijkl} \dot{D}_{kl} - \frac{\mathcal{K}^1_{ijpq} \psi_{pq} \frac{\partial \phi}{\partial \sigma_{mn}} \mathcal{K}^1_{mnkl} \dot{D}_{kl}}{h + \frac{\partial \phi}{\partial \sigma_{mn}} \mathcal{K}^1_{mnkl} \psi^D_{kl}} \tag{16}$$

for loading only where ϕ is the yield function, ψ the direction of the plastic strain and h the hardening modulus

4 An application
of local elasto-plastic second gradient model

4.1 The problem to be solved

In this section some analytical solutions for a boundary value problem are derived (a more general study can be seen in [3]). We consider an infinite layer of geomaterials bounded by two parallel plane corresponding to $x = 0$ and $x = l$. z is the direction normal to the studied part of plane. This means that plane strains are assumed in this direction. The velocity field is defined by only two variables namely u in the x direction and v in the y direction. Moreover it is assumed that u and v are functions of x only. So in the following derivatives with respect to x are denoted by a' which is here unambiguous. The second gradient part obeys equation 13 and here we need only $\mathcal{K}^2_{111111} = a^{12345}$ and $\mathcal{K}^2_{211211} = a^{34}$

The present stress state is homogeneous and we search solutions of the rate problem. The rates of boundary conditions are known namely $\dot{u} = 0$, $\dot{v} = 0$ and $\dot{P}_i = 0$ for $x = 0$ and $\dot{p}_i$ and $\dot{P}_i$ are given for $x = l$. The body forces are assumed to be constant. The only non vanishing components of h_{ijk} are $\dot{h}_{111} = \dot{u}''$ and $\dot{h}_{211} = \dot{v}''$

The balance equations 9 read:

$$\dot{\sigma}'_{11} - \dot{\chi}''_{111} = 0 \tag{17}$$

$$\dot{\sigma}'_{21} - \dot{\chi}''_{211} = 0. \tag{18}$$

The boundary conditions (equation 10) becomes for $x = l$:

$$\dot{\sigma}_{11} - \dot{\chi}'_{111} = \dot{p}_1 \quad \dot{\sigma}_{21} - \dot{\chi}'_{211} = \dot{p}_2 \tag{19}$$

$$\dot{\chi}_{111} = \dot{P}_1 \quad \dot{\chi}_{211} = \dot{P}_2. \tag{20}$$

Taking into account the boundary conditions, the balance equations can be integrated once, this yields:

$$\dot{\sigma}_{11} - \dot{\chi}'_{111} = \dot{p}_1 \tag{21}$$

$$\dot{\sigma}_{21} - \dot{\chi}'_{211} = \dot{p}_2 \tag{22}$$

Using now the constitutive equations yields

$$\mathcal{K}^1_{1111}\dot{u}' + \mathcal{K}^1_{1112}\dot{v}' - a^{12345}\dot{u}''' = \dot{p}_1 \tag{23}$$

$$\mathcal{K}^1_{2111}\dot{u}' + \mathcal{K}^1_{2121}\dot{v}' - a^{34}\dot{v}''' = \dot{p}_2 \tag{24}$$

4.2   Partial solutions

Solutions of such a set of equations depend on the values of the constitutive coefficients. As we work with isotropic models, only the stress orientation influences the values of the plastic constitutive coefficients. The solution depends only on the orientation of the stress tensor with respect to the chosen axis. Our problem can be seen as a general shear band analysis where the shear band orientation is assumed but the stress orientation is free. It is a shear band analysis in the spirit of Vermeer [11] but for an elasto-plastic second gradient model.

Solutions of equations 23 and 24 are depending on the roots of the characteristic equation 25.

$$A(s)^4 - B(s)^2 + C = 0 \tag{25}$$

where

$$A = a^{12345}a^{34}$$

$$B = a^{12345}\mathcal{K}^1_{2121} + a^{34}\mathcal{K}^1_{1111} \tag{26}$$

$$C = \mathcal{K}^1_{2121}\mathcal{K}^1_{1111} - \mathcal{K}^1_{1112}\mathcal{K}^1_{2111}$$

Denoting $S = (s)^2$, equation 25 reads:

$$A(S)^2 - BS + C = 0 \tag{27}$$

Δ the discriminant of equation 27 reads: $\Delta = (a^{12345}\mathcal{K}^1_{2121} - a^{34}\mathcal{K}^1_{1111})^2 + 4a^{12345}a^{34}\mathcal{K}^1_{1112}\mathcal{K}^1_{2111}$. It is positive for elastic moduli and also for elasto-plastic moduli corresponding to materials obeying the normality rule. For other elasto-plastic materials as C is usually decreasing as the material is loaded, Δ is positive and the roots of equation 27 are necessarily real.

Finally, solutions of equations 23 and 24 have the following forms.

1. If the two roots of equation 27 are positive, then solutions read:

$$\begin{bmatrix} \dot{u}' \\ \dot{v}' \end{bmatrix} = \begin{bmatrix} \dot{U}'^0 \\ \dot{V}'^0 \end{bmatrix} + \begin{bmatrix} \dot{U}'^1 \\ \dot{V}'^1 \end{bmatrix} (\lambda^{11} cosh(\eta^1 x) + \lambda^{12} sinh(\eta^1 x))$$

$$+ \begin{bmatrix} \dot{U}'^2 \\ \dot{V}'^2 \end{bmatrix} (\lambda^{21} cosh(\eta^2 x) + \lambda^{22} sinh(\eta^2 x)) \qquad (28)$$

where $\dot{U}'^0$ and $\dot{V}'^0$ are solutions of equation:

$$\begin{bmatrix} \mathcal{K}^1_{1111} & \mathcal{K}^1_{1112} \\ \mathcal{K}^1_{2111} & \mathcal{K}^1_{2121} \end{bmatrix} \begin{bmatrix} \dot{U}'^0 \\ \dot{V}'^0 \end{bmatrix} = \begin{bmatrix} \dot{p}_1 \\ \dot{p}_2 \end{bmatrix} \qquad (29)$$

and $(\eta^1)^2 = S^1$ and $(\eta^2)^2 = S^2$ are the two roots of equation 27. Moreover $\dot{U}'^i$ and $\dot{V}'^i$, $i \in \{1,2\}$ have to meet:

$$\begin{bmatrix} \mathcal{K}^1_{1111} - S^i a^{12345} & \mathcal{K}^1_{1112} \\ \mathcal{K}^1_{2111} & \mathcal{K}^1_{2121} - S^i a^{34} \end{bmatrix} \begin{bmatrix} \dot{U}'^i \\ \dot{V}'^i \end{bmatrix} = \begin{bmatrix} 0 \\ 0 \end{bmatrix} \qquad (30)$$

2. If one root of equation 27 denoted S^1 is positive and the other one denoted S^2 negative, then solutions read:

$$\begin{bmatrix} \dot{u}' \\ \dot{v}' \end{bmatrix} = \begin{bmatrix} \dot{U}'^0 \\ \dot{V}'^0 \end{bmatrix} + \begin{bmatrix} \dot{U}'^1 \\ \dot{V}'^1 \end{bmatrix} (\lambda^{11} cosh(\eta^1 x) + \lambda^{12} sinh(\eta^1 x))$$

$$+ \begin{bmatrix} \dot{U}'^2 \\ \dot{V}'^2 \end{bmatrix} (\mu^{21} cos(\omega^2 x) + \mu^{22} sin(\omega^2 x)) \qquad (31)$$

where $\dot{U}'^0$ and $\dot{V}'^0$ are solutions of equation 29. We have $(\eta^1)^2 = S^1$ and $(\omega^2)^2 = -S^2$ and $\dot{U}'^i$ and $\dot{V}'^i$, $i \in \{1,2\}$ have to meet equation 30.

3. If one root of equation 27 vanishes which implies that $C = 0$ and the other denoted S^1 is positive, then solutions read:

$$\begin{bmatrix} \dot{u}' \\ \dot{v}' \end{bmatrix} = \begin{bmatrix} \dot{U}'^1 \\ \dot{V}'^1 \end{bmatrix} (\lambda^{11} cosh(\eta^1 x) + \lambda^{12} sinh(\eta^1 x))$$

$$+ \begin{bmatrix} \mathcal{K}^1_{1112}\dot{p}_2 - \mathcal{K}^1_{2121}\dot{p}_1 \\ \mathcal{K}^1_{2111}\dot{p}_1 - \mathcal{K}^1_{1111}\dot{p}_2 \end{bmatrix} \frac{(x)^2}{2B} + \begin{bmatrix} -\mathcal{K}^1_{1112} \\ \mathcal{K}^1_{1111} \end{bmatrix} k^1 \frac{x}{B} \qquad (32)$$

$$+ \begin{bmatrix} \dfrac{a^{12345}\mathcal{K}^1_{1112}\dot{p}_2 + a^{34}\mathcal{K}^1_{1111}\dot{p}_1}{B(\mathcal{K}^1_{1111} + \mathcal{K}^1_{1112})} + \dfrac{B\mathcal{K}^1_{1112}}{a^{12345}\mathcal{K}^1_{1112}\dot{p}_2 + a^{34}\mathcal{K}^1_{1111}\dot{p}_1} k^2 \\[3ex] \dfrac{a^{12345}\mathcal{K}^1_{1112}\dot{p}_2 + a^{34}\mathcal{K}^1_{1111}\dot{p}_1}{B(\mathcal{K}^1_{1111} + \mathcal{K}^1_{1112})} - \dfrac{B\mathcal{K}^1_{1111}}{a^{12345}\mathcal{K}^1_{1112}\dot{p}_2 + a^{34}\mathcal{K}^1_{1111}\dot{p}_1} k^2 \end{bmatrix}$$

where we have $(\eta^1)^2 = S^1$; $\dot{U}'^i$ and $\dot{V}'^i$, $i \in \{1,2\}$ have to meet equation 30.

It is not necessary to study other solutions for the characteristic equation (see [3])

4.3   Patch conditions and full solutions

Given a stress state, depending on whether loading or unloading condition is assumed, two partial solutions are possible. A part of the body where the loading solution is used is called soft part. Similarly, a part of the body where unloading solution is used is called hard part. Given a stress state, there are two different partial possible solutions. Each solution correspond to formulae chosen among the three possible ones depicted in the previous section, depending on the value of the constitutive parameters, on the stress and on the orientation of the stress with respect to the chosen axes. Let us notice that each partial solution depends on 4 independent constants.

The solution of a given problem is then a patch of partial solutions. The body is split in hard parts or soft parts. Let us denote N the number of parts. To get a complete solution, it is necessary to find 5 unknowns per part, 4 ones corresponding to the independent constants and the fifth one to the width of the part. Globally, we have $5N$ unknowns. At each of the $N-1$ junction points it is necessary to write the continuity of $\dot{u}'$ and $\dot{v}'$. Moreover as a junction point belongs to a hard part as well as to a soft part, at this junction point the strains have to correspond to neutral loading. For instance if we choose as the classical part of the model, the 2D Mohr-Coulomb constitutive equations of Vardoulakis and Sulem ([12] chapter 6), then the neutral loading is characterized by equation 33.

$$(\sigma_{11} - \sigma_{22})G\dot{u}' + 2\mu\tau K\dot{u}' + 2\sigma_{12}G\dot{v}' = 0 \qquad (33)$$

where K is the bulk modulus, G the shear modulus, μ the friction angle, τ the stress deviator.

It is necessary to write the balance equation at a junction point. The ones corresponding to the classical term are necessarily met as equations 17 and 18 have been integrated into equations 19. On the contrary continuity of $\dot{\chi}_{111}$ and $\dot{\chi}_{211}$ at a junction point has to be enforced. Finally every junction point gives us 5 equations, which mean an amount of $5(N-1)$ equations.

Moreover it is necessary to write the boundary conditions on the two sides of the body which means that $\dot{\chi}_{111}$ and $\dot{\chi}_{211}$ are known for $x = 0$ where they vanishes and for $x = l$ where they are given by equation 20. Finally the sum of the width of the part has to be equal to l. We end up with a system of $5N$ equations of $5N$ unknowns which generally give us $\dot{u}'$ and $\dot{v}'$ as a function of x. It is then necessary to check that unloading condition holds in hard part whereas loading condition holds in soft part. At the end it is possible to get of $\dot{u}$ and $\dot{v}$ as we know that for $x = 0$: $\dot{u} = 0$ and $\dot{v} = 0$.

This is the generalization of the method given in [2] for a one dimensional case. We can build up basic solutions for the rate problem of a local elasto plastic second gradient model. This is useful to check the numerical codes. It is also interesting to discuss a little bit the conditions of apparition of the diverse solutions depicted in section 4.2.

4.4   Discussion

It is worth noticing first that as they correspond to an elastic behaviour, the two parameters a^{12345} and a^{34} are positive, so $A > 0$ (equation 27) is always positive and C has the same sign as the product of the two roots of equation 27. Moreover B has the opposite sign of the two roots sum.

An other important remark is that in fact we can write:

$$C = det(n_i \mathcal{K}^1_{ijkl} n_l) \tag{34}$$

where n_i is the normal to the chosen boundaries (here $n_1 = 1$ and $n_2 = 0$). This means that C is in fact the determinant of the acoustic tensor of the underlying first gradient model, involved in the classical shear band analysis ([9]). If our medium behaves elastically then $C > 0$ and $B > 0$. In this case the only partial solution is solution 1. In this case, $\mathcal{K}^1_{1112} = \mathcal{K}^1_{2111} = 0$, the two roots of the characteristic equation are positive and we get: $\eta^1 = \frac{\mathcal{K}^1_{2121}}{a^{34}}$ and $\eta^2 = \frac{\mathcal{K}^1_{1111}}{a^{12345}}$. Finally there is no coupling between the two differential equations and the solutions can be written:

$$\dot{u}' = \dot{U}'^0 + \lambda^{11} cosh(\eta^1 x) + \lambda^{12} sinh(\eta^1 x) \tag{35}$$

$$\dot{v}' = \dot{V}'^0 + \lambda^{21} cosh(\eta^2 x) + \lambda^{22} sinh(\eta^2 x) \tag{36}$$

The first terms of equations 35 and 36 describe an homogeneous straining of the whole body.

For moderate plastic part the behavior is similar to an elastic one, and only parts obeying equation 28 are available.

Let us examine now what happens if the medium behaves more and more plastically. First one of the root of equation 27 vanishes while the other one remains positive. In this case, equation 33 becomes available. Then this root becomes negative and solution 31. which involves a *cosine* and consequently a localized structure appears. This structure has in this case a clear internal length $2\pi/\omega^2$. It can be concluded first that the threshold of possible localization in a second gradient model like the one studied here is the same as the threshold of localization for the corresponding underlying first gradient model. It is important to emphasize that the localization criterion of studied second gradient models involves only the first gradient part of the model. However as localized solutions are linked with a characteristic length which decreases as the medium is more and more plastic and which is infinite at the threshold of localization the appearance of a localized band depends on the geometry and on the boundary conditions of the whole problem. Practically this means that appearance of localization can be somewhat delayed for a second gradient media.

If we are dealing with a materials for which normality holds and if the chosen axes are the principal directions of the stress then $\mathcal{K}^1_{1112} = \mathcal{K}^1_{2111} = 0$ and similarly to elastic solutions, equations for $\dot{v}'$ and $\dot{v}'$ are uncoupled.

For every case seen above the solution of the characteristic equation varies continuously with respect to the orientation of the stress. Consequently like for classical model (i.e. first gradient model) localization can be possible for a fan of stress orientation (with respect to the shear band). However once more for the second gradient model only, it is possible that details in the geometry of the body and/or of the boundary conditions inhibit the localization phenomenon. Some solutions got by using the previous method can be seen in [3].

5   Equations with Lagrange multipliers

Another way to deal with second gradient models is keeping the equations of a continuum with microstructure simplifying them only with the first equality of equation 7 and using the second equality as a mathematical constraint. Then Lagrange multipliers are introduced, which correspond to every component of v_{ij} and are thus denoted λ_{ij}. The following equations are got.

$$\int_{\Omega}[\sigma_{ij}\frac{\partial u_i^*}{\partial x_j} + \chi_{ijk}\frac{\partial v_{ij}^*}{\partial x_k} + \lambda_{ij}(\frac{\partial u_i^*}{\partial x_j} - v_{ij}^*)]dv = \int_{\Omega} G_i u_i^* dv \int_{\partial\Omega}(p_i u_i^* + P_i Du_i^*)ds \tag{37}$$

which hold for any kinematic admissible fields u_i^* and v_{ij}^*, and

$$\int_{\Omega}\lambda_{ij}^*(\frac{\partial u_i}{\partial x_j} - v_{ij})dv = 0 \tag{38}$$

which holds for any λ_{ij}^*. This way which seems less natural than the previous one is in fact very useful in finite element applications because it is difficult to get shape functions meeting the necessary continuity conditions. A two dimensional application of this way of working in finite element method for incompressible second gradient elastic solids have been done by Shu et. al. ([10]).

Here we want to develop a finite element method to solve problems involving an elasto plastic second gradient model. As usual it is necessary to discretize the time in order to be able to numerically solve the problem. In fact the problem is not solved for every time t but for a sequence of time Δt, $2\Delta t$, $3\Delta t$, and so on up to $N\Delta t = T$. In the next section we will consider the following time step problem. The solution of the general problem are known up to the time $t - \Delta t$, and we are looking for the values of u_i^t and v_{ij}^t at time t which then have to meet equations 37, and 38 at time t, the values of G_i, p_i and P_i of being assumed to be known. In order to solve this problem we will build up in the next section an iterative procedure based on a Newton- Raphson method.

6   Equations for the iterative procedure

Let us detail here the central point of the iterative procedure corresponding to a given time step. Let us assume that we know an approximation of the solution for time t. This means that we know a configuration called Ω^{t_n}. We are now looking for a better (unknown) approximation called Ω^{t_n+1}. First of all for every kinematically admissible field, it is possible to define residuals denoted $R^{t^*_n}$ and $Q^{t^*_n}$ with the following equation:

$$\int_{\Omega^{t_n}} \left[\frac{\partial u_i^*}{\partial x_l^{t_n}} \sigma_{il}^{t_n} + \frac{\partial v_{ij}^*}{\partial x_l^{t_n}} \chi_{ilk}^{t_n} - \frac{\partial u_i^*}{\partial x_l^{t_n}} \lambda_{il}^{t_n} + v_{ij}^* \lambda_{ij}^{t_n} \right] d\Omega^{t_n} - P_e^{t^*_n} = R^{t^*_n} \qquad (39)$$

$$\int_{\Omega^{t_n}} \lambda_{ij}^* \left(\frac{\partial u_i^{t_n}}{\partial x_j^{t_n}} - v_{ij}^{t_n} \right) d\Omega^{t_n} = Q^{t^*_n} \qquad (40)$$

Let us notice that $R^{t^*_n}$ and $Q^{t^*_n}$ are linear forms with respect to u_i^* and v_{ij}^* for the first one and with respect to λ_{ij}^* for the second one. Then similar equations are written in the unknown configuration $\Omega^{t_{n+1}}$ assuming that this new configuration is the solution, which means assuming that the corresponding residuals vanish. Writing both set of equation in configuration Ω^{t_n} and substracting the ones corresponding to configuration Ω^{t_n} from the ones corresponding to configuration $\Omega^{t_{n+1}}$ yields:

$$\int_{\Omega^{t_n}} \left[\frac{\partial u_i^*}{\partial x_l^{t_n}} (\sigma_{ij}^{t_{n+1}} \frac{\partial x_l^{t_n}}{\partial x_j^{t_{n+1}}} detF - \sigma_{il}^{t_n}) + \frac{\partial v_{ij}^*}{\partial x_l^{t_n}} (\chi_{ijk}^{t_{n+1}} \frac{\partial x_l^{t_n}}{\partial x_k^{t_{n+1}}} detF - \chi_{ilk}^{t_n}) \right.$$
$$\left. - \frac{\partial u_i^*}{\partial x_l^{t_n}} (\lambda_{ij}^{t_{n+1}} \frac{\partial x_l^{t_n}}{\partial x_j^{t_{n+1}}} detF_{ij} - \lambda_{il}^{t_n}) + v_{ij}^* (\lambda_{ij}^{t_{n+1}} detF - \lambda_{ij}^{t_n}) \right] d\Omega^{t_n}$$
$$- (P_e^{t^*_{n+1}} - P_e^{t^*_n}) = -R^{t^*_n} \qquad (41)$$

$$\int_{\Omega^{t_n}} \lambda_{ij}^* \left[(\frac{\partial u_i^{t_{n+1}}}{\partial x_k^{t_n}} \frac{\partial x_k^{t_n}}{\partial x_j^{t_{n+1}}} detF - \frac{\partial u_i^{t_n}}{\partial x_j^{t_n}}) - (v_{ij}^{t_{n+1}} detF - v_{ij}^{t_n}) \right] d\Omega^{t_n} = -Q^{t^*_n} \qquad (42)$$

where $detF$ is the Jacobian of $F_{ij} = \partial x_i^{t_{n+1}} / \partial x_j^{t_n}$ and obeys the following equation:

$$d\Omega^{t_{n+1}} = detF d\Omega^{t_n} \qquad (43)$$

Let us denote:

$$du_j^{t_n} = x_j^{t_{n+1}} - x_j^{t_n}$$
$$dv_{ij}^{t_n} = v_{ij}^{t_{n+1}} - v_{ij}^{t_n}$$
$$d\sigma_{il}^{t_n} = \sigma_{il}^{t_{n+1}} - \sigma_{il}^{t_n}$$
$$d\chi_{ikl}^{t_n} = \chi_{ikl}^{t_{n+1}} - \chi_{ikl}^{t_n}$$
$$d\lambda_{il}^{t_n} = \lambda_{il}^{t_{n+1}} - \lambda_{il}^{t_n} \qquad (44)$$

Using a Taylor expansion, assuming that $du_j^{t_n}$, $dv_{ij}^{t_n}$, $d\sigma_{il}^{t_n}$, $d\chi_{ikl}^{t_n}$, and $d\lambda_{il}^{t_n}$ are on the same order and discarding terms of degree greater than one yields after some mathematics to the following linearized equations

$$
\int_{\Omega^{t_n}} \left[\frac{\partial u_i^\star}{\partial x_l^{t_n}} \, \frac{\partial v_{ik}^\star}{\partial x_l^{t_n}} \, v_{il}^\star \right] \left(\begin{bmatrix} d\sigma_{il}^{t_n} \\ d\chi_{ikl}^{t_n} \\ 0 \end{bmatrix} + \begin{bmatrix} \sigma_{il}^{t_n} - \lambda_{il}^{t_n} \\ \chi_{ikl}^{t_n} \\ \lambda_{il}^{t_n} \end{bmatrix} \frac{\partial du_m^{t_n}}{\partial x_m} \right.
$$

$$
\left. + \begin{bmatrix} -\sigma_{ij}^{t_n} + \lambda_{ij}^{t_n} & -1 \\ -\chi_{ikj}^{t_n} & 0 \\ 0 & 1 \end{bmatrix} \begin{bmatrix} \frac{\partial du_j^{t_n}}{\partial x_i^{t_n}} \\[2mm] d\lambda_{il}^{t_n} \end{bmatrix} \right) d\Omega^{t_n} = (P_e^{t_{n+1}^\star} - P_e^{t_n^\star}) - R^{t_n^\star} \tag{45}
$$

$$
\int_{\Omega^{t_n}} \lambda_{ij}^\star \left(\left(\frac{\partial u_i^{t_n}}{\partial x_j^{t_n}} - v_{ij}^{t_n} \right) \frac{\partial du_m^{t_n}}{\partial x_m^{t_n}} + \frac{\partial du_i^{t_n}}{\partial x_j^{t_n}} - \frac{\partial u_i^{t_n}}{\partial x_k^{t_n}} \frac{\partial du_j^{t_n}}{\partial x_k^{t_n}} - dv_{ij}^{t_n} \right) d\Omega^{t_n} = -Q_{t_n}^\star \tag{46}
$$

The term $P_e^{t_{n+1}^\star} - P_e^{t_n^\star}$ in eq.(45) depends on f_i^t, p_i^t and P_i^t. Usually f_i^t are position independent. For simplicity it is assumed in the sequel that $P_i^t = 0$. So if p_i^t are dead loading forces which means that they are position independent too, then:

$$
P_e^{t_{n+1}^\star} - P_e^{t_n^\star} = 0 \tag{47}
$$

This is assumed in the following. Adding the corresponding terms in the case of following forces (for p_i^t) such as pressure originated ones is straightforward and can be done as in classical (without gradient effects) analysis.

In the following for simplicity only two dimensional problem are studied. So now the indices $i, j, \ldots \in \{1, 2\}$.

Let us define the following vector with 20 variables:

$$
[dU^{t_n}]^T \equiv \left[\frac{\partial du_1^{t_n}}{\partial x_1^{t_n}} \, \frac{\partial du_1^{t_n}}{\partial x_2^{t_n}} \, \frac{\partial du_2^{t_n}}{\partial x_1^{t_n}} \, \frac{\partial du_2^{t_n}}{\partial x_2^{t_n}} \, \frac{\partial dv_{11}^{t_n}}{\partial x_1^{t_n}} \, \frac{\partial dv_{11}^{t_n}}{\partial x_2^{t_n}} \, \frac{\partial dv_{12}^{t_n}}{\partial x_1^{t_n}} \, \cdots \, \frac{\partial dv_{22}^{t_n}}{\partial x_2^{t_n}} \right.
$$

$$
\left. dv_{11}^{t_n} \, dv_{12}^{t_n} \, dv_{21}^{t_n} \, dv_{22}^{t_n} \, d\lambda_{11}^{t_n} \, d\lambda_{12}^{t_n} \, d\lambda_{21}^{t_n} \, d\lambda_{22}^{t_n} \right] \tag{48}
$$

$[U^\star]^T$ is defined similarly for the corresponding virtual quantities. After some mathematics eq.(45) and (46) yields:

$$
\int_{\Omega^{t_n}} [U^\star]^T [E^{t_n}] [dU^{t_n}] d\Omega^{t_n} = -R^{t_n^\star} - Q^{t_n^\star} \tag{49}
$$

where, $[E]^{t_n}$ is a 20×20 elements matrix.

7 Finite Element Method

7.1 Shape functions

Equation 49 is now discretized using the finite element method. The quadrilateral elements used are detailed hereafter. They have 8 nodes for u_i, 4 nodes for v_{ij} and one node for λ_{ij}. The parent element is defined in a ξ η spaced by ξ $(-1 < \xi < 1)$

and η $(-1 < \eta < 1)$. Finally all the variables are described with respect to these parent coordinates.

The usual quadratic serendipity shape function denoted ϕ are used for u_i. The linear serendipity one denoted ψ are used for v_{ij}, and λ_{ij} is constant and can be seen as known in the center of the element:

$$u_i^{t_n}(\xi, \eta) = [\phi_1(\xi, \eta) \ \phi_2(\xi, \eta) \ \dots \ \phi_8(\xi, \eta)] \begin{bmatrix} u_i^{t_n}\big|_{(\xi=-1, \eta=-1)} \\ u_i^{t_n}\big|_{(0,-1)} \\ \vdots \\ u_i^{t_n}\big|_{(-1,0)} \end{bmatrix} \quad (50)$$

$$v_{ij}^{t_n}(\xi, \eta) = [\psi_1(\xi, \eta) \ \psi_2(\xi, \eta) \ \psi_3(\xi, \eta) \ \psi_4(\xi, \eta)] \begin{bmatrix} v_{ij}^{t_n}\big|_{(\xi=-1, \eta=-1)} \\ v_{ij}^{t_n}\big|_{(1,-1)} \\ v_{ij}^{t_n}\big|_{(1,1)} \\ v_{ij}^{t_n}\big|_{(-1,1)} \end{bmatrix} \quad (51)$$

$$\lambda_{ij}^{t_n}(\xi, \eta) = \lambda_{ij}^{t_n}\big|_{(\xi=0, \eta=0)} \quad (52)$$

The mapping function from $(x_1^{t_n}, x_2^{t_n})$ to (ξ, η), are the same as the shape function for $u_i^{t_n}(\xi, \eta)$ (isoparametric element).

7.2   Element stiffness matrix

The transformation matrices, $[T^{t_n}]$ and $[B]$, which connect $[dU^{t_n}]$ with the nodal variables are obtained through the common finite element procedure:

$$[dU^{t_n}] = [T^{t_n}][dU_{(\xi,\eta)}^{t_n}] \quad (53)$$

$$[dU_{(\xi,\eta)}^{t_n}] = [B][dU_{node}^{t_n}] \quad (54)$$

where,

$$[dU_{(\xi,\eta)}^{t_n}]^T =$$
$$\left[\frac{\partial du_1^{t_n}}{\partial \xi} \ \dots \ \frac{\partial du_2^{t_n}}{\partial \eta} \ \frac{\partial dv_{11}^{t_n}}{\partial \xi} \ \dots \ \frac{\partial dv_{22}^{t_n}}{\partial \eta} \ dv_{11} \ \dots \ dv_{22} \ d\lambda_{11} \ \dots \ d\lambda_{22} \right] \quad (55)$$

$$[dU_{node}^{t_n}]^T =$$

$$
\begin{aligned}
&\Big[du_1^{t_n}{}_{(-1,-1)} \; du_2^{t_n}{}_{(-1,-1)} \; dv_{11}^{t_n}{}_{(-1,-1)} \; dv_{12}^{t_n}{}_{(-1,-1)} \; dv_{21}^{t_n}{}_{(-1,-1)} \; dv_{22}^{t_n}{}_{(-1,-1)} \\
&du_1^{t_n}{}_{(-1,0)} \quad du_2^{t_n}{}_{(-1,0)} \\
&du_1^{t_n}{}_{(-1,1)} \quad du_2^{t_n}{}_{(-1,1)} \; dv_{11}^{t_n}{}_{(-1,1)} \; dv_{12}^{t_n}{}_{(-1,1)} \; dv_{21}^{t_n}{}_{(-1,1)} \; dv_{22}^{t_n}{}_{(-1,1)} \\
&du_1^{t_n}{}_{(0,-1)} \quad du_2^{t_n}{}_{(0,-1)} \\
&d\lambda_{11}^{t_n}{}_{(0,0)} \quad d\lambda_{12}^{t_n}{}_{(0,0)} \; d\lambda_{21}^{t_n}{}_{(0,0)} \; d\lambda_{22}^{t_n}{}_{(0,0)} \\
&du_1^{t_n}{}_{(0,1)} \quad du_2^{t_n}{}_{(0,1)} \\
&du_1^{t_n}{}_{(1,-1)} \quad du_2^{t_n}{}_{(1,-1)} \; dv_{11}^{t_n}{}_{(1,-1)} \; dv_{12}^{t_n}{}_{(1,-1)} \; dv_{21}^{t_n}{}_{(1,-1)} \; dv_{22}^{t_n}{}_{(1,-1)} \\
&du_1^{t_n}{}_{(1,0)} \quad du_2^{t_n}{}_{(1,0)} \\
&du_1^{t_n}{}_{(1,1)} \quad du_2^{t_n}{}_{(1,1)} \; dv_{11}^{t_n}{}_{(1,1)} \; dv_{12}^{t_n}{}_{(1,1)} \; dv_{21}^{t_n}{}_{(1,1)} \; dv_{22}^{t_n}{}_{(1,1)} \Big]
\end{aligned}
\tag{56}
$$

Finally the element stiffness matrix $[k^{t_n}]$ is obtained from eq.(49) as follows:

$$
\int_{\Omega_{elem}^{t_n}} [U^*]^T [E^{t_n}][dU^{t_n}] d\Omega^{t_n} =
$$

$$
[U_{node}^*]^T \int_{-1}^{1} \int_{-1}^{1} [B]^T [T^{t_n}]^T [E^{t_n}][T^{t_n}][B] det J^{t_n} d\xi d\eta [dU_{node}^{t_n}]
$$

$$
\equiv [U_{node}^*]^T [k^{t_n}][dU_{node}^{t_n}]
\tag{57}
$$

where,

$$
det J^{t_n} =
\begin{vmatrix}
\dfrac{\partial x_1^{t_n}}{\partial \xi} & \dfrac{\partial x_1^{t_n}}{\partial \eta} \\[2ex]
\dfrac{\partial x_2^{t_n}}{\partial \xi} & \dfrac{\partial x_2^{t_n}}{\partial \eta}
\end{vmatrix}
\tag{58}
$$

It is worth mentioning here that in the following applications, the integration procedure uses a Gauss method. In order to avoid possible locking in the plastic range, the elements are under integrated, which means here that only four Gauss points are used to compute the element stiffness matrix. Obviously the same method holds for the computation of the residual terms detailed in the following.

7.3   Element residual terms

The residual terms $R^{t_n^*} + Q^{t_n^*}$ in eq.(49) can be computed for one element as follows, defining the elementary out of balance forces $[f_{HE}^{t_n}]$:

$$
(-R^{t_n^*} - Q^{t_n^*})_{elem} =
$$

$$
P_e^{t_n^*} - [U_{node}^*]^T \int_{-1}^{1} \int_{-1}^{1} [B]^T [T^{t_n}]^T [\sigma^{t_n}] det J^{t_n} d\xi d\eta
$$

$$
\equiv [U_{node}^*]^T [f_{HE}^{t_n}]
\tag{59}
$$

where,

$$[\sigma^{t_n}] = \left[\sigma_{11}^{t_n} - \lambda_{11}^{t_n} \cdots \sigma_{22}^{t_n} - \lambda_{22}^{t_n} \chi_{111}^{t_n} \cdots \chi_{222}^{t_n} \lambda_{11}^{t_n} \cdots \lambda_{22}^{t_n} \right.$$

$$\left. \frac{\partial u_1^{t_n}}{\partial x_1^{t_n}} - v_{11}^{t_n} \cdots \frac{\partial u_2^{t_n}}{\partial x_2^{t_n}} - v_{22}^{t_n} \right] \tag{60}$$

The external virtual power, $P_e^{t_n^\star}$, consists of two terms; the body force term and the boundary surface force term.

7.4   Global matrices

Adding the equations 57 and 59 for each element and eliminating the virtual displacement vector, $[U_{node}^\star]$, the global finite element equations are finally obtained:

$$[K^{t_n}][\delta U_{node}^{t_n}] = -[F_{HE}^{t_n}] \tag{61}$$

where $[K^{t_n}]$ is the global stiffness matrix obtained by assembling the element stiffness matrices $[k^{t_n}]$ (see equation 57, $[F_{HE}^{t_n}]$ is the global out-of-balance force vector obtained by assembling the elementary out of balance forces $[f_{HE}^{t_n}]$ (see equation 59), and $[\delta U_{node}^{t_n}]$ is the global matrix of the corrections of the nodal displacement values. They are the unknown of the nth iteration of the time step from $t - \Delta t$ to t. After solving this auxiliary linear system, the current configuration is actualized and the following iteration starts. In order to clarify the various aspects detailed above, an algorithm is detailed in table 1.

8   Applications: two dimensional elasto-plastic constitutive relation

As far as the first gradient part, is concerned, the model implemented in this study is based on the Prandtl-Reuss elasto-plastic model as follows:

$$\dot{\sigma} = 3K\dot{e} \tag{62}$$

$$\overset{\triangledown}{s}_{ij} = \begin{cases} 2G_1 \dot{\varepsilon}_{ij} & (\|\varepsilon\| \le e_{lim}) \\ 2G_1 (\dot{\varepsilon}_{ij} - \frac{G_1 - G_2}{G_1} \frac{s_{kl}\dot{\varepsilon}_{kl}}{\|s\|^2} s_{ij}) & (\|\varepsilon\| > e_{lim}) \end{cases} \tag{63}$$

where, $\overset{\triangledown}{s}_{ij}$ is the Jaumann deviatoric stress rate tensor, $\dot{\varepsilon}_{ij}$ is deviatoric strain rates, $\dot{\sigma}$ is the mean stress rate and $\dot{e}$ is the mean strain rate. K, G_1 and G_2 are the bulk modulus, the shear moduli before peak and after peak, respectively.

For the second gradient part, the more general isotropic linear relation is used. A six parameters model with five independent coefficients was derived by

Table 1. The iterative algorithm for one loading step

1. Initial configuration :
   stress $\sigma^{t-\Delta t}$, double stress $\chi^{t-\Delta t}$, coordinates $x^{t-\Delta t}$
2. Assumption on the final configuration
   update coordinates : x^{tn} for $n = 1$
3. Beginning of the iteration n :
4. For each element :
   - For each integration point :
     - compute the strain rate, the rotation rate and the second gradient rate
     - compute $\Delta\sigma_n$ and $\Delta\chi_n$ using the constitutive equations
     - update the stress and the double stress
       $\sigma^{tn} = \sigma^{t-\Delta t} + \Delta\sigma_n$, $\chi^{tn} = \chi^{t-\Delta t} + \Delta\chi_n$,
     - compute the consistent stiffness matrices by means of perturbation methods
   - Compute the element stiffness matrix $[k^{tn}]$
   - Compute the element out of balance forces $[f_{HE}^{tn}]$
5. Compute the global stiffness matrix $[K^{tn}]$
6. Compute the global out of balance forces $[F_{HE}^{tn}]$
7. Compute $[\delta U_{node}^{tn}]$ by solving $[K^{tn}][\delta U_{node}^{tn}] = -[F_{HE}^{tn}]$
8. Check the accuracy of the computed solution
   - if convergence : go to 9
   - if no convergence : update the new assumed final configuration, $n = n + 1$, go to 3
9. End of the step

Mindlin in this case ([7], [6]). We used a particular case depending only on one parameter as can be seen in the following equation.

$$[h]\begin{bmatrix} \overset{\nabla}{\chi}_{111} \\ \overset{\nabla}{\chi}_{112} \\ \overset{\nabla}{\chi}_{121} \\ \overset{\nabla}{\chi}_{122} \\ \overset{\nabla}{\chi}_{211} \\ \overset{\nabla}{\chi}_{212} \\ \overset{\nabla}{\chi}_{221} \\ \overset{\nabla}{\chi}_{222} \end{bmatrix} = \begin{bmatrix} D & 0 & 0 & 0 & 0 & D/2 & D/2 & 0 \\ 0 & D/2 & D/2 & 0 & D/2 & 0 & 0 & D/2 \\ 0 & D/2 & D/2 & 0 & D/2 & 0 & 0 & D/2 \\ 0 & 0 & 0 & D & 0 & D/2 & D/2 & 0 \\ 0 & D/2 & D/2 & 0 & D & 0 & 0 & 0 \\ D/2 & 0 & 0 & D/2 & 0 & D/2 & D/2 & 0 \\ D/2 & 0 & 0 & D/2 & 0 & D/2 & D/2 & 0 \\ 0 & D/2 & D/2 & 0 & 0 & 0 & 0 & D \end{bmatrix} \begin{bmatrix} \frac{\partial \dot{v}_{11}}{\partial x_1} \\ \frac{\partial \dot{v}_{11}}{\partial x_2} \\ \frac{\partial \dot{v}_{12}}{\partial x_1} \\ \frac{\partial \dot{v}_{12}}{\partial x_2} \\ \frac{\partial \dot{v}_{21}}{\partial x_1} \\ \frac{\partial \dot{v}_{21}}{\partial x_2} \\ \frac{\partial \dot{v}_{22}}{\partial x_1} \\ \frac{\partial \dot{v}_{22}}{\partial x_2} \end{bmatrix} \quad (64)$$

The model presented so far has been already used in computations and gave good results. Mesh independence, scale effects, influence of the large strain assumption are satisfactory got by the Finite Element Method described above [6].

9 Conclusions

It is possible within the continuum mechanics framework to take into account the influence of the microstructure on to the overall behavior of a body by using enhanced models. Amongst others, we have here pointed out that the local second gradient continua is able to take into account the micro scale. Analytical solutions as well as numerical procedures detailed in this paper which increases the confidence we can have in such model to solve properly problems involving the microstructure like rupture.

References

1. F. Calvetti, G. Combe, J. Lanier: Mechanics of Cohesive Frictional Materials **2**, 121 (1997)
2. R. Chambon, D. Caillerie, N. El Hassan: Eur. J. Mech. A/Solids **17**, 637 (1998)
3. R. Chambon, D. Caillerie, T. Matsushima: 'Plastic continuum with microstructure, local second gradient theories for geomaterials, localization studies'. (submitted for publication) (2000)
4. P. Germain: SIAM J. Appl. Math. **25**, 556 (1973)
5. T. Matsushima, R. Chambon, D. Caillerie: C. R. A. S-Série II b **328**, 179 (2000)
6. T. Matsushima, R. Chambon, D. Caillerie: 'Media with microstructure, a two dimensional large strain numerical analysis'. (in preparation) (2000)
7. R. D. Mindlin: Arch. Rational Mech. Anal. **16**, 51 (1964)
8. R. D. Mindlin: Int. J. Solids Structures. **1**, 417 (1965)
9. J. Rice: 'The localization of plastic deformation'. In: *International Congress of Theoretical and Applied Mechanics*, ed. by W. T. Koiter, (North Holland Publishing Comp. 1976) pp. 207–220
10. J. Shu, W. King, N. Fleck: Int. J. Numer. Meth. in Engineering **44**, 373 (1999)
11. P. A. Vermeer: 'A simple shear band analysis using compliances'. In: *IUTAM conf. Def. Fail. Gran. Media DELFT*, ed. by P. A. Vermeer, H. J. Luger (A. A. Balkema Rotterdam 1982) pp. 493–499
12. I. Vardoulakis, J. Sulem: *Bifurcation Analysis in Geomechanics*, (Blackie Academic and Professional, Glasgow 1995)

Damage in a composite material under combined mechanical and hygral load

H. Sadouki, F. H. Wittmann

Institute for Building Materials, Swiss Federal Institute of Technology, Zürich, Switzerland

Abstract. Endogenous shrinkage and drying shrinkage are phenomena of hardened cement paste which may provoke micro-cracking. Microcracking may have several consequences on the physical and mechanical properties and the durability of cement-based composite materials. This paper presents results of a numerical study related to crack formation induced by endogenous and/or drying shrinkage. The analyses are performed on the mesostructural level of concrete-like composite materials. This means that the material is considered to be a composite consisting of two fundamental phases, i.e. coarse aggregates embedded in a porous hardened cement paste or mortar matrix. The model called Numerical Concrete has been used to perform these analyses.

First of all, time-dependent moisture distributions in cement-based composite materials such as normal and high performance concrete subjected to endogenous and external drying have been simulated. The hardening binding cement matrix with low water-cement ratio such as in high performance concrete is generally subject to strong self-desiccation inducing endogenous drying. These inner hygral strains of the matrix are restrained by the presence of the generally stiff aggregates. As consequence, eigenstresses will develop in the composite material. If in addition, the composite material is exposed to a dry atmosphere differential shrinkage between the inner and outer layers of the specimen will occur. This hygral strain gradient will induce additional eigenstresses. In this contribution, the eigenstresses induced by both endogenous and drying shrinkage in normal and high performance concrete are analyzed. As it will be shown the generated tensile stresses often exceed the relatively low tensile strength of the cement matrix and thus leading to damage and microcracking. Crack formation and crack growth are simulated by means of the fictitious crack concept. Special attention has been paid to the risk of crack formation due to endogenous shrinkage in the composite structure of the high performance concrete. Finally, it will be shown how an applied sustained load on a concrete specimen influences crack formation. If drying takes place in a uniaxially loaded specimen the result will be a pronounced anisotropy of the composite material.

1 Introduction

In order to understand the distinction which is made in this contribution between endogenous (or autogenous) shrinkage and drying shrinkage, it is necessary to recall the basis of both phenomena [1].

(i) Endogenous shrinkage in a cement-based material with a low water-cement ratio starts at an early age of the material and is mainly the consequence of dissolution of Portland cement and self-desiccation. Self-desiccation can occur in a sealed or non-sealed cement-based material, as long as the hydration process proceeds. The amount of free water in hardening cement paste gradually decreases due to water consumption by the hydration process of the clinker minerals. The degree of self-desiccation is more pronounced for mortar or concrete mixtures with low water-cement ratios typically used for high performance concrete (HPC). In this case, the relatively small amount of water used for mixing is rapidly consumed by early age hydration of cement. The gradual moisture loss in the rigid porous matrix then will induce a reduction of volume. This time-dependent phenomenon is most often called endogenous shrinkage. Although endogenous shrinkage of concrete is known for many years [1], no attention has been paid to the influence on short and long term mechanical behaviour and time-dependent deformation i.e. creep and shrinkage of the material. In concrete, the endogenous shrinkage of the hydrating binding matrix will be restrained by the embedded, generally stiff aggregates. This restraining action will undoubtedly induce eigen-stresses. One of the main purposes of this contribution is to evaluate, by means of the model called Numerical Concrete Model [2], the tensile stresses in the matrix generated by restrained endogenous shrinkage and consequently to assess the risk of early micro-cracking in the composite material.

(ii) As for endogenous drying, drying shrinkage of cement-based materials is caused by a time-dependent decrease of the moisture held in the pores of the matrix. The mechanism of the moisture loss, however is not due to hydration but to moisture migration (diffusion) from the specimen towards the outer dry environment. In contrast to the case of endogenous drying, considerable attention has been paid to drying shrinkage especially with respect to crack formation [3–7]. Nevertheless, the phenomenon will also be treated in this contribution at the mesostructural level of the composite material.

2   Generation of numerical concrete

Concrete is a multiscale particle composite. In this contribution, the structural mesolevel of concrete will be modeled by the numerical concrete [2]. Concrete is considered as a composite material, composed of two major phases, namely coarse aggregates (inclusions) embedded in a hydraulic cement paste matrix. The inclusions, randomly distributed in the porous binding matrix, are assumed to follow a predetermined grain size distribution. The volume content of the aggregates and their shape can be chosen according to predefined requirements in the computer-generated composite structures. The generated composite structures are then subdivided into finite elements (FE), in order to perform the intended numerical analysis, such as for example heat and moisture transfer or crack formation. Fig. 1 shows FE idealization of two different computer-generated composite structures. In one case hexagonal aggregates (crushed) and in the other spherical aggregates (river gravel) have been approximated. More complex geometries have also been taken into consideration. At this moment a project to simulate three dimensional aggregates is on its way.

Fig. 1. Computer-generated composite structures and their finite element idealization

3   Drying process and self-desiccation

3.1   Basic elements and equations governing the processes

In order to predict the evolution of the endogenous and drying shrinkage strains, first, spatial moisture variations in the material must be known for any time. As it has been mentioned in the introduction, drying of concrete is due to moisture migration and moisture loss. In the porous material, moisture gradients or corresponding moisture potentials are then the driving force of this process. In reality, moisture flow in porous materials like concrete is a very complex process including capillary suction, capillary condensation and subsequent evaporation, gas phase flow and surface diffusion [8]. In the present approach, all phenomena are globally described by means of a single non-linear flow equation. This means that the involved different mechanisms are not separated. Mathematically, the drying process can be expressed by the following second-order partial differential equation:

$$\frac{\partial h}{\partial t} = \mathrm{div}(D(h,\tau) \cdot \mathrm{grad}\, h) \tag{1}$$

where h is the moisture potential, τ is the age of the material, $D(h,\tau)$ is the moisture transfer coefficient. If the material is subjected to self-desiccation, a sink term $q(\tau)$, representing the rate of the moisture loss in the system, must be added to the right side of (1). The global equation governing drying and self-desiccation then reads as follows:

$$\frac{\partial h}{\partial t} = \mathrm{div}(D(h,\tau) \cdot \mathrm{grad}\, h) + q(\tau) \tag{2}$$

It must be underlined, that both parameters ($D(h,\tau)$ and $q(\tau)$) are strongly dependent on the composition of the cement-based material. If the specimen is exposed to a given surrounding atmosphere with a relative humidity h_{ext}, then the moisture flux due to an exchange between the system and the environment can be expressed as follows:

$$q = \beta \cdot (h_s - h_{ext}) \cdot \Gamma \tag{3}$$

in which β is the hygral convection coefficient and h_s the humidity on the exposed surface Γ of the specimen. If the specimen is totally sealed, then the moisture flux in (3) becomes zero.

In this simplified approach, the influence of thermal effects, such as the rise of temperature induced by the hydration reaction, is not taken into consideration. Coupled heat and mass transfer can be simulated, however, in a similar way.

3.2   Material parameters

The hygral material parameters in (3) ($D(h, \tau)$ and $q(\tau)$) are strongly dependent on the composition of the cement-based materials. The moisture transfer coefficient can easily be determined from drying specimens as follows: after a given curing time τ_0, specimens are exposed to different ambient relative humidities at constant temperature. The moisture loss is measured periodically by determining the weight of specimens as function of time. By means of an adequate inverse analysis, the moisture transfer coefficient can be evaluated [9]. Typical moisture transfer coefficients are shown in Fig. 2 for two different mortars [7]. In this figure, curve (A) represents the moisture transfer coefficient for a standard mortar with a water/cement ratio 0.50. Curve (B) is obtained for a high performance mortar with the following mix (in kg/m3): w= 135, c= 375, sand= 1620 and as admixtures (in % of cement weight): polymer dispersion= 15%, superplasticiser= 1.2 and air-entraining agents= 0.2.

Moisture loss due to self-desiccation is measured by means of hygrometric sensors embedded in sealed specimens [10, 11]. The humidity in the specimen is then recorded as function of time (age of the material). Fig. 3 [10] shows the decrease of the humidity as function of age, for two types of concrete and two types of hardened cement paste. The composition of the four materials is given in Table 1.

The moisture sink term $q(\tau)$ (endogenous drying) in (3) can be derived from curves as shown in Fig. 3. In all simulations performed in this study, aggregate particles are assumed not to absorb or contain any water. This assumption is not valid for a series of special concretes. The non-linear FE Code DIANA, has been used to perform the numerical analyses [12].

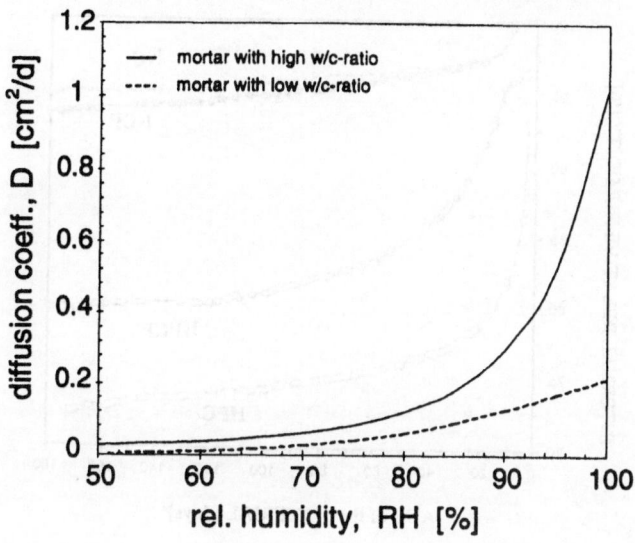

Fig. 2. moisture transfer coefficient for two different mortars

Table 1. Composition of four cement-based materials. The endogenous drying of these materials is shown in Fig. 3, after Baroghel-Bouny [10]

Type of material	Abbreviation	Water-cement ratio	Cement content [kg/m³]	Aggregates and sand [kg/m³]	Micro silica [kg/m³]	Plasticiser [kg/m³]
Normal Concrete	NC	0,487	353	1936	-	-
High Performance concrete	HPC	0,267	421	1917	112,3	7,59
Normal Hardened Cement Paste	NCP	0,348	-	-	-	-
High Performance Hardened Cement paste	HPCP	0,196	1757	-	175,7	31,7

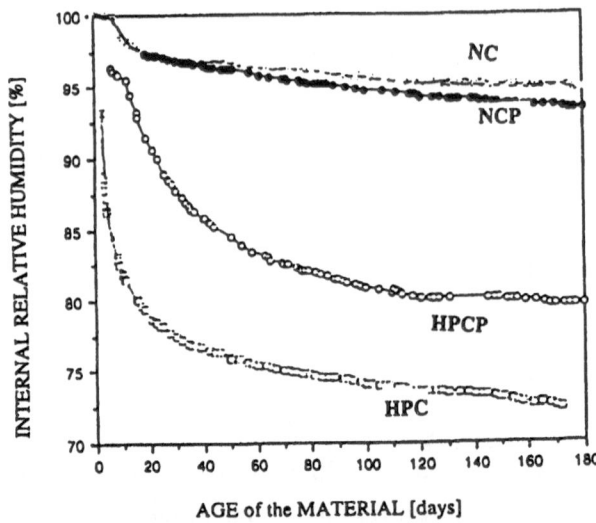

Fig. 3. Kinetics of self-desiccation as function of age of the material: 2 different hardened cement pastes and 2 different concretes have been studied [10]

As it can be seen from Fig. 3, materials characterized by a low w/c-ratio and a high cement content undergo a remarkable self-desiccation.

3.3   An example of simulation of drying

This example deals with the drying process in normal concrete idealized by a computer-generated composite structure consisting of 62% coarse non-porous aggregates embedded in a porous mortar matrix. The matrix is characterized by a relatively high moisture transfer coefficient $D(h, \tau)$ (curve (B) in Fig. 2). The self-desiccation term $q(\tau)$ in (3) is neglected in this case, because of the relatively high w/c-ratio. This last assumption means that the hydration process does not contribute to the decrease of the relative humidity in the porous structure. After a given curing time (7 days for example), the specimen is exposed to a dry atmosphere with 60% RH. Fig. 4 shows the moisture distribution of a specimen which i.e. a thickness of 32mm at two drying times, 7 and 30 days. The lightest grey shades in the figures indicate a low RH (60%), the darkest grey a RH=100%. The Figures reveal a gradual drying of the composite, starting from the exposed surfaces. As it can be seen from Fig. 4b, even after an advanced drying time of 30 days, the humidity of the inner parts is still close to 100% RH, but the hygral state of the system progressively converges to the equilibrium with the surrounding atmosphere.

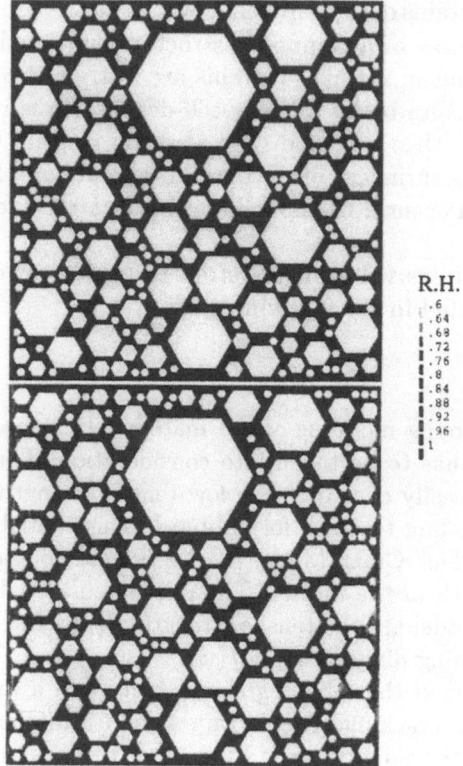

R.H.
.6
.64
.69
.72
.76
.8
.84
.88
.92
.96
1

Fig. 4. moisture distribution in a drying NC specimen at 2 drying times, (a) 7d and (b) 30d

4   Endogenous and drying shrinkage

4.1   General concept

Moisture decrease, induced either by self-desiccation or by drying or both has been considered to be the single driving force for hygral shrinkage strains. Further, it has been assumed, that the free shrinkage strain $\Delta\varepsilon_{sh}$ generated by an infinitesimal decrease of humidity Δh is given by the following relationship:

$$\Delta\varepsilon_{sh} = \alpha_{sh} \cdot \Delta h \qquad (4)$$

in which α_{sh} is the so-called coefficient of linear hygral shrinkage. This coefficient depends not only on the matrix composition by also on other state variables such as moisture content, temperature and degree of hydration. In this contribution, for a quantitative evaluation of shrinkage strains and shrinkage-induced stresses, this coefficient has been assumed to be constant for a given type of concrete.

If the shrinkage strains of the hardening porous matrix are restrained, stresses will be built up. In case of a composite structure subjected to self-desiccation only, i.e. sealed specimen, shrinkage strains are restrained by the inert stiff aggregates. If the structure under ongoing self-desiccation is in addition exposed to a dry atmosphere, then a second type of strain restrain will appear, due to the differential drying shrinkage of the outer drier and inner layers. A third type of strain restrain can occur if the boundaries of the structure are not allowed to deform freely.

These strain restrains will generate stresses. In a linear elastic analysis, the stresses σ are computed in the following way:

$$\sigma = E \cdot \alpha_{\mathrm{sh}} \cdot \Delta h \qquad (5)$$

in which E is the Young modulus of the material. In a more realistic analysis creep of the matrix has to be taken into consideration. The shrinkage-induced tensile stresses can easily overcome the low tensile strength of the porous matrix consequently leading to crack formation. In this contribution, the smeared version of the Fictitious Crack Model [12–14] is used to simulate crack formation and crack growth in the matrix. The relevant material parameters of this softening fracture model are the tensile strength f_t, the fracture energy G_f and the stress-crack-opening diagram $\sigma_c = f(w)$.

It has been assumed that the aggregates behave in a linear elastic manner without possibility of cracking. The above mentioned mechanical properties are strongly dependent on many parameters, such as composition, degree of hydration, curing time, moisture content and temperature. But as a first approximation, in this contribution, which is aimed to be a qualitative evaluation of the risk of crack formation under self-desiccation and drying in NC and HPC, the material parameters are assumed to be constant for a given type of concrete. The chosen values for the mechanical parameters will be given for each case which has been investigated. More information can be found in [7, 11]

4.2   Shrinkage in normal and high performance concrete

4.2.1 Material parameters used in numerical experiments

In order to evaluate the contribution of the endogenous and drying shrinkage to crack formation in NC and in HPC, the following numerical simulations have been performed. Both concrete types are represented by computer-generated composite structures. In case of NC, material parameters $D(h, \tau)$ and $q(\tau)$ in (3) are given by the curve (B) in Fig. 2 and the curve (NC) in Fig. 3, respectively. In case of HPC, these parameters are given by the curve (A) of the Fig. 2 and the curve (HPC) in Fig. 3, respectively. In the first analysis, drying shrinkage in drying NC-specimens is analyzed, while in the second analysis, the endogenous shrinkage in sealed HPC -specimens is investigated.

The aggregates with a Young's modulus of 60 GPa and a Poisson's ratio of 0.2 are considered to behave linear elastic without the possibility of cracking.

The mechanical material parameters of the matrices at an age of seven days are given in Table 2. The tensile stress f_t of the mortar matrix is considered to be a statistical variable following a Gauss distribution, the standard deviation is given in Table 2 in parenthesis.

Table 2. Mechanical material parameters of the matrix in NC and HPC

Matrix in:	E [GPa]	ν [-]	α_{sh} [1/h]	f_t [MPa]	G_f [N/m]
NC	16.00	0.2	0.0025	4.0 (2.0)	30
HPC	17.86	0.2	0.0035	5.0 (2.5)	60

4.2.2 Stresses and crack formation induced by endogenous and drying shrinkage

a) Normal concrete

If the system is sealed, that means the material is subjected to self-desiccation only, the highest induced tensile stress around the stiff aggregates remains very low as compared to the tensile strength f_t of the matrix (4MPa) even in the matured state. This means that the tensile stresses induced by self-desiccation in NC cannot provoke noticeable damage.

In the following analysis, the composite structure of NC (see Fig. 4) is exposed to a dry atmosphere with 60%RH. Initially, the porous matrix is assumed to be moisture saturated (100%RH). Stress and crack evolutions are then analyzed as function of drying time. Some hours after the beginning of exposure, tensile stresses start to be built up in the outer drying layers of the composite structure near both exposed surfaces (left and right edges). As the drying process proceeds, tensile stresses in some FE of the outer layers reach the tensile strength of the matrix. This means that the micro-cracking process starts to develop. Figure 5a shows the resulting crack pattern after a drying time of 2.5 days. As the drying process continues, cracks become wider and the front of the cracked zone progresses towards the center of the specimen. If, during drying, a compressive sustained load is applied on the top surface of the specimen (normal to the drying direction), then cracking is reduced, because the number of cracks normal to the load direction decreases. This effect is more pronounced for high compressive load. Figure 5b shows the resulting crack pattern after a drying time of 2.5 days but with an applied load of 5 MPa. Figure 6 shows the effective drying shrinkage of the composite structure as function of drying time. In this figure, the upper line marked by circles is the "hypothetical" effective shrinkage if the matrix is assumed to behave linear elastical (no crack formation). In reality, due to high tensile stresses, the matrix undergoes micro-cracking, the resulting effective drying shrinkage is then given by the line marked by triangles. This curve is actually the observed shrinkage in usual experiments. In fact, it is a superposition of the real shrinkage and the induced micro-cracking. If a sustained compressive load is applied on top of the specimen, the effective observed shrinkage is higher as compared with unloaded specimens (see Fig. 6, lines with + symbols). If the load level is increased, the observed shrinkage increases, this is shown by the line marked with X-symbols in Fig. 6 (compressive load 10 MPa). If the compressive applied load is further increased, then the observed shrinkage is almost the same as the hypothetical case without crack formation (see line marked with circles in Fig. 6). Similar results based on a combination of a micro-structural FE model [15] and a physical model [16–18] have been obtained in the case of unloaded and loaded micro-mortar specimens.

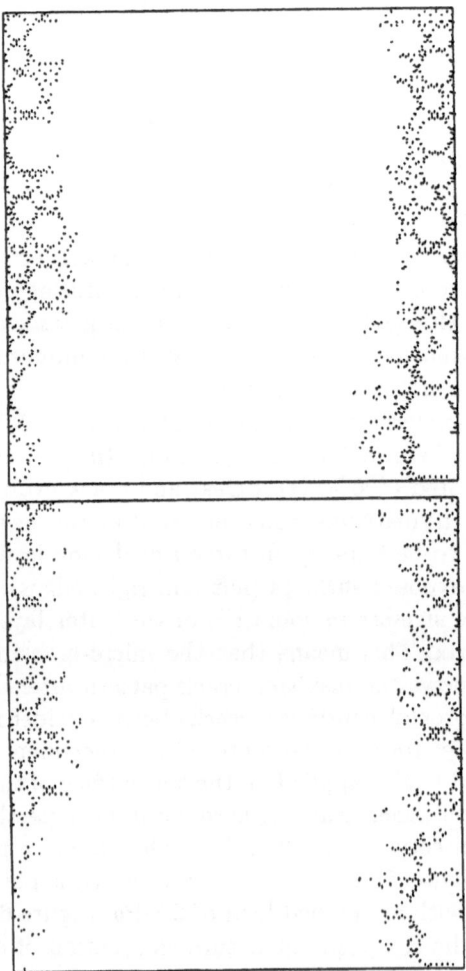

Fig. 5. Micro-crack patterns in the matrix of a NC-composite structure subjected to external drying at (a) unloaded specimen (b) with a compressive load of 5 MPa

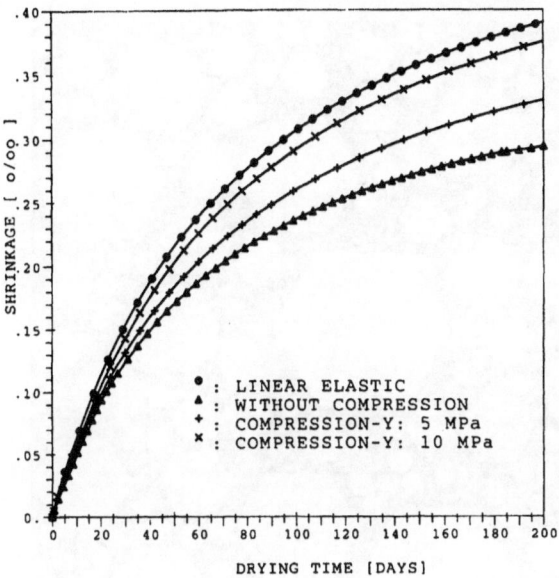

SHRINKAGE [o/oo]

DRYING TIME [DAYS]

● : LINEAR ELASTIC
▲ : WITHOUT COMPRESSION
+ : COMPRESSION-Y: 5 MPa
× : COMPRESSION-Y: 10 MPa

Fig. 6. Effective drying shrinkage as function of drying time in a NC-composite structure with and without an external load

b) High performance concrete

Stresses induced by the endogenous drying in HPC are much higher than in NC. In this analysis a HPC specimen is idealized by a composite structure consisting of 57% of aggregates embedded in a mortar matrix with a low w/c-ratio. All faces of the specimen are sealed that means that the moisture decrease in the porous matrix is due to self-desiccation only. The induced stresses and micro-cracking are analyzed. Figures 7a and b show the distribution of the principal stresses induced in the matrix and in the stiff inclusions, respectively, at an age of 1.5 days. As it can be seen from Fig. 7a, the endogenous shrinkage induces tensile stresses in the matrix. The shrinkage of the matrix is strongly restrained by the presence of the stiff aggregates. On the other hand, the aggregates are in a confined state (compression) as shown in Fig. 7b. The highest tensile stresses in the matrix are predominantly localized around the inclusions. These tensile stresses increase rather rapidly during the hardening period, this is a consequence of the rapid self-desiccation (see Fig. 3). Tensile stresses in the matrix overcome quickly the tensile strength consequently micro-cracking starts. The resulting crack patterns are given in Fig. 8a and b respectively for an age of 2 and 4 days. The cracks are mostly distributed radialy in the regions surrounding the stiff inclusions. Few of them will develop to real cracks only. It is obvious that micro-cracking induced by endogenous drying must affect the mechanical performance of the material. The severity of the induced damage can be estimated by the reduction of the effective stiffness of the material.

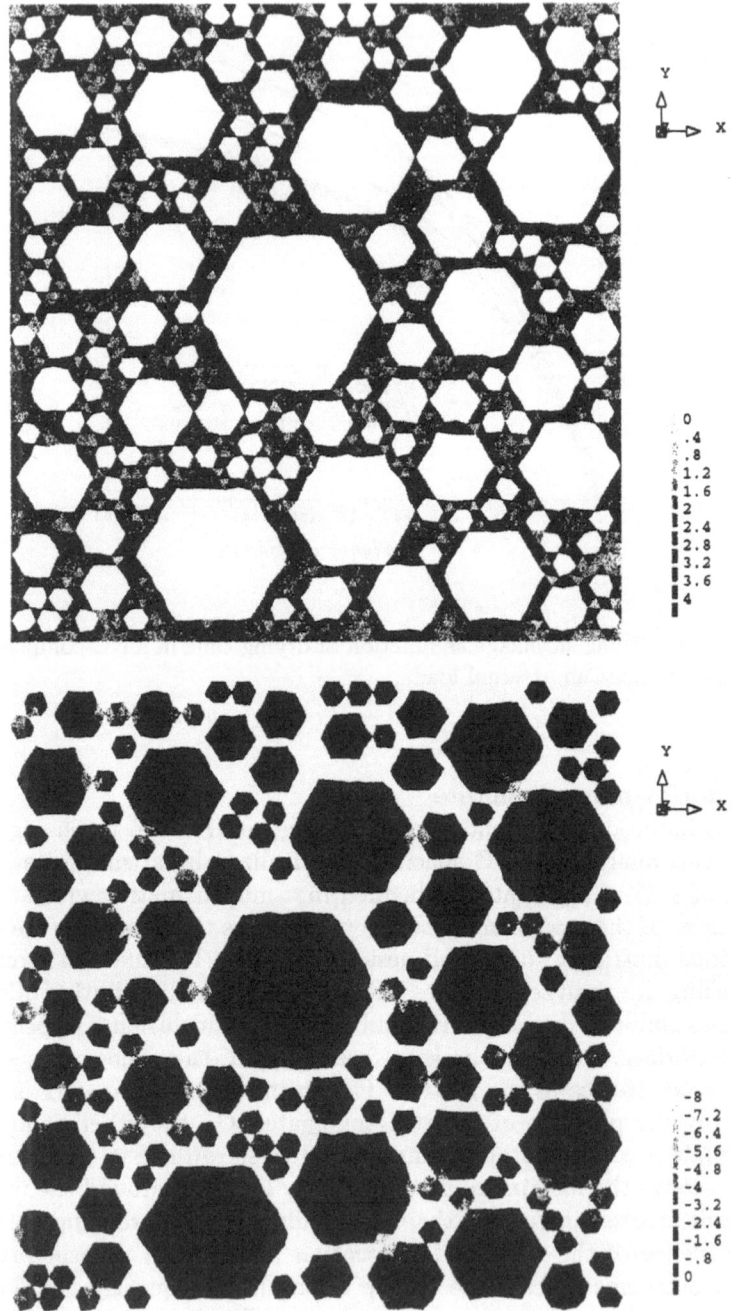

Fig. 7. Distribution of principal stresses (a) in the matrix and (b) in the aggregates of a HPC-composite structure subjected to endogenous drying at an age of 1.5 day

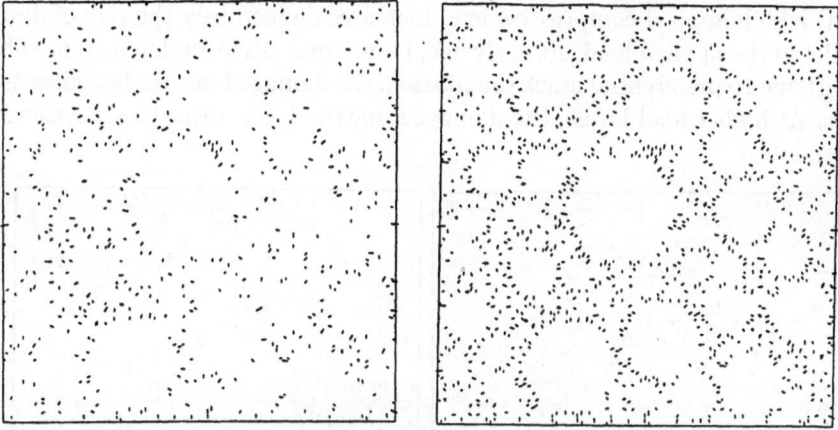

Fig. 8. Micro-crack patterns in a HPC-composite structure subjected to endogenous drying at (**a**) 2 and (**b**) 4 days

c) High performance concrete under sustained load

As it has been shown in the previous section, the restrained endogenous shrinkage in HPC induces micro-cracks in the porous matrix. The performed numerical experiments showed that the micro-cracks are mostly oriented normal to the circumference of the stiff inclusions. In other words, the microcracking in the matrix is isotropic (see Fig. 8). The following numerical experiments will show how an applied sustained load influences the orientation of micro-cracks induced by the endogenous shrinkage and thus imposes anisotropy to the composite material.

The same composite structure as in Sect. 4.2.2 b has been used to idealize a HPC specimen. In this case the system is assumed to be totally sealed. This means the matrix is subjected to self-desiccation only. The mechanical parameters of the matrix are given in Table 2, while the kinetics of the self-desiccation is given by the curve (HPC) in Fig. 3. In the performed mechanical analyses, the origin of time (time= 0) coincides with an age of 1 day of the material. A sustained compressive load is applied to the upper edge of the composite structure during the self-desiccation process. The behavior under two different compressive load levels is analyzed, namely 10 and 20 N/mm².

Figure 9a and b shows the resulting crack patterns at an age of 4 days in unloaded and loaded specimen, respectively. The following conclusions can be drawn from the performed analyses:

(i) an applied compressive load favors formation of micro-cracks running parallel to the direction of the applied load. The number of horizontal cracks (normal to the load) is reduced while the number of vertical cracks (parallel to the load) increases. This phenomenon of selective orientation becomes more pronounced as the load level increases.

(ii) a high applied compressive load increases dangerously the risk of damage parallel to the applied load. At early age, failure may occur under sustained load.

(iii) due to preferential crack orientation, the damaged matrix becomes anisotropic. At higher load levels, the damaged matrix degenerates to an orthotropic material.

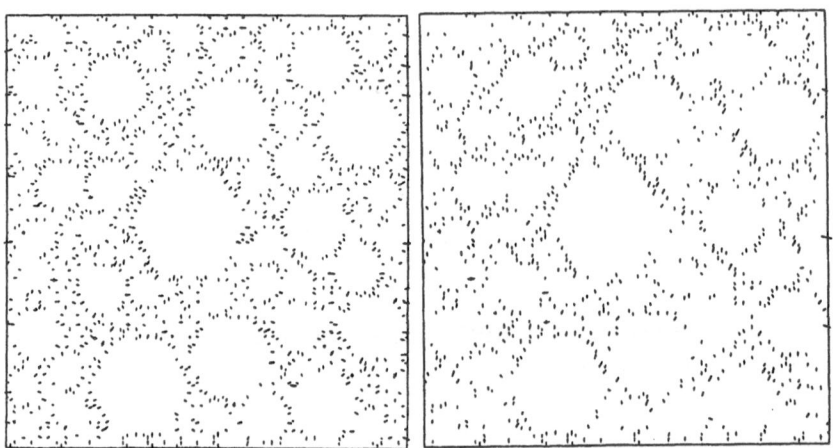

Fig. 9. Micro-crack patterns of a HPC-composite structure subjected to endogenous drying at an age of 4 days. (a) unloaded, (b) with an external load of $10 \, \text{N/mm}^2$

5   Conclusions

The heterogeneous structure of the composite material concrete can be represented by a numerical model called Numerical Concrete. By means of this model self-desiccation and drying of normal and high performance concrete can be simulated in a realistic way.

The observed drying shrinkage of normal concrete increases if the specimen is subjected to a compressive load applied normal to the drying direction. As a consequence drying shrinkage under uniaxial load becomes anisotropic.

Endogenous shrinkage plays a major role with respect to crack formation in high performance concrete, while this phenomenon is negligible in normal concrete with a comparatively high water-cement ratio. The micro-cracked matrix in high performance concrete is statistically isotropic.

If high performance concrete, under endogenous drying, is subjected to a sustained compressive load in addition, the resulting micro-cracks are preferentially oriented parallel to the direction of load. The damaged matrix can not be considered to be a homogenous material. Damage induced by simultaneously acting endogenous shrinkage and uniaxial external load leads to an anisotropic material. The effective mechanical and physical properties of the damaged composite material are anisotropic.

References

1. E. Tazawa, Autogenous Shrinkage of Concrete, *Proc. of the Int. Workshop organized by the Japan Concrete Institute*, Hiroshima, Japan, E. Tazawa editor, 1998.
2. P. E. Roelfstra, H. Sadouki and F. H Wittmann, Le béton Numérique, *Materials and Structures*, 18:327–336, 1985.
3. Z. P. Bazant, Material Models for Structural Analysis, *Mathematical Modelling of Creep and Shrinkage*, Bazant Z.P. editor, J. Wiley and Sons Ltd., 1988.
4. P. Haardt, Zementgebundene und kunstoffvergütete Beschichtungen auf Beton, PhD Thesis, Schriftenreihe des Instituts für Massivbau und Baustofftechnologie der Universität Karlsruhe, 13, Germany, 1991.
5. A. M. Alvaredo, Drying Shrinkage and Crack Formation, PhD Thesis, Swiss Federal Institute of Technology, also published in *Building Materials Reports*, No. 5, Aedificatio Publishers, 1994.
6. G. Martinola, H. Sadouki and F.H. Wittmann, Numerisches Modell zur Beschreibung der Eigenspannungen und der Rissbildung in Beschichtungenssystemen, Proc. of the 4th Int. Colloq. On Materials Science and Restoration, Aedificatio Verlag, Freiburg, Germany, Vol. 1, pages 393–407, 1996.
7. G. Martinola and H. Sadouki, A numerical Model to Prevent Shrinkage Induced Cracking in Repair Systems, in Proc. of the 5th Int. Worshop on Material Properties and Design, Durable Reinforced Concrete Structures, P. Schwesinger and F.H. Wittmann editors, Aedificatio Publishers, pages 161–190, 1998.
8. D. Quenard and H. Sallee, Water vapour adsorption and transfer in cement-based materials: a network simulation, *Materials and Structures*, 25:515–524, 1992.
9. X. Wittmann, H. Sadouki and F. H. Wittmann, Numerical Evaluation of Drying Test Data, Trans. 10th Int. Conf. On Struct. Mechanics in Reactor Technology, Vol. Q, pages 71–79, 1989.
10. V. Baroghel-Bouny, Caractérisation des Pâtes de Ciment et des Bétons, thèse de doctorat de l'Ecole Nationale des Ponts et Chaussées, 1994.
11. G. Martinola, Zum Verhalten zementgebundener Beschichtungen auf Beton, PhD Thesis, Swiss Federal Institute of Technology, to appear in *Building Materials Reports No. 2*, Aedificatio Publishers, 2000.
12. DIANA, Users manual-Release 7.0, Diana Analysis BV, Delft, Netherlands, 1998.
13. A. Hillerborg, M. Modéer and P. E. Petersson, Analysis of Crack Formation and Crack Growth in Concrete by Means of Fracture Mechanics and Finite Elements, *Cement and Concrete Res.*, 6:773–782, 1976.
14. Z. P. Bazant and B. H. Oh, Crack Band Theory for Fracture of Concrete, *Materials and Structures*, 16:155–172, 1983.
15. L. Cedolin, M. C. Contriono and F. Civetta, A Microstructural Model For Drying Shrinkage of Concrete, *Werkstoffe im Bauwesen*, Theorie und Bauwesen, Festschrift Reinhardt, ibidem, pages 273–285, 1999.
16. F. H. Wittmann, Grundlagen eines Modells zur Beschreibung charakteristischer Eigenschaften des Betons, *Deutscher Ausschuss für Stahlbeton*, Report Nr. 292, Wilhelm Ernst und Sohn, Berlin, 1976.
17. F. H. Wittmann, Structure of Concrete with respect to Crack Formation, *in Fracture Mechanics of Concrete*, F. H. Wittmann editor, Elsevier Science Publishers, pages 43–74, Amsterdam, 1983.
18. F. H. Wittmann, Drying and Shrinkage of Hardened Cement Paste, *in Pore Solution in Hardened Cement Paste*, M. J. Setzer editor, Aedificatio Publishers, 2000.

Printing: Weihert-Druck GmbH, Darmstadt
Binding: Buchbinderei Schäffer, Grünstadt

Lecture Notes in Physics

For information about Vols. 1–528 please contact your bookseller or Springer-Verlag

Monographs

For information about Vols. 1–24
please contact your bookseller or Springer-Verlag